Green Village and Town Construction in China

Xiaoming Wang · Hong Hua

Green Village and Town Construction in China

Science Press
Beijing

Springer

Xiaoming Wang
School of Civil and Hydraulic Engineering
Huazhong University of Science
and Technology
Wuhan, China

Hong Hua
School of Environmental Science
and Engineering
Huazhong University of Science
and Technology
Wuhan, China

ISBN 978-981-16-2097-3 ISBN 978-981-16-2098-0 (eBook)
https://doi.org/10.1007/978-981-16-2098-0

Jointly published with Science Press
The print edition is not for sale in China (Mainland). Customers from China (Mainland) please order the print book from: Science Press.

This Springer imprint is published by the registered company Springer Nature Singapore Pte Ltd.
The registered company address is: 152 Beach Road, #21-01/04 Gateway East, Singapore 189721, Singapore

Preface

In the 21st century, China's village and town construction is undergoing rapid development and tremendous changes that have never before been seen over thousands of years of Chinese history. The construction activity of villages and towns, which is aimed at rural revitalization and green development, will not only enable rural farmers in China to realize a comprehensively moderately prosperous life and share the civilized progress of urban development but will also leave a wealth of valuable rural construction and rural cultural heritage for future Chinese generations.

Since the reform and opening up in 1978, China has made great achievements in urbanization but has also paid a heavy price in areas such as out-of-control city scale, loss of ecological landscape, monotonous architectural style, damage to historical and cultural heritage, severe traffic congestion, pollution fog and haze, frequent flooding disasters, and other common phenomena. The differences in natural geographical and socioeconomic conditions between villages and towns in different regions are large, so construction cannot simply copy the urban construction mode, nor can it leave an unsustainable ecological environment for future generations. Green development is an inevitable choice for village and town construction.

To achieve the goal of green and sustainable development in village and town construction, it is necessary to answer the major question of how to realize the sustainable green transformation of the village and town construction from the perspective of combining theory and practice in the course of development over the next 10–30 years.

The construction and sustainable development of green villages and towns is a systematic innovation project that benefits both contemporary and future society, and there are four basic questions that must be answered. First, why should China build green villages and towns? The construction evolution of villages and towns and the urgent need for green development in China must be analyzed. Second, how should green villages and towns be built? The planning method and key content of green village and town construction should be mastered. Third, how should green village and town construction be managed? The key point is to select a suitable construction development pattern and build a construction community in green villages and towns. Fourth, how can the benefits of green village and town construction be evaluated? The effects of planning implementation and construction in green villages and towns should be scientifically evaluated. Therefore, the overall requirements of green village and town construction are to achieve good planning, good construction, good management, and good benefits.

The theoretical research and practical innovation of green village and town construction in China originated from small-town construction research and practice that started in the 1990s and have undergone more than 20 years of development to the present day. The authors are fortunate to have participated in the exploration process and to have successively pursued 6 national research subjects and 3 provincial village and town construction research projects; investigated more than 100 village and town construction and development demonstration projects in different regions across the country; and visited Yanhe Village, a green village and town construction demonstration site, more than 80 times. The demonstration project of Yanhe Ecological Village construction in Hubei Province, which was completed under the direction of the authors, was selected as the national major scientific and technological achievement during the 11th Five-Year Plan period. The authors have published 30 monographs, which have laid a solid foundation for the completion of this book.

This book is not only the authors' research summary on the theory and practice of green village and town construction in China over the years but also a systematic approach to the future green development and innovation of Chinese village and town construction. The relevant information and results provided in this book are also of great reference value for understanding and studying the evolution and development of Chinese village and town construction.

The Chinese government has clearly listed "widespread formation of green production and lifestyle", "prioritize the development of agricultural and rural areas and comprehensively promote rural revitalization", "promote green development and promote harmonious coexistence between man and nature" as goals in the "Five-Year Plan (2021–2025) for National Economic and Social Development" and "Development and Long-Range Objectives Through the Year 2035". It is believed that vigorously implementing and promoting green village and town construction will benefit the green development of rural areas and the green practices of farmers.

The completion of this book has received strong support from many departments and collaborators. This support was obtained from government departments, research institutions, experts, construction enterprises, grassroots construction organizations, and many villagers and also included the continuous participation of more than 100 doctoral and postgraduate students under the guidance of the authors' team. The authors thank doctoral graduates: Li Ming, Wang Xu; current doctoral graduate students: Li Xiaokang and He Nanjun for their assistance in writing this book, Chap. 4 of this book was completed by the authors in cooperation with Dr. Wang Xu. And thanks to master graduates: Yu Jian, Zhang Yujing, Chen Chong, Li Wen, Luo Di, Yang Haoliang, Rao Wei etc; and current graduate students: Li Dingqing, Ma Bingqian, Zhao Runfeng, Zhang Kaining, Yang Yang etc.

As this book goes to press, special thanks are owed to the following departments and supporters.

Thanks to the Rural Science and Technology Department and the Social Development Science and Technology Department of the Ministry of Science and Technology (MST), the China Rural Technology Development Center, the Science and Technology Department of Building Energy Conservation, and the Science and Technology Department of Village and Town Construction of the Ministry of Housing and Urban-Rural Development for long-term scientific research projects. Many thanks to the Housing and Urban-Rural Development Department of Hubei Province, the Science and Technology Department of Hubei Province, and the Science and Technology Department of Hainan Province for long-term support. Thanks to Wushan Town Government and Gucheng County Government of Hubei Province and Fushan Town Government and Chengmai County Government of Hainan Province for their strong support of green village and town construction demonstration projects.

Thanks to China Architectural Design and Research Institute Co., Ltd, Rural Development Institute of Chinese Academy of Social Sciences, Beijing Green Cross Culture Communication Center, Wuhan University of Technology, Harbin Institute of Technology, and Huazhong University of Science and Technology for their strong support of the authors' research and engineering demonstration. Special thanks to Prof. Cheng Zhenhua, Prof. Wan Yanhua, Prof. Yang Zhiyong, Prof. Liu Yi, Director Sun Jun, Prof. Du Zhixiong, Prof. Li Hong, Prof. Xiong Yan, Prof. Liu Jing, Mr. Gan Shijun, and Ms. He Gehua.

Special thanks to Dr. Wang Wei for her support and important contribution to the writing of this book.

Thanks to the village committee and all villagers of Yanhe Village for their full support of the planning and implementation of the national green ecological village construction demonstration base project over the past 10 years. Special thanks to Secretary Min Hongyan, Director Ren Anlun, Director Ji Dacui, Mr. Yang Shiquan, Mr. Fang Hongjun, and Mr. Liu Xiaoyun. Thanks for the participation and support of many cooperative enterprises that are members of the green village and town construction community.

Human beings share one earth. As China is the largest developing country and one of the largest agricultural countries in the world, its theoretical exploration, development path, and practical experience in green village and town construction have global significance and can serve as a model for other developing countries. Therefore, the book is published in both Chinese and English for readers domestically and abroad. The authors aim to introduce the theory and practice of China's construction development of green villages and towns to the world to enhance communication and cooperation between China and the rest of the world in the construction development of green villages and towns.

Owing to the limitations of our knowledge, some mistakes and errors doubtless remain in the book. Suggestions from readers would be appreciated.

Wuhan, China
October 2020

Xiaoming Wang
Hong Hua

Brief Introduction

This book systematically summarizes the authors' pioneering research results on the construction theory, construction methods, evaluation technology and application of demonstration projects in China's green villages and towns in the past 20 years. It adopts the research method of combining theory and practice, construction and evaluation, using photographs and illustrations as well as rich cases. The book is divided into three parts and eleven chapters. Part I is about the theory and development of green village and town construction, including theory and innovation, the evolution and development, the patterns and mechanisms, and the construction community of green villages and towns. Part II is about the planning and construction methods of green villages and towns, including the plan compilation, construction of the environmental infrastructure, and construction and renovation of green buildings in villages and towns. Part III is about the evaluation of the planning and construction of green villages and town, including the evaluation of plans, the evaluation of environmental infrastructure construction, the evaluation of green building construction, and the comprehensive evaluation of the planning and construction of green village and town.

This book is suitable as a reference for professionals engaged in theoretical research on green village and town planning and construction, engineering technology and construction management; teachers and students of engineering; and technical personnel but can also be used as a teaching and reference resource for college and university courses in related majors.

Contents

About the Authors

Prof. Xiaoming Wang of School of Civil and Hydraulic Engineering, Huazhong University of Science and Technology, director of Institute of Urban Construction and Human Settlements Environment. He is the council member of the China Society for Sustainable Development, the member of the Expert Committee of the National Sustainable Development Experimental Area, the member of the China Rural Green Building and Comprehensive Disaster Prevention Committee. He has presided over more than 20 China national research projects, more than 30 technical research and application projects and 4 international cooperation research projects. The green village construction research and demonstration program leads by him was selected in the "China National Major Science and Technology Achievement Exhibition". He has published more than 180 papers and 4 books.

Research interests: sustainable city and sustainable building, green urban and rural construction, protection and renewal of historic and cultural cities.

Associate Professor Hong Hua of School of Environmental Science and Engineering, Huazhong University of Science and Technology. She has presided over and participated in 10 China national research projects related to village and town construction, more than 10 technical research and application projects and 2 international cooperation research projects. She won the second prize of "Science and Technology Award of "Huaxia Construction" for the low-carbon construction and demonstration program in 2016. As the key member, she participated in the green village construction research and demonstration program was selected in the "China National Major Science and Technology Achievement Exhibition". She has published more than 30 papers.

Research interests: sustainable improvement of living environment, green and ecological urban and rural construction, environmental quality evaluation, urban low-carbon development.

Part I
Theory and Development of Green Village and Town Construction

Chapter 1
Theory and Innovation of Green Village and Town Construction

Abstract Green village and town construction in China not only responds to an urgent need to implement the national rural revitalization strategy and rural green development but also is an inevitable way to improve the quality of life of farmers in villages and towns, protect and improve the ecological environment, and promote the economic growth of villages and towns. In this chapter, the development system and legal guarantee of green village and town construction are proposed, and the research progress, innovative practices, and theoretical system of green village and town construction are discussed.

During the 71 years since the founding of the People's Republic of China in 1949, China has achieved world renown in village and town construction. Since the 21st century, China has entered a new period of development with the commitment to realizing the goals of a prosperous rural society and rural revitalization. Carrying out green village and town construction and establishing a theoretical system of China's rural characteristics that is suitable for green village and town construction, are necessary for promoting the coordinated development and green development of China's rural areas.

1.1 Green Villages and Towns and Green Village and Town Construction

1.1.1 Basic Content of Village and Town Construction

1. Definition of village and town

Village and town is general name for all kinds of villages and towns in rural areas. A village is a settlement where rural residents live and produce within the jurisdiction, and a town is the location of the township government and the administrative, economic, cultural and living service centre of the jurisdiction.

X. Wang and H. Hua, *Green Village and Town Construction in China*,
https://doi.org/10.1007/978-981-16-2098-0_1

Villages are places where rural residents live and work. They can be further divided into central villages and basic-level villages. A central village is a village with public facilities serving the surrounding villages, and a basic-level village is a village outside the central village.

The administrative region of a village refers to the area managed by the village committee.

A town is established with the approval of the provincial people's government. Towns can be divided into central towns and general towns. A central town is a town that plays a central role in the economic, social and spatial development of each district of a county town planning system. A general town is a town outside the central town in the county town planning system.

The administrative region of the town refers to the area under the administration of the town government.

2. Village and town construction

Village and town construction refers to the general activities involved in transforming the living environment of villages and towns and constructing various material facilities in the system of villages and towns based on the development plan. The key tasks of rural construction include infrastructure construction, housing construction and transformation, ecological environmental protection and village and town management.

The purpose of village and town construction is to create a good living environment for residents to ensure a normal life for them through planning and construction and to serve the economic and social development of villages and towns.

3. Content of town construction

Town construction should be carried out in accordance with a construction plan based on the construction contents of the *Standard for Planning of Town* GB 50188-2007, including county-rural system and population forecasts; town land classification and calculation; and planning standards for construction land, residential land, land for public facilities, production facilities and storage land use, road traffic, utilities facilities, disaster prevention and mitigation, environmental needs, and historical and cultural protection.

The construction content of the town is divided into six main categories.

(1) Construction land includes residential land in town areas and land for public facilities, roads and squares, and public green spaces.
(2) Construction of public works facilities includes facilities for water supply, drainage, power supply, communications, gas, heating, and engineering pipelines.
(3) Housing construction in town areas includes the construction of residential buildings; administrative office buildings; cultural, educational and sports buildings; medical and health buildings; commercial service buildings; and bazaars and markets.

(4) Disaster prevention and reduction. include fire control, flood control, earthquake prevention and wind mitigation in town areas.
(5) Environmental construction includes production pollution prevention and control, environmental sanitation, and environmental greening and landscaping in town areas.
(6) Historical and cultural protection includes the historical spatial pattern and traditional architectural style of the town, mountains, water systems, terrain, ground objects, ancient and famous trees and other elements closely related to the historical culture and other immovable historical relics reflecting the historical style, essence of folk customs, and sites and fixed facilities of traditional celebration activities.

4. Content of village construction

A village construction plan should be based on the *Guidelines for the Construction of Beautiful Village* GB/T 32000-2015 and other contents in terms of construction plans, adjusting measures to local conditions, the principle of villagers' participation, rational layout, and economical use of land and combined with the superior planning requirements of villages and towns.

The content of village construction is divided into 7 main categories.

(1) Construction land includes mainly village residential land, public service facility land, production facilities and storage land, green land and square land.
(2) Road construction includes mainly village roads and roads between villages.
(3) Public facilities and engineering construction includes water supply engineering, drainage engineering, power supply and communication, gas and heating facilities, environmental protection and environmental sanitation facilities.
(4) Construction of village farmhouses includes the construction, reconstruction, expansion and renovation of farmhouses, the renovation of old and dangerous farmhouses, and the construction of green farmhouses.
(5) Disaster prevention and mitigation includes fire protection, flood prevention and drainage, earthquake prevention and disaster reduction, geological disaster prevention, and meteorological disaster prevention.
(6) Environmental construction includes the maintenance and restoration of the ecological environment of the village, landscape and environmental greening, and domestic sewage treatment.
(7) Historical and cultural protection and special features includes the continuation of the traditional pattern of the village, the protection of the historical buildings of the village and the maintenance of historical and cultural heritage, and the protection of ancient and famous trees in the village.

1.1.2 Basic Concepts of Green Village and Town Construction

1. Green village and town

Green village and town refers to sustainable villages and towns that fully implement land savings, water savings, energy savings, material savings and environmental protection and achieve the coordinated development of the economy, society and ecological environment.

The goals of green villages and towns, guided by the coordinated development of the economy, society and ecological environment, are to promote the efficient use of resources, the harmonious coexistence of humans and nature, and the full integration of technology and nature. The key is to fully implement land savings, water savings, energy savings, material savings and environmental protection to maximize the productivity and creativity of villages and towns and ensure that the residents of can live and work safely.

2. Green village and town construction

Green village and town construction is the general term for the green construction and transformation activities of the living environment and various material facilities in the village and town system to promote green development as the construction goal, with green development planning for guidance and the systematic application of green construction engineering technology.

3. Terms related to green villages and towns

(1) Environmental infrastructure of green villages and towns is the general term for the material engineering facilities for environmental public services provided to ensure the green development of village and town production and residents' lives. It includes water source protection and water saving facilities, sewage reduction and re-use facilities, garbage classification and harmless treatment facilities, clean energy utilization facilities, greening and landscaping facilities, green transportation facilities, disaster prevention and mitigation facilities and early-warning facilities.
(2) Green building refers to a high-quality building that can save resources; protect the environment; reduce pollution; provide healthy, applicable and efficient use spaces for people; and maximize the harmonious coexistence of humans and nature.
(3) Green rural housing refers to healthy and comfortable new rural housing that is safe and practical, saves energy and reduces waste, and is economical and beautiful [1].
(4) Green energy, also known as clean energy, usually refers to a low degree of pollutant emissions and energy that can be directly used for production and living needs. It includes "renewable energy" and "non-renewable energy", with

renewable energy referring to energy production sources such as hydropower, wind power, solar energy, biomass (biogas), geothermal energy (including geothermal and water sources), and tidal energy.

1.1.3 Components of the Construction System of Green Villages and Towns

1. Development goal of green village and town construction

The development goal of green village and town construction is to adopt various green construction measures according to local conditions to meet the development requirements "industry prosperity, ecological liveability, civilized civilization, effective governance, and affluent life" and promote the coordinated development of a liveable green lifestyle, green economic growth and green ecological environment in villages and towns.

2. Main tasks of green village and town construction

The main tasks of green village and town construction are reasonable selection of the construction development model, establishment of the construction community, preparation of the construction development plan, construction of a green environmental infrastructure, construction of green buildings and green renovation of existing farmhouses. The effects of constructing green villages and towns should be monitored and evaluated, and construction demonstration projects should be promoted. Three types of tasks can be summarized: "good plan, good construction and good management".

3. Development system of green village and town construction

Green village and town construction is a complex system that involves scope, a long implementation cycle, and rigorous innovation requirements. Compared with general village and town construction, green village and town construction is characterized by the multi-objective coordination of construction planning, diversified selection of construction modes, multi-subject participation in construction projects, application of new technologies in construction projects, monitoring and evaluation of construction effects, and replication and promotion of construction experience.

The evolution and development direction of the green village and town construction system involves promoting the implementation of national rural revitalization strategies and the sustainable development of villages and towns.

The development system of green village and town construction in China is shown in Fig. 1.1.

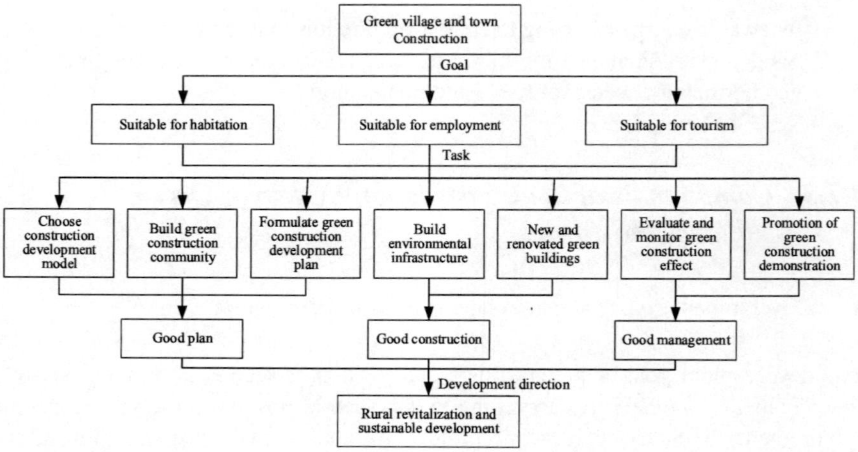

Fig. 1.1 Development system of green village and town construction in China

1.1.4 Legal System of Green Village and Town Construction

China initially formed a legal system for planning, constructing and evaluating green villages and towns that included relevant national laws, standards and technical documents.

1. National law related to green village and town construction

The main laws include *Urban and Rural Planning Law of the People's Republic of China, Construction Laws of the People's Republic of China, Environmental Protection Law of the People's Republic of China, Land Administration Law of the People's Republic of China, Energy Conservation Law of the People's Republic of China, Law of the People's Republic of China on Prevention and Control of Water Pollution, Law of the People's Republic of China on the Prevention and Control of Environmental Pollution Caused by Solid Waste, Law of the People's Republic of China on the Prevention and Control of Atmospheric Pollution*, and *Law of the People's Republic of China on Environmental Impact Assessment*.

2. Standard specifications related to green village and town construction

The main standards are *Standard for Planning of Towns* GB50188-2007, *Guidelines for the Construction of Beautiful Villages* GB/T 32000-2015, *Technique Code for Village Rehabilitation* GB 50445-2008, *Hygienic Specification for Township-village Planning* GB 18055-2012, *Standards for Drinking Water Quality* GB 5749-2006, *Integrated Wastewater Discharge Standard* GB 8978-2002, *Discharge Standard of Pollutants for Municipal Wastewater Treatment Plants* GB 18918-2002, *Technical Standard for Domestic Wastewater Treatment Engineering of Rural Areas* GB 51347-2019, *Code for Fire Protection and Prevention of Rural Areas* GB 50039-2010, *Code for Design of Civil Buildings* GB 50352-2019, *Code for Seismic Design of Buildings* GB 50011-2010, *Assessment Standard for Green Buildings* GB/T 50378-2019, *Code for Green Construction of Buildings* GB/T 50905-2014, *Technical Specifications of Domestic Pollution Control for Town and Village* HJ 574-2010, *Guideline for Town-level General Land Use Planning* TD/T 1025-2010, and *Technical Specifications for Village Landscape Environmental Engineering* CECS 285-2011.

3. Technical documents related to green village and town construction

The main documents are *National Ecological Village Construction Standards (Trial), National Ecological Township Construction Indicators (Trial), National Ecological Civilization Construction Demonstration Village and Town Indicators (Trial)*, and *Green Rural Housing Construction Guidelines (Trial).*

1.2 Research Progress in Green Village and Town Construction

Theoretical research on green village and town construction in China has gradually developed since the beginning of the 21st century. It is concentrated mainly in areas such as development planning, construction techniques and construction evaluation.

1.2.1 Research on Development Planning of Green Villages and Towns

Jin Zhaosen et al. proposed that ecological villages and towns inherit and develop ecological villages and are complex socioeconomic ecosystems with internal functions of production, life and ecology and external functions of tourism, education and demonstration [2]. Implementation was the focus of the project and included four aspects: ecological industry, ecological human settlement, ecological environment and ecological culture. Peng Yigang analysed the influence of natural factors and social factors on traditional village and town settlement forms and proposed a type of composition of traditional village and town settlement forms, landscape analysis elements, and the extension and regeneration of rural architectural culture [3]. Wu Liangyong created a generalized architectural theory from the multi-dimensional perspectives of settlement theory, regional theory, cultural theory, scientific and technological theory, political and legal theory, business theory, education theory, and art theory and methodology and proposed the framework of a theoretical research system on urban and rural construction and governance centred on the improvement of the living environment [4]. The FAO (Food and Agriculture Organization of the United Nations) proposed that in the process of rural rejuvenation, the concept of "regional development" should be promoted from a planning perspective, and green development should be achieved by creating long-term self-sufficient development and employment, but only if local stakeholders are able to fully participate in decision-making and planning [5].

Zhang Xiaoshan et al. systematically analysed the construction and development of villages and towns since China's reform and opening up in 1978; proposed the main content of the development of future towns, new rural communities and villages; and analysed the development trends of improving infrastructure construction, reforming the use of construction land, establishing a construction and development investment and financing system, and innovating the construction management system [6]. Li Xiaohong researched the evolution and innovation process of China's ancient and modern rural governance methods and analysed the role played by village committees in the governance of rural construction [7]. Sun Jun pointed out that farmers were the subject of new rural construction, and farmers' participation was the key to the success of new rural construction [8].

Li Xiaokang et al. quantitatively analysed the influence and mechanism of structural changes in 8 types of rural land (cultivated land, garden land, forestland, pasture grassland, other agricultural land, residential land, independent industrial and mining land, transportation land, and water conservancy facility land) in relation to carbon emissions, taking the monitoring and evaluation of green land for villages and towns as the object [9]. Li Ming et al., based on the principles of ecology, pointed out that the rural village construction pattern should be optimized in the stages of planning and design, construction implementation, operational management, demolition and transformation. Optimized strategies include for the agricultural circular economy model, ecological agricultural development model and industrial development model [10]. Li Ming et al. researched the grey correlation degree coupling model of a rural ecological settlement construction system and established a construction composite system and a construction system coupling measurement model based on construction as the main component and a rural ecological settlement with construction, resources, economy, environment and society as sub-systems [11]. Wang Xiaoming et al. proposed the concept and evolutionary process of green agricultural housing construction partnerships, analysed the main participants in green agricultural housing construction and their participating behaviours, and constructed a network analysis model of the community [12].

In 2010, "The European 21st Century Village Declaration" was issued in Germany and proposed that rural areas have renewable resources and a granary production base, maintain ecological balance, and provide multiple functional spaces for rural trade, culture, leisure, residence and life. In rural construction, to create efficient infrastructure and practice and create jobs, the natural ecological system should be protected, and the full participation of residents should be ensured.

In 2015, the United Nations' "Transforming Our World: The 2030 Agenda for Sustainable Development" called for supporting positive economic, social and environmental linkages between urban, suburban and rural areas by strengthening national and regional development programmes. The aim was to ensure access to proper, safe and affordable housing and basic services for all; rehabilitate slums; and provide safe, inclusive, accessible and green public spaces for all by 2030.

In 2016, "The New Urban Agenda" adopted at the United Nations Conference on Housing and Sustainable Urban Development (Habitat III) proposed supporting sustainable regional infrastructure development, promoting sustainable economic productivity, and promoting the equitable development of integrated urban and rural areas.

1.2.2 Research on Construction Techniques of Green Villages and Towns

Guo Zhiwei et al. analysed the development trend of China's urbanization and rural construction technology, including key technologies such as village and town planning, land use change monitoring and ecological carrying capacity evaluation, environmental monitoring and evaluation, infrastructure construction, water resource utilization and management, energy utilization and energy savings, village and town ecological construction and environmental protection, rural communities and residential technology [13]. Li Bingdi proposed adopting an integrated technology system to promote rural green development and residential environmental management, implementing village planning with low-cost green technology system guarantees, implementing low-impact development and construction modes, and adopting non-engineering measures to solve rural infrastructure construction problems [14]. Di Qing et al. pointed out that the construction of a green infrastructure system was the key to solving the construction of green liveable villages and towns [15].

Liu Xianjue et al. systematically researched the theory of green buildings and pointed out that the understanding of modern green buildings should include the system perspective, ecological perspective, space-time perspective, economic perspective, and social perspective [16]. Brian Edwards et al. proposed that components of sustainable buildings should include building products and professional services, energy conservation and pollution prevention, water conservation, environmental impact analysis, environmental health and safety, recycling, ecology, and sustainable development [17]. Jerry Yuelson believed that the green transformation of existing buildings should include the planning of existing buildings, a system for grading and evaluating green buildings, the value of existing buildings, the improvement of green energy, green construction management, and the greenness of buildings [18].

Ma Suzhen pointed out that green building integration technology should include saving land and the outdoor environment, energy saving and energy utilization, water saving and water resource utilization, material saving and material resource utilization, indoor environmental quality, construction management, operational management and other types [19]. Lu Hongyi discussed the improvement of traditional building technology and case practice in combination with the construction of green agricultural houses in a beautiful countryside [20]. Zhao Shiyong et al. integrated the key technologies of rural green ecological residential area planning, green suitable house construction, green energy development and utilization, and garbage and sewage treatment [21]. Mei Hongyuan et al. proposed planning and design strategies such as spatial layout, ecological protection and land use of green villages and towns, energy-saving building strategies such as roof, wall, door and window energy savings, and infrastructure strategies such as water resources, garbage treatment and clean energy utilization [22].

Yang Zhiyong et al. proposed the practical technology of seismic design for rural houses, especially the seismic technological construction and seismic reinforcement method of original ecological buildings such as stone houses, soil houses and wood houses. In addition, they proposed a practical approach to integrating the technology of energy-saving design and construction of rural houses [23]. Zhang Kui, et al. proposed key technologies for the energy conservation transformation of existing agricultural buildings [24]. Li Yue et al. discussed planning innovation based on the intensive use of land, energy innovation based on the rational use of natural resources, material innovation based on the use of green building materials, and design innovation based on the enhanced adaptability of residential functions in village and town construction [25]. Yang Fan et al. analysed the construction of ecological energy-saving houses in villages and towns, demonstrated the main components of science and technology, and established a construction demonstration model of ecological energy-saving houses in villages and towns [26].

The key research project "technological innovation of green liveable villages and towns" launched by the Ministry of Science and Technology of China in 2018 [27] found that research on the construction technology of green villages and towns in China had entered the stage of systematic development. The project fell into two broad categories.

(1) Common key technologies of green liveable villages and towns

There are 12 key techniques: the system of calculating the resource environmental bearing capacity of village and town construction, the development pattern and technical path of village and town construction, the digital simulation and evaluation model of settlement space in villages and towns, the regional village building disaster mechanism and appropriate disaster prevention and reduction system, the key technologies of the rural "toilet revolution", living garbage high-value utilization and secondary pollution-control technology equipment in villages and towns, low-cost clean energy heating and heat storage technology in villages and towns, planning technology for spatial development intelligent control and function, key technologies of rural residential design and construction, ecological building materials research and components in villages and towns, research and integrated innovation of rural human settlement environmental improvement technology, and research on the intelligent investigation of key technologies of village and town land.

(2) Construction application technologies of green liveable villages and towns

There are 16 application technologies: key technology research and equipment for improving drinking-water quality in villages and towns, equipment for sewage treatment and recycling in villages and towns, key technology and equipment for domestic waste mobile miniaturization treatment in villages and towns, key technology for rural ecological landscape construction, key technology for community environmental monitoring and rehabilitation, key technology for clean energy and high-value utilization of biomass resources in villages and towns, planning technology for protecting and transforming characteristic villages and towns, optimizing and laying out community spaces in villages and towns, key technology for protection of the utilization and modern inheritance of traditional villages, the ecological structural system of prefabricated housing in villages and towns, the technology for improving community public services based on "Internet +" in villages and towns, key technology and applications for bio-based building materials in villages and towns, key technology for land intelligence surveys in villages and towns, key technology for improving the integrated disaster prevention capability of existing buildings in villages and towns, key technology for community infrastructure construction in villages and towns, and digital rural development technology.

1.2.3 Research on Construction Evaluation of Green Villages and Towns

Su Xing screened the environmental impact index factors of green ecological villages and towns, constructed the environmental index system and evaluation method for green ecological villages and towns, and established two environmental impact evaluation models of energy utilization and waste recycling in villages and towns [28]. Wu Jianguo proposed the landscape ecology theory—the theoretical analytical method of patterns, processes, scale and hierarchy [29], which has important reference value for the construction of green village landscape patterns and environmental restoration planning evaluation. Hua Hong et al. analysed and measured the energy-saving efficiency of a rural adobe house and proposed data for the energy consumption per unit of materials commonly used in village and town buildings [30]. Hua Hong et al. established the overall goal of low-carbon village construction based on controlling carbon sources and established an evaluation benchmark and level of low-carbon village construction under the mode of energy-saving village construction, energy-saving practices of farmers, energy-saving production and construction, and green land use [31]. Yang Jun et al. established a benefit analysis and calculation model of energy savings, land savings, water savings, material savings and waste reduction in village and town houses [32]. Yang Jun et al. researched the connotations of the durability of village and town houses and established an economic analysis and benefit evaluation model of a durable design scheme [33]. Cen Qianting et al. proposed a sustainable evaluation method for infrastructure construction in a green village [34].

Wang Yang et al. proposed a method to evaluate the integration and comprehensive benefits of low-carbon construction technology in rural buildings in areas with hot summers and cold winters, focusing on building energy conservation and technology economy [35].

The "*Assessment Standard for Green Building* GB/T 50378-2019" provided an important basis for evaluation research on green village and town construction; vigorously advocated a simple and moderate, green and low-carbon lifestyle; and carried out actions such as creating energy-saving agencies, green families, green schools, green communities and green travel. It defined green performance as the comprehensive performance related to building safety and durability, health and comfort, convenience of life, resource savings (land, energy, water, and material savings) and environmental liveability. It established five indexes of green buildings: security and durability, health and comfort, convenience, resource savings, and environmental habitability.

The "*Evaluation for the Construction of Beautiful Village* GB/T 37072-2018" put forward principles, contents, procedures and calculation methods applicable to the comprehensive evaluation of the construction of beautiful villages.

1.3 Innovative Exploration of Green Village and Town Construction

Green village and town construction in China is an innovative exploratory process that combines practice and theory. From 2002 to 2018, the authors conducted a field survey of typical villages and towns in more than 100 counties and cities in different regions of the country and successively undertook seven national research projects on village and town construction: "Research on and demonstration of monitoring and evaluation technology of planning and construction of green ecological villages and towns", "Integrated application of residential construction technology in a mountainous area", "Research on and demonstration of construction technology of ecological buildings in different regions", "Research on and demonstration of key technologies of ecologically energy-saving community construction in Huazhuang Town", "Investigation and research of residential protection planning and remediation in Qingyan Ancient Town, Guizhou Province", "Research on Evaluation Index System and Key Technology of Small Town Construction", and "Research on innovation mechanism of demonstration project for small-town construction pilots".

The authors also presided over the completion of more than ten village and town construction planning technology research and development and application projects, including "Construction planning of a green development demonstration town in Fushan Town, Chengmai County, Hainan Province (2016–2020)", "Planning of low-carbon town construction in Shilong Town, Dongguan City, Guangdong Province (2011–2015)", "Planning of economic and social development in Mulan

Township, Huangpi District, Wuhan City (2006–2020)", "Planning research of immigrant village construction with Jijiagou Village, Danjiangkou City, as an example", "Demonstration research of modernization construction and informatization technology of villages and towns in Hubei Province", and "Demonstration of a construction project of aseismic and energy saving in Yanhe Village, Gucheng County", as well as international cooperative research such as "Co-research on and demonstration of eco-village construction in China and Australia".

These research achievements are an important basis for the content of this book, and they reflect the research and development and innovation process of green village and town construction in China.

1.3.1 Research on the Innovation Mechanism of the Demonstration Project for Small-town Construction Pilots

This project was a national scientific and technological research topic, and the research period was 2002–2005.

1. Research method

The subject was the construction demand and sustainable development process of small towns in China at the beginning of the 21st century, combined with pilot demonstration projects. Experts were organized from multiple disciplines, including town planning, urban construction, environmental engineering, project management and information systems, to jointly tackle the key problems and explore the theory and innovation mechanisms of the sustainable and coordinated development of small-town construction.

2. Research achievements

1) The sustainable development innovation mechanism system of small-town construction

The research considered four elements of the sustainable development of China's small towns: sustainable utilization of resources, sustainable growth of the industrial economy, sustainable improvement of the ecological environment, and sustainable fairness of urban and rural society, and three innovative approaches: urbanization innovation support, industrialization innovation guidance and ecological innovation guarantees.

2) The dynamic planning method and elastic coordination mechanism for sustainable construction of small towns

The research considered methods and promotion mechanisms of town coordinated development planning, established methods and promotion mechanisms of coordinated development planning in small towns and townships, and established the dynamic elastic planning method and promotion mechanism for the coordinated development of small towns from "time limit planning" to "scale planning" and from a "total conclusion" fixed pattern to a "non-conclusion" elastic pattern.

3) The integration pattern and regulation mechanism of sustainable utilization of resources in small-town construction

A mathematical model was constructed to evaluate the utilization efficiency of construction resources of small towns, and the value composition and evaluation method of construction resources of small towns were established.

4) Analysis method and early-warning mechanism of ecological environmental carrying capacity in small-town construction

A quantitative analysis method for the ecological environmental carrying capacity of small towns was established, and an evaluation model of the ecological comprehensive status and early-warning level of small towns was established.

5) Innovation and adaptive adjustment mechanism of the small-town construction development model

A new model of small-town construction and the development of "resource capitalization-integration-marketization-characterization" was established, and an innovative mechanism for the operation of the policy securities market for small-town construction was proposed.

6) Dynamic coordination system for sustainable development of small-town construction

An interactive development coordination mechanism based on three types of subsystems, urbanization, industrialization and ecologicalization, was proposed, the model diagram and dynamic equations of the dynamic coordination system for small-town construction were established, and simulation analysis and policy experimental applications were carried out.

The sustainable development system of China's small-town construction is shown in Fig. 1.2.

The above research results are the authors' preliminary foundation for discussion on the theoretical system and construction development system of green villages and towns.

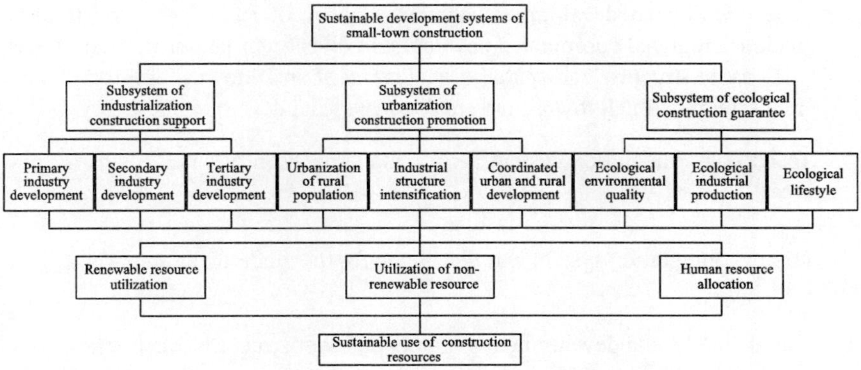

Fig. 1.2 Sustainable development system of China's small-town construction

1.3.2 Research on Evaluation Index System and Key Construction Technology of Small Towns

This project was conducted by a research group of the National Spark Program, and the research period was 2002–2005.

1. Research method

The aim was to meet the needs of the sustainable development of rural urbanization in China in the early 21st century and to solve the three rural issues; study the construction goals, construction standards, technical methods, evaluation systems and management models of modern small-town rural modernization technology demonstration actions; establish a system of technological innovation and construction technological guidance for the construction of small towns in line with the characteristics of China's rural areas; and develop evaluation and promotion methods for the demonstration of science and technology in the modernization of small towns in rural areas.

2. Research achievements

1) Theory of sustainable and coordinated development of rural small-town construction

(1) The theory of small-town construction and sustainable development in rural areas includes small-town construction and rural sustainable development, small-town construction and rural economic and social coordination, and small-town construction to promote the coordinated development of rural areas.

(2) Small-town economic agglomeration and economic development theory includes the stratification of the small-town cluster economy, small-town construction resources and resource endowment, and small-town regional economic development model selection.

(3) The coordinated development planning theory of small-town construction includes regional coordinated development division of labour in small towns, the spatial structure and spatial organization of small towns, differential land rent effects in small towns, and small-town social development theory.

2) Technology integration system for the modernization of small towns in rural areas

A technology integration system was developed for the modernization of rural small towns, including:

(1) Rural small-town development planning and infrastructure construction guidance technology;
(2) Rural small-town industrialization construction guidance technology; and
(3) Rural small-town ecological environmental protection guidance technology.

3) Evaluation index system of demonstration towns for the modernization of small towns in rural areas

The evaluation indexes are composed of 8 types of first-level evaluation indexes:

(1) Basic conditions for small-town construction;
(2) Rationale of overall planning for small towns;
(3) Rationale of infrastructure and ecological environmental protection;
(4) Small towns' impact on the regional economy and the promotional effect of development;
(5) The promotional effect of small cities and towns on social development;
(6) The promotional effect of small cities and towns on agricultural industrialization and township and village enterprises;
(7) The technological innovation ability of small cities and towns; and
(8) The implementation management capabilities of construction projects in small cities and towns.

4) Technology integration system software for rural small-town construction.

The aim was to establish a human-machine interactive selection platform for rural small-town construction technology application; establish a basic database for the modern small-town modernization construction technology integration system; form an interactive and friendly advanced and suitable construction technology selection and matching platform; realize an open and continuous construction technology integration system that supplements and updates functions; and encourage the application and promotion of modernization construction techniques in small rural towns.

5) Dynamic coordination system for rural small-town construction

The aim was to establish a dynamic model of a small-town construction coordination system based on the sustainable development of construction → urbanization planning and regulation → industrialization model and operation → ecological guarantee and policy, establish a dynamic coordination simulation technology method for small-town construction in rural areas, and propose a rural small-town modernization development promotion mechanism focusing on urbanization support, industrialization guidance, ecological protection, and overall urban-rural development.

6) A case study on the creation of a demonstration town for the modernization of small rural towns

The construction and economic and industrial development of 34 demonstration towns in different regions of China were investigated. Yangxunqiao Town in Zhejiang Province in the east, Chengui Town in Hubei Province in the central region, and Tianhui Town in Sichuan Province in the west were selected.

This research provided theoretical and technological support for the development model of green village and town construction.

1.3.3 Research on and Demonstration of Construction Technology of Ecological Buildings in Different Regions

This project was a national key scientific and technological research project, and the research period was from 2004 to 2006.

1. Research method

Guided by the theory of the sustainable development of urban and rural construction and sustainable building technology and focusing on the establishment of application systems and management models of ecological building construction technology in different regions of villages and towns, ecological building construction technology demonstration bases were established in national sustainable development communities in different regions.

2. Research achievements

1) Compilation and application of eco-building construction planning guidelines

The guidelines include new ideas, new approaches and new mechanisms for the construction of ecologically energy-efficient residential areas; action guidelines for the construction of ecologically energy-efficient residential areas; guidelines for the construction of forest areas and scenic spots; old village reconstruction and green volunteer home planning; and ecological environmental planning for green open spaces in residential areas.

2) Research and application of key technologies of ecological building construction

The technologies include key technologies for the construction of ecologically energy-efficient residential areas, complete technologies for ecological buildings in rural residential areas, energy-saving and material-saving technologies for ecological village buildings, green energy-saving technologies for ecological buildings in scenic spots, construction technologies for ecologically energy-efficient grass-brick buildings, and integrated system software for key technologies for ecologically energy-efficient residential areas.

3) Research and application of green construction technology and patterns of ecological buildings

The technology and patterns include green construction guides for ecological building construction, green construction management models for ecological buildings, construction codes and standards for energy-saving straw-brick houses, and research and demonstration guides for eco-village construction and green volunteer home projects.

4) Research and application of the environmental improvement evaluation index of the ecological community

Evaluation index systems were created for residents of ecologically energy-saving residential areas, ecological living environments in rural residential areas, ecological village dwellings in severe cold areas, the energy efficiency of ecological residential buildings, etc.

3. Demonstration achievements

The project resulted in the completion of 9 village and town ecological building construction demonstration projects, the building of 12 demonstration projects, and the formation of 15 demonstration sites. Significant land savings and energy savings, ecological and environmental protection, and economic and social benefits were achieved. See Table 1.1.

The above research provides a practical basis for establishing green village and town planning and construction technology systems and demonstration and promotion mechanisms.

Table 1.1 Demonstration achievements of eco-building construction technology in different regions of villages and towns

Number	Demonstration project location	Demonstration project theme
1	Kaifayuan Community, Huazhuang Town, Wuxi City	Eco-environmental planning and construction of green open spaces in community
2	Shuiqingmuhua Community, Huazhuang Town, Wuxi City	Comprehensive solar energy utilization technology in residential area
3	Luoxiayuan Community, Huazhuang Town, Wuxi City	Digital construction and management of new rural community
4	Wuxi Yiduo Environmental Thermal Power Company	Rendering harmless and reduction of domestic garbage in small towns
5	Wuxi Wangxing Curtain Wall Decoration Engineering Company	New technology for green energy utilization of photovoltaic curtain wall
6	Zhuanghu Village, Bohai Town, Huairou District, Beijing	Demonstration construction of eco-village and green volunteer home
7	Hong'an New Village, Hongtuya Town, Baishan City	Construction of ecological energy-saving straw-brick houses in the frigid regions of northeast China
8	Weihu Mountain Scenic Area in Hailin City, Heilongjiang Province	Ecological planning of forest region scenic spots and application of building green energy-saving technology
9	Huahong Century Garden, Zhouzhuang Town, Jiangyin City	Construction of rural ecological community and improvement of human settlement environment in developed areas

1.3.4 Integrated Application of Residential Construction Technology in Mountainous Areas

This project was part of the National Science and Technology Support Program, and the research and demonstration base was constructed from 2008 to 2012.

1. Research method

The aims were to promote the overall level of rural green ecological village and town housing construction through the integration and promotion of suitable housing ecological village construction technology; build a suitable new green ecological building demonstration community through the demonstration project of suitable housing construction in Yanhe Ecological Village in Hubei Province; enable the community to achieve the goal of resource-saving, environmentally friendly housing suitable for farmers; and provide a national-level demonstration base for improving the living environment of villages and towns and the quality of life of residents.

2. Research achievements

The suitable ecological village residential construction technology integration system was composed of four sub-systems: planning and design technology, construction technology, living environment improvement and function improvement technology, and new community management technology, including 16 groups of specialized technologies. See Fig. 1.3.

3. Demonstration achievements

The location of the demonstration base is Yanhe Village, Wushan Town, Gucheng County, Hubei Province, which is a national ecological village and a suitable green village. After five years of construction, 50,000 m^2 of suitable construction and renovation for farmers were completed, and four groups of construction technology

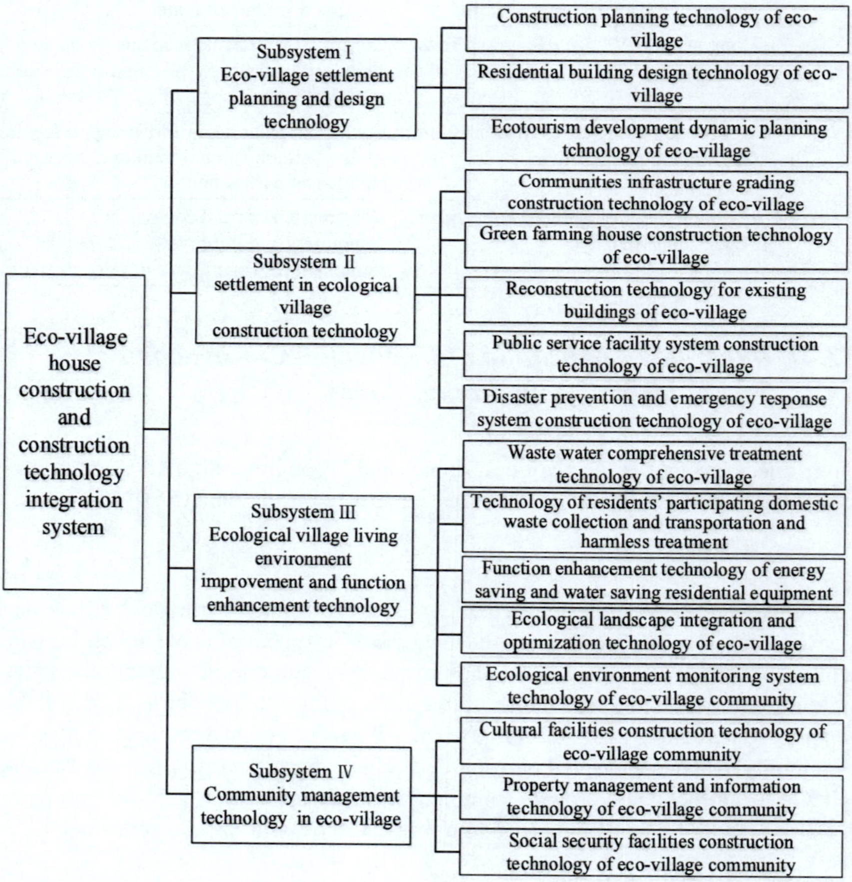

Fig. 1.3 Integration system of eco-village construction technology

application demonstrations were completed: planning and design technologies of suitable residential areas; construction technologies of suitable residential areas; improvement of living environment and residential functions; and new community management technology.

Through the implementation of the construction demonstration project, the human settlement environmental quality and infrastructure construction level of Yanhe Village were comprehensively improved. A green ecological village style was formed with the characteristics of western Hubei, making Yanhe Village a demonstration base for the construction of suitable housing in Hubei Province. The project won the honour of being designated a national AAA tourist scenic spot and a national ecological village. The demonstration result in 2011 was that the "Hubei Yanhe Ecological Village Construction" was selected for the National Eleventh Five-Year National Exhibition of Major Scientific and Technological Achievements.

The annual per capita income of farmers in Yanhe Village increased from 3600 yuan per person before the implementation of the project to 8300 yuan per person in 2011, with an average annual growth rate of 15%. The Gucheng County government vigorously promoted the construction model and technological achievements of suitable housing in Yanhe Village throughout the county and built 23 suitable ecological new peasant villages based on Yanhe Village.

The above research provided technical support for the establishment of the planning and construction system of green villages and towns.

1.3.5 Research on and Demonstration of Monitoring and Evaluation Technology of Planning and Construction of Green Ecological Villages and Towns

This project was part of the National Science and Technology Support Program, and the research and demonstration base were constructed from 2014 to 2018.

1. Research method

Taking the key technologies for evaluation and monitoring of green village and town planning and construction as the object, the research focused on the evaluation technologies of green village and town planning and infrastructure construction and rural housing construction and aimed to integrate the key technologies of green village and town planning and construction evaluation and monitoring to establish a green village and town planning and construction evaluation monitoring platform. Yanhe Village, Hubei Province, was selected as the demonstration base for construction technology research and application.

2. Research achievements

The key technologies for green village planning and construction evaluation are divided into three categories, with a total of 19 evaluation technologies.

1) Evaluation technology of green village and town planning

The technology has 7 categories: a green village and town evaluation index system for planning and construction, land location and function partition optimization evaluation technology, ecological niche and bearing capacity evaluation technology, sewage treatment and utilization of water resources planning and evaluation technology, solid waste disposal plan evaluation technology, clean energy utilization evaluation technology, and ecological landscape and environmental restoration planning evaluation technology.

2) Evaluation technology of infrastructure green village and town construction

This technology has 6 categories of green village and town evaluation technology: water conservation and water-saving facility construction, sewage reduction and reuse facility construction, separate waste collection and harmless treatment facility construction, clean energy facility construction, greening and landscape facility construction, and disaster prevention and mitigation and early-warning facility construction.

3) Evaluation technology of green village and town farmhouse construction

This technology has 6 categories of evaluation technology of green farmhouses: green construction, energy-saving efficiency, water-saving facility effectiveness, indoor environmental quality, and low-carbon operational management.

3. Demonstration achievements

1) Application demonstration of monitoring technology

The application demonstration of the construction monitoring technology of green villages and towns has two categories: planning implementation and farmhouse construction.

The first category, planning implementation, includes site selection and functional area monitoring and monitoring of ecological carrying capacity and environmental quality factors. A total of 7 monitoring technologies were applied and demonstrated.

The second category, farmhouse construction, includes monitoring the energy-saving efficiency of farm buildings, indoor environmental quality testing of farm buildings, and safety testing of existing farm buildings. A total of 7 monitoring technologies were applied and demonstrated.

2) Application demonstration of evaluation technology

The application demonstration of the evaluation technology of green village and town construction has five categories: planning and implementation, infrastructure construction, clean energy utilization, farmhouse construction and reconstruction engineering, and ecological landscape and environmental restoration.

The first category, construction planning and implementation evaluation technology, has five categories: site selection and functional zoning evaluation, ecological niche planning and environmental carrying capacity evaluation, water resource utilization and sewage treatment planning evaluation, solid waste treatment planning monitoring evaluation, and ecological landscape and environmental restoration planning evaluation. A total of 11 demonstration techniques were applied and demonstrated.

The second category, environmental construction infrastructure, has four categories: wastewater reduction and re-use facility construction evaluation, waste separation collection and harmless treatment facility construction evaluation, disaster prevention and mitigation and early-warning facility construction evaluation, and water conservation and water-saving facility construction evaluation. A total of four demonstration techniques were applied and demonstrated.

The third category, clean energy utilization, has two categories: clean energy utilization planning evaluation and clean energy facility construction evaluation. A total of three demonstration techniques were applied and demonstrated.

The fourth category, green farmhouse construction and renovation, has three categories: energy-saving efficiency evaluation for newly built farmhouses, energy-saving efficiency evaluation for farmhouse renovation, and indoor environmental quality evaluation for farmhouses. A total of five demonstration techniques were applied and demonstrated.

In the fifth category, ecological landscape and environmental restoration, one demonstration technique was applied and demonstrated.

The above research provides technical support for the establishment of planning and construction evaluation systems for green villages and towns.

1.4 Theoretical System of Green Village and Town Construction

Establishing a theoretical system for green village and town construction is an important foundation for meeting the needs of green and sustainable development in rural China.

1.4.1 Theoretical Research Foundation of Green Village and Town Construction

The theoretical research foundation of green village and town construction includes sustainable development, construction engineering, and new urbanization.

1. Sustainability development theory

Research on green village and town construction aims to achieve sustainable development and green development for villages and towns, and the theoretical system should be guided by sustainable development theory. Green development is an important component of sustainable development theory, and it focuses on how to use environmental protection as a new development method to achieve sustainable development under the constraints of ecological environmental capacity and resource-carrying capacity.

2. Construction engineering theory

The engineering research on green village and town construction focuses on the planning, construction and management of villages and towns and is an expansion and innovation of contemporary Chinese architectural engineering theory. Since the beginning of modern times, construction engineering theoretical research in China has relied on the introduction and absorption of Western architectural theories as a mainstay, and the foundation for rural construction theoretical research has been weak. Since the reform and opening up in 1978, China has gradually attached importance to researching and establishing an urban-rural architectural engineering theory system with Chinese characteristics. The generalized architecture theory established by Professor Wu Liangyong and the soft engineering design theory established by Professor Wang Guangyuan are outstanding representations of innovation.

1) Integrated architecture theory

Professor Wu Liangyong proposed the theory of integrated architecture in 1989 [4], expanding architecture from the traditional sense of the knowledge of building houses to the study of settlement, starting with the analysis of temporal and spatial views of humanity, nature, geography, technological economy, culture and art. He established the integrated architecture theoretical system by means of a multi-dimensional analytical method and pointed out the development direction of multi-domain comprehensive planning theory. This approach has been of enlightening significance in the establishment of the theoretical system of green village and town construction based on multi-dimensional construction theory.

2) Soft design in engineering theory

Professor Wang Guangyuan proposed the theory of soft design in engineering in 1992 [36], analysed the trend from "determinism" to "selection" in natural and technical sciences, and pointed out the softening tendency of traditional "hard" science and the shift from "soft" science, which previously could not be calculated, to "hardening". He believed that the active human participation and intervention in construction activities was the core of engineering choice theory and proposed the global optimization of civil and architectural engineering systems and the development direction of the theory of diversified construction engineering. This is significant in the establishment of a theoretical system of green village and town construction based on multi-disciplinary integration.

3. New urbanization theory

Research on green village and town construction is based on new urbanization and urban-rural integration development as the foundation and driving force and is closely connected with new urbanization development theory. China's National New Urbanization Plan (2014–2020) was the first to prioritize the promotion of urban-rural development integration in the national urbanization development plan and proposed three goals and tasks for the construction of a new socialist countryside: first, improving the planning and management of villages and towns; second, strengthening the construction of rural infrastructure and service networks; and third, accelerating the development of rural social undertakings. China's tremendous achievements in urbanization over the years have provided an important foundation for comprehensive research on and practice of green village and town construction.

1.4.2 Theoretical System Establishment of Green Village and Town Construction

Based on the analysis of the theoretical attributes and development of green village and town construction, the theoretical system consists of three levels.

1. Theoretical foundation of green village and town construction

(1) Sustainable development theory focuses on studying how to achieve sustainable utilization of village and town resources, sustainable growth of village and town economies, sustainable improvement of village and town environments, and sustainable progress of village and town societies.

(2) Construction engineering theory focuses on researching development planning theory of green villages and towns, environmental infrastructure construction and green building construction theory of villages and towns, and green management theory of village and town construction.

(3) New urbanization theory focuses on theories of urban and rural green and low-carbon overall development planning, the joint development of urban and rural environmental infrastructure construction, and collaborative management of urban and rural green community construction.

2. Engineering theory of green village and town construction

(1) The planning theory of green villages and towns includes the planning of village and town construction, the overall development of urban and rural areas, and new rural construction and development.
(2) The construction theory of green villages and towns includes engineering theory, environmental theory, and sociology theory of village and town construction.
(3) The management theory of green villages and towns includes the development pattern of village and town construction, construction project management patterns, and construction community of villages and towns.
(4) The evaluation theory of green villages and towns includes planning implementation evaluation of green villages and towns, environmental infrastructure construction evaluation of green villages and towns, and green building construction evaluation of villages and towns.

3. Application of the theory of green village and town construction

The key points of the application of the theory of green village and town construction are building suitable habitations for villages and towns to achieve a green lifestyle; building suitable employment facilities for villages and towns to promote green production; and building suitable tourism facilities for villages and towns to protect green ecology. The overall goal of the application of green village and town construction theory is to guide and promote the green and sustainable development of villages and towns in China.

The composition of the system of green village and town construction theory is shown in Fig. 1.4.

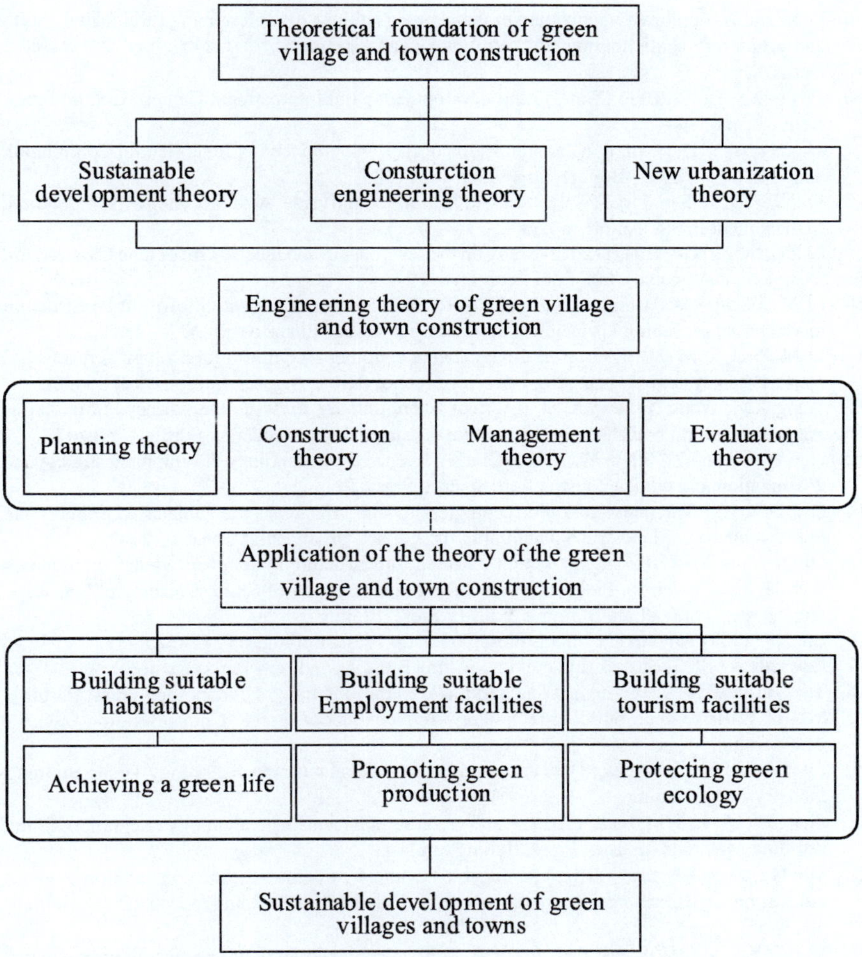

Fig. 1.4 Composition of the system of green village and town construction theory

References

1. Ministry of Housing and Urban-Rural Development, Ministry of Industry and Information Technology (2013) Notice of the Ministry of Housing and Urban-Rural Development of the Ministry of Industry and Information Technology on the development of green rural housing construction. http://www.mohurd.gov.cn/wjfb/201312/t20131225_216676.html. Accessed 18 Nov 2013
2. Jin ZS, Lu WG (2011) Planning of village and town. Southeast University Press, Nanjing, pp 288–289
3. Peng YG (2017) Landscape analysis of traditional village and town settlements, 2nd edn. China building Industry Press, Beijing, pp 1–2
4. Wu LY (2011) Generalized architecture. Tsinghua University Press, Beijing

5. Food and Agriculture Organization of the United Nations (2018) Realizing rural revitalization and promoting youth employment. http://www.fao.org/3/mx457zh/mx457zh.pdf. Accessed 5 Oct 2018
6. Zhang XS, Li Z (2011) China's rural development path. International Chinese Culture Press, Beijing, pp 85–94
7. Li XH (2012) The evolution and innovation of rural governance in China. Central Compilation and Translation Press, Beijing, pp 1–3, pp 185–188
8. Sun J (2013) Rural road—the road to a new countryside with no shortcuts. Economic Management Press, Beijing, pp 330–333
9. Li XK, Wang XM, Hua H (2018) Study on the relationship and mechanism of land use structure change on carbon emission. Ecol Econ 1:14–19
10. Li M, Wang XM, Wang X, et al (2010) Research on the optimization of rural construction model based on ecology principle. China's Popul Resour Environ (20):377–382
11. Li M, Wang XM (2010) Grey relational degree coupling model of rural ecological residential construction system. J Huazhong Univ Sci Technol (Urban Science Edition) 6(27):54–58
12. Wang XM, Wang X, Wu JX, et al (2016) A preliminary study on the partnership model of green agricultural housing construction. Real Estate China (Academic Edition) 7:36–42
13. Guo ZW, Wang Z, Zhang YL (2010) China's science and technology development strategy for urbanization and village construction. Science Press, Beijing
14. Li BD (2019) The construction of ecological livable villages as the subjects to promote the implementation of land space planning in rural areas. Urban Rural Const 17:26–27
15. Di Q, Wang YC (2014) Construction of green infrastructure system for livable communities in village and town. In: Proceedings of 2014 annual meeting of Chinese society of landscape architecture, vol I. China Building Industry Press, Beijing
16. Liu XJ (2009) Arcology. China Building Industry Press, Beijing, pp 317–320
17. Edwards B (2003) Sustainable building. China Building Industry Press, Beijing
18. Yuelson J (2014) Greening existing buildings. Harbin Institute of Technology Press, Harbin
19. Ma SZ (2016) Green building technology implementation guide. China Building Industry Press, Beijing
20. Lv HY (2018) Improvement and case study of traditional building technology. China Building Industry Press, Beijing
21. Zhao SY, Qiang WM, Fu SJ (2017) Rural green suitable housing technology integration. China Building Materials Industry Press, Beijing, pp 1–4
22. Mei HY, Zhou LJ, Yin Q (2015) Applicable technical guidelines for the construction of green village and town in severe cold areas of northeast China. China Building Industry Press, Beijing, pp 16–55
23. Yang ZY, Chen JD, Wang XM, et al (2013) Practical technology of seismic energy saving in rural building construction. Wuhan University of Technology Press, Wuhan
24. Zhang K, Wang XM, Wang ZZ (2012) Key technology of building energy saving transformation of existing farmhouses. Const Technol 6:82–83
25. Li Y, Wang XM, Cheng S (2009) Inheritance and innovation—the road of rural housing construction and development in China. Small Town Const 9:10–14
26. Yang F, Wang XM, Li M (2009) The participation mode of village and town eco-energy house construction and science and technology demonstration. Hubei Civil 1:62–64
27. Ministry of Science and Technology of the People's Republic of China (2018) Guidelines for 2018 annual project application of the key special project of "Technological Innovation of Green Livable Village and Town". http://www.most.gov.cn/xxgk/xinxifenlei/fdzdgknr/qtwj/qtwj2018/201807/t20180706_140465.html. Accessed 6 July 2018
28. Su X (2017) Environmental index system and evaluation criteria of green ecological village and town. Tongji University Press, Shanghai, pp 147–178
29. Wu JG (2007) Landscape ecology—pattern, process, scale and hierarchy. Higher Education Press, Beijing
30. Hua H, Li W, Wang XM, et al (2017) Comparative study on energy consumption measurement and energy saving efficiency of adobe building. J Civil Eng Manag 6:17–22

31. Hua H, Wang XM, Peng WJ (2012) Village low-carbon construction and carbon emission evaluation. J Civil Eng Manag 1:20–24
32. Yang J, Wang XM, Hua H (2011) Benefit analysis of village house greening. Hydropower Energy Sci 8:203–206
33. Yang J, Wang XM, Wang Y (2011) Economic analysis of the durability of village houses. J Civil Eng Manag 2:82–86
34. Cen QT, Wang XM, Yang XJ (2012) Evaluation on the sustainability of eco-village infrastructure construction. Const Technol 6:88–89
35. Wang Y, Wang XM, Zhu HP (2010) Study on integration and benefit evaluation of rural low-carbon construction technology in hot summer and cold winter areas. J Civil Eng 43(S2):410–415
36. Wang GY (1992) Engineering soft design theory. Science Press, Beijing, pp i–iii

Chapter 2
Evolution and Development of Green Village and Town Construction

Abstract Green village and town construction in China is a new rural construction mode produced over the long-term process of evolution that requires a systematic understanding of the evolutionary process and developmental characteristics of village and town construction in China. This chapter discusses the origin and evolution of village and town construction in China, analyses the organizational system formation and the developmental stage of village and town construction in contemporary China, and studies the influence of rural development and change on green village and town construction by collecting and analyzing 20 items of construction and development statistics during the 40 years from 1978 to 2017 about Chinese village and town.

Planning and green village and town construction should not only inherit and carry forward the traditions of village and town construction in China but also combine modern rural development goals and the actual needs of village and town construction and adapt to the profound impact of developmental demands and rural social and economic changes in different regions on green village and town construction.

2.1 Origin and Evolution of Village and Town Construction in China

The evolutionary history of village and town construction in China can be generally divided into three periods: ancient origin, modern evolution and contemporary development.

X. Wang and H. Hua, *Green Village and Town Construction in China*,
https://doi.org/10.1007/978-981-16-2098-0_2

2.1.1 Origin of Village and Town Construction in Ancient Times

1. Birth of villages and towns

In the agricultural society, agricultural production was the basis for human survival and development, and land was the basic resource for agricultural production. Farmers from one generation to the next engaged in agricultural production activities on the land on which they lived, settled nearby for convenience in farming and reproduction, and thus built and formed numerous primitive villages.

In the feudal society, with the development of socially productive forces, to adapt to trade in surplus agricultural products and the increased demand for production tools and daily necessities, market trade gradually came into existence, and the construction of "market", "street" and "town" then evolved into the forms of "village" and "town" settlement.

2. Origin of the village

According to Chinese historical records, the original character for village, "*Cun*", appeared in the *Qin* and *Han* dynasties. Before the middle of the *Eastern Han* dynasty, there were many early forms, such as "*Lu*", "*Qiu*" and "*Ju*". (These are all terms for small-scale settlements in ancient China.)

Villages, also known as rural settlements, are the earliest form of settlement units built by humans. The evolution of village construction in different historical periods reflects the development and evolution of residential life and labour production in various stages of human society, from gathering and hunting in the agrarian age to primitive agriculture, traditional agriculture, and industrialization.

3. Origin of the township

The original meaning of "town" is "the first mountain of a party", that is, the highest and largest mountain in a certain area. A military town was set up in a northwest frontier fortress from 422 to 484 A.D., serving as a local administrative institution to resist the invasion of the enemy.

In the early years of the *Northern Song* dynasty, towns gradually formed as certain areas of jurisdiction and population, with regular trading markets and fixed streets and shops. These areas were gradually divided into streets and lanes. After the *Ming* dynasty, with the continuous development of local handicraft industries and commercial trade, the town's functions as the center of rural commercial trade and a commodity distribution center gradually improved.

A town that functions as a commercial market is also called a "*Shizhen*", a place where farmers in rural areas exchange necessities and tools of production. "*Shi*" is the most concentrated construction in the city and thus is also known as "*Dushi*". (These words are different but have the same meaning.) The *Song* dynasty generally refers to small commercial markets below the county level. During the *Ming* and

Qing dynasties, China's urban construction and development entered their heyday [1]. The core function of town construction is that of a market [2], meaning a public place for buyers and sellers to gather for commodity trading at a fixed place and time. The town has a certain scale of permanent population, with fixed shops over a long term. A town usually has a market, but a market does not have to be in a town.

By the *Qing* dynasty, the administrative functions of the town had begun to weaken, and the county government usually set up administrative units, such as "*Jia*", "*Li*", "*Tu*", "*Fang*" and "*Xiang*", according to the needs to levy taxes and exact corvée. (*Jia, Li, Tu, Fang* and *Xiang* are forms of small-scale settlements in ancient China, similar to villages.) By the end of the *Qing* dynasty, towns had begun to take shape as units of the basic administrative system. In 1909, the regulations on *Local Autonomy of Towns and Townships* [3] promulgated by the *Qing* government, based on the principle of the division of the urban and rural systems, stipulated that government offices, prefectures and counties should be established as cities, while villages and towns outside cities should be established as towns if they had a population of over 50,000 and otherwise as townships.

2.1.2 Evolution of Village and Town Construction in Modern Times

The division of China's modern history began with the Opium War in 1840 and ended with the founding of the People's Republic of China in 1949. The construction of villages and towns in modern China mainly underwent three stages of evolution.

1. Period of turbulent and declining rural construction (1840–1919)

During the feudal society, for thousands of years, the construction of villages and towns in China experienced steady and gradual development consistent with rural self-sufficiency and social and economic development. However, after the Opium War launched by Britain against China from 1840 to 1842, the construction of villages and towns in China entered a period of turmoil and decline, and the serious poverty and backwardness of villages became major problems affecting the survival of the country and the development of the nation.

The 1911 Revolution in China overthrew feudal rule and established the Republic of China (1912–1949) with the capital in Nanjing. Faced with a huge gap between rural construction and the development of Western developed countries in the same period, a group of insightful people began to ponder and explore ways to build and rejuvenate villages and towns that were suitable for China's national conditions. Rural construction in China started the difficult process of transforming from the traditional closed type to the transformation and rejuvenation type.

2. Period of the rise of the rural construction movement (1920–1930)

The rural construction movement of the Republic of China from 1920 to 1930 was an important enlightenment period for the reform and innovation of rural construction in modern China. By the early 1930s, more than 600 institutions had participated in rural educational experiments, and more than 1,000 experimental areas had been established [4]. At this stage, the rural construction movement sprang up in various places, with the rural reconstruction experiment initiated by Mr. Yan Yangchu, the rural construction theory put forward by Mr. Liang Shuming, and the rural construction demonstration implemented by Mr. Lu Zuofu as typical representations. During this period, the Communist Party of China first put forward the programme of rural land system reform, and the national government in Nanjing initially constructed the system of township governance. Since its founding in 1921, the Communist Party of China has been exploring and solving the basic problem of equitable development in urban and rural areas through land system reform.

Yan Yangchu was a famous educationist and sociologist. He believed that the realization of the mission of "national reconstruction" and the creation of the conditions of "rural construction" required a stage of rural education [5]. The rural reconstruction experiment was carried out in Ding County, Hebei Province.

Liang Shuming was long engaged in theory and experimentation on rural construction. He carried out an experiment in village governance in Henan Province in 1928 and then went to Zouping County in Shandong Province to carry out a rural construction campaign. He wrote *The Theory of Rural Construction*, *The General Idea of Rural Construction* and other books. In 1930, he founded the Beijing Institute of Rural Construction, established the Research Department of Rural Construction and the Training Department of Rural Service Personnel, and cultivated various rural construction talents.

Lu Zuofu was a famous industrialist and pioneer of early modernization in China. He devoted much effort to creating the Beibei experimental area in Sichuan Province for the demonstration of rural construction [6]. His main contributions were improving the rural ecological environment, vigorously promoting various rural economic undertakings, and establishing rural culture and social public services.

During this period, the Communist Party of China (CPC) first put forward the programme of rural land system reform. As early as 1924, the CPC put forward *The Restrictions on Urban Rent Increase and Construction of Housing for Working Civilians*. In the Resolution on Land in 1927, the CPC explicitly proposed the immediate and free confiscation of the land and property of the nobility and landlord class; the confiscated land would be disposed of by the peasant representative conference (Soviet Union) and allocated to peasants who had no land or little land to use.

In June 1929, the government in Nanjing issued the revised *County Organization Law*, followed by the *Law on the Implementation of Township Autonomy* and other regulations. *The County Organization Law* clearly divided counties into the following four levels: district, township (town), "*Lv*" and "*Lin*". Each county was divided into 4 to 10 districts. Each district had jurisdiction over 10 to 50 townships. A village with more than 100 households was called a township, and a market community with

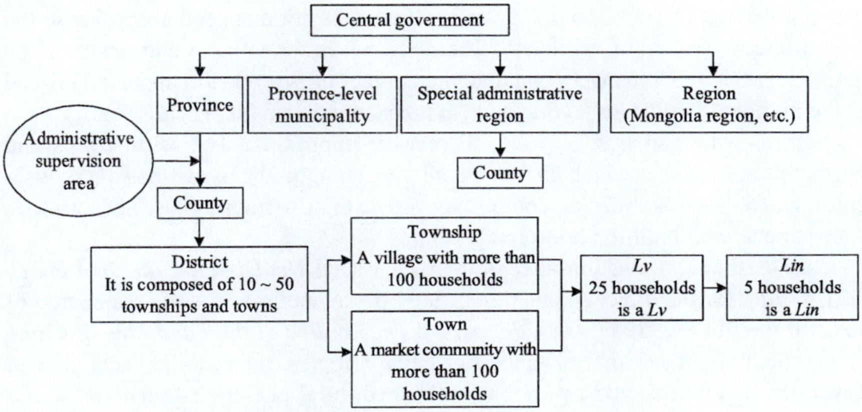

Fig. 2.1 Administrative division and township management systems of China in 1929

more than 100 households was called a town. Additionally, the law stipulated that a township with 25 households was a "*Lv*", and a township with 5 households was a "*Lin*" [7]. ("*Lv*" and "*Lin*" are forms of small-scale settlements.)

The administrative division and township management systems of the Republic of China in 1929 are shown in Fig. 2.1.

3. Period of slow development of rural construction (1931–1949)

From the outbreak of the War of Resistance Against Japanese Aggression on September 18, 1931, to the surrender announced by Japan on August 15, 1945, the War of Liberation lasted for more than three years until the founding of the People's Republic of China in 1949. During this period, the construction of villages and towns in China was in a stage of slow development or even stagnation caused by war.

The outstanding representative of the field of rural construction research during this period was Fei Xiaotong. As one of the earliest Chinese scholars to study the theory of rural economy and the development of small towns, he travelled to Zouping County, Shandong Province, in 1933 to participate in rural construction. In 1936, he went to Kaixungong Village, Suzhou City, Jiangsu Province, to conduct research. In 1938, he completed his doctoral dissertation [8], named *Peasant Life in China*, at the London School of Economics and Political Science [9]. He investigated the planning, residential area, transportation, environment, public facilities, industrial layout and landownership of Kaixungong Village and performed an in-depth analysis of the relationships among households, neighbourhood production cooperatives, social groups and the village government that is still of reference value in present-day research on the construction of villages and towns.

In 1934, the national government in Nanjing introduced the "*Bao-jia*" system in all provinces. This system defined 10 households as a "*Jia*", 10 "*Jia*" as a "*Bao*", and 10 "*Bao*" as a township and assigned governors for each "*Bao*" and "*Jia*". *The Organizational Outline of Counties at All Levels* [10] issued in 1939 specified the composition of the "*Bao-jia*" system, which was not managed according to the organizational system of "villages". Therefore, when the villages and towns set up in the Republic of China were counted, the number of *Bao* and *Jia* appeared instead of the number of "villages". (*Bao* and *Jia* are also units of small settlements.)

According to statistics, in 1946, there were approximately 54,156 towns and townships in China, of which 631,587 were *Bao*, composed of more than 100 households, and 6,499,670 were *Jia*, composed of more than 10 households. The total was approximately 650 million households [11].

In 1946, the Central Committee of the CPC issued *The Directive on Land Policy* [12], calling for the direct acquisition of land from landlords and the realization of land for the tiller. In 1947, the CPC passed *The Outline of the Land Law of China* [12], which stipulated that all ancestral halls, temples, monasteries, schools and government bodies should be abolished. Instead, rural peasant associations would take over the livestock, farming tools, houses, grain and other property of landlords and give them to peasants and other poor people who lacked such property. The government under the leadership of the CPC had already implemented land reform in the liberated areas for more than 100 million peasants before 1949 [13], thus driving the gradual recovery and development of rural construction.

2.2 Development Stage of the Organizational System and Construction of Villages and Towns in China

After the founding of the People's Republic of China in 1949, the construction of villages and towns in China entered a period of recovery and development. The establishment of villages and towns underwent seventy years of evolution along with the development of rural areas and agriculture, forming the social development foundation for the current green village and town construction in China.

2.2.1 Formation and Evolution of the Administrative System of Villages and Towns in China

As the basic administrative level in rural areas, towns are the political, economic and cultural centres of the smallest administrative regions. Towns are usually the seats of county organs or residential areas with more industry and commerce in rural areas. The village is the settlement point where peasants live and engage in all kinds of production.

In 1954, the *Constitution of the People's Republic of China* was promulgated and stipulated that "counties and autonomous counties are divided into townships, national townships and towns" to clarify the administrative status of towns. In 1955, the State Council promulgated *The Decision on the Establishment of City and Town*

Organizations, which further stipulated that the town is the concentration of industry, commerce and handicraft production.

From 1954 to 1958, the state started pilot agricultural production primary cooperatives in rural areas and initiated the collectivization of mutual aid groups based on villages. The agricultural cooperative movement launched in 1956 upgraded all primary rural cooperatives into advanced cooperatives, gradually forming a basic system of rural construction in which public buildings were built uniformly by villages and towns and collectively owned, and farmers built their own houses on allocated homesteads. This system has been in use ever since. In 1958, the original high-level cooperatives were transformed into people's communes in rural areas across the country.

From 1959 to 1961, China's village and town construction was in a period of stagnation. In 1963, the Central Committee of the CPC and the State Council issued *The Instructions on Adjusting Urban Structure and Shrinking Suburbs*, calling for the compression and adjustment of the urban structure.

From 1966 to 1977, China experienced a period of transformation of villages and towns. People's communes and production brigades generally set up revolutionary committees to directly administer rural areas and the construction of villages and towns. Due to the state's implementation of the planned economy and the development policy of "first produce, then live" and "first urban and then rural", the development gap between rural and urban construction expanded.

In 1978, China entered a new era of reform, opening up and development. In 1979, the Fifth National People's Congress passed *The Resolution on Amending Several Decisions Concerning the Constitution of the People's Republic of China*, in which local revolutionary committees were replaced by local people's governments. This change gradually restored the vigour of the rural construction management system.

In 1984, the State Council issued *The Report on Adjusting the Standards for the Establishment of towns*, relaxed the standards for the establishment of towns in accordance with the principle of actively developing small towns and implemented the system of village and town management.

The report stipulated that all county-level state organs should establish an administrative town: townships could be established with a total population of less than 20,000 and a resident non-agricultural population of more than 2,000. In a township with a total population of more than 20,000, the resident non-agricultural population of the township government could account for more than 10% of the township population. Towns could also be established in areas inhabited by ethnic minorities, sparsely populated remote areas, mountainous areas, small-scale mining areas, small ports, scenic spots, border ports, etc. with a non-agricultural population of less than 2,000.

In 1987, the Standing Committee of the Sixth National People's Congress passed *The Organic Law of Villagers' Committees.* It stipulated that the administrative village was the management scope of villagers' self-governance carried out by village committees and was the mass autonomous unit at the grassroots level in rural China.

In 1993, the State Council issued *Regulations on the Administration of Village and Town Planning and Construction*, which pointed out that "village" referred to rural villagers living and engaging in production in a variety of settlements. "Town" referred to the location of people's governments of townships and national townships and was confirmed by the people's government at the county level by the market development of certain areas as rural economy, culture and life service centres.

In the 1990s, the state began to fully implement the small-town construction strategy as an important strategic measure to realize the coordinated development of urban and rural areas, solve the problems of agriculture, rural areas and farmers, accelerate rural economic development, increase farmers' income, improve the rural consumption environment, improve the quality of rural and township construction, narrow the gap between urban and rural areas, and realize the coordinated development of urban and rural social economies.

Since 2005, China has entered a new period of building new countryside and rejuvenating rural areas. Rich theoretical innovations and practical achievements have been made in the management mode and innovation mechanism of village and town construction, and green village and town construction has developed simultaneously.

2.2.2 Development Stage of Village and Town Construction in China

Since the founding of the People's Republic of China, the construction and development of villages and towns can be divided into four stages.

1. Period of land reform to the systematic transformation of villages and towns (1949–1977)

This stage covers the period of land reform in rural areas, the period of rural collectivization and the period of systematic transformation of villages and towns.

1) Period of land reform in rural areas (1949–1953)

During this period, the state initially established property and landownership in village and town construction. From the promulgation of *The Land Reform Law of the People's Republic of China* in 1950 to the basic completion of the land reform in 1952, landownership by shrines, temples, monasteries, schools and government organizations was transferred to rural collectives, and landlords' houses were distributed to poor farmers. These changes improved the living conditions of farmers and laid the foundation for the establishment of a system of collective ownership of public buildings in villages and towns and a distribution system of homesteads.

2) Period of rural collectivization (1954–1965)

During this period, the construction of villages and towns was macroscopically guided by the planned economy. Farmers built their own houses and collectively constructed public buildings, aiming to meet the basic needs of their production and daily lives. Landownership was transformed from peasant private ownership to collective ownership. In 1958, people's communes were established in the countryside, replacing the original functions of the rural system. The construction of villages and towns in the early 1960s was basically maintained.

3) Period of systematic transformation of villages and towns (1966–1977)

During this period, people's communes and brigades in rural areas throughout the country generally established revolutionary committees to directly govern the countryside and manage the construction of villages and towns, which led to the weakening of the construction management functions of villages and towns. It was not until the fifth National People's Congress in 1979 that local revolutionary committees were replaced by local people's governments at all levels and the management system for village and town construction was gradually improved and perfected.

2. Period of economic reform and opening up and small-town construction (1978–2004)

The main characteristics of this stage were the rural economic system reform and the construction of small towns to promote the development of rural construction.

1) Reform of the rural economic system promoted the rapid development of rural construction

In 1978, China entered a new period of reform and opening up, and the construction of villages and towns also encountered unprecedented new development opportunities. In 1979, the State Construction Commission, the State Agricultural Commission, the Ministry of Agriculture, the Ministry of Building Materials and the State Administration of Construction Industry jointly held the first national conference on rural housing construction in Qingdao City, Shandong Province, marking the beginning of the transformation of rural construction from traditional spontaneous construction to modern cooperative construction [14]. With the continuous development of rural economic collectives and the increase in farmers' incomes, enthusiasm for the construction of villages and towns across the country was unprecedentedly high, making up for the serious lag in the construction and development of villages and towns.

Reform practices had proved that the development of the rural economic industry was not only the development foundation but also the development goal of green village and town construction.

2) Construction of small towns led to the constant improvement of the level of rural construction

The construction and development strategy of small towns implemented by the state in the 1990s was a significant innovation stage in the construction and development mode of villages and towns. In 1993, the Ministry of Construction held the first national conference on the construction of villages and towns in Suzhou City, Jiangsu

Province, defining the goal of taking the construction of small towns as the center, driving the construction of villages and towns and promoting the comprehensive development of the rural economy.

In 2000, the central committee of the CPC and the State Council issued *Several Opinions on Promoting the Healthy Development of Small Towns*, which pointed out the direction for the development of small towns. In 2004, the Ministry of Housing and Urban-Rural Development began to promote the construction demonstration of national key towns, famous towns with distinctive landscapes, famous towns with historical culture and green and low-carbon key small towns and put forward construction requirements for green small towns.

During this period, theoretical research and practices of small-town construction were carried out throughout the country, laying a solid foundation for the formation of the theoretical and technological system of green village and town construction with Chinese characteristics. This was also the budding period of green village and town construction development.

3. Period of new countryside and beautiful countryside construction (2005–2016)

The strategic decision on the construction of new countryside, formally proposed during the fifth plenary session of the 16th Central Committee of the CPC in 2005, and *Guidelines for the Construction of Beautiful Villages*, issued in 2015, were the important landmarks of this stage.

1) Urbanization theory and village construction demonstration led new rural construction

The theory and technology of modern village and town construction in China originated from the development of urbanization. In 2005, the state first included urbanization and urban development as a complete field in *The Outline of the National Plan for Medium and Long-term Scientific and Technological Development (2006–2020)*. Afterward, the Ministry of Science and Technology, the Ministry of Housing and Urban-Rural Development, the former Ministry of Land and Resources, the State-owned Assets Supervision and Administration Commission, the Ministry of Education, Chinese Academy of Sciences and other departments organized a group of pioneering science and technology research projects and demonstration village construction engineering [15]. A total of 524 national universities, scientific research units and related enterprises and 5078 researchers took part in the great project of science and technology. Such a large-scale comprehensive study of the development history of village and town construction was previously rare worldwide.

In 2014, the Central Committee of the CPC and the State Council promulgated *The National Plan for New Urbanization (2014–2020)*, which promoted the integration of urban and rural development as a national strategy for rural development and put forward clear goals and key tasks for building a new socialist countryside: first, improve the planning and management level of villages and towns; second, strengthen rural infrastructure and service network construction; and third, accelerate the development of social programmes in rural areas.

2) Construction of beautiful villages promoted the transformation of villages and towns through green development

The Guideline for Building Beautiful Villages, a national standard issued in 2015, pointed out that the essence of a beautiful village is to build a liveable and business-friendly countryside with sustainable development. It stipulated the requirements for village planning and construction, ecological environment, economic development, public services, rural customs and civilization, primary organizations and long-term management of beautiful rural areas. In addition, it outlined the development direction of green village construction.

In 2016, the State Council issued several opinions on deepening the supply-side structural reform of agriculture and accelerating the cultivation of new drivers of agricultural and rural development, further pointing out that agricultural and rural development should shift towards the pursuit of green ecology and sustainability.

4. Period of implementation of strategic plan for rural revitalization (2017–)

This phase started in 2017. The strategy for rural revitalization proposed at the 19th CPC National Congress in 2017 and *Strategic Plan for Rural Revitalization (2018–2022)* issued by the Central Committee of the CPC and the State Council in 2018 were important landmarks.

The strategic plan for rural revitalization clearly put forward the key tasks of "promoting green development", advocated a simple and moderate green and low-carbon lifestyle, and carried out actions such as creating green families, green schools, green communities and green travel. The strategic planning required that people respect, comply with and protect nature; coordinate the management of mountains, rivers, forests, fields, lakes and grasslands; accelerate the implementation of a green development model for rural areas; promote the revitalization of rural ecology; and build beautiful, ecologically liveable villages with clean and attractive living environments, stable and healthy ecosystems, and harmonious coexistence between humans and nature.

The *Three-Year Action Plan for Rural Living Environment Improvement* issued by the General Office of the Central Committee of the CPC and the General Office of the State Council in 2018 proposed six key construction tasks:

(1) Promote the treatment of rural household garbage;
(2) Carry out toilet manure pollution control;
(3) Establish progressive treatment of rural domestic sewage;
(4) Improve the appearance of the village and build a green ecological village;
(5) Strengthen the management of village planning; and
(6) Improve construction, management and protection mechanisms.

As discussed above, modern Chinese rural construction from 1920 to 2020, a period of one hundred years, experienced six major evolutionary stages: the rural construction movement, the slow development of rural areas, land reform and the systematic

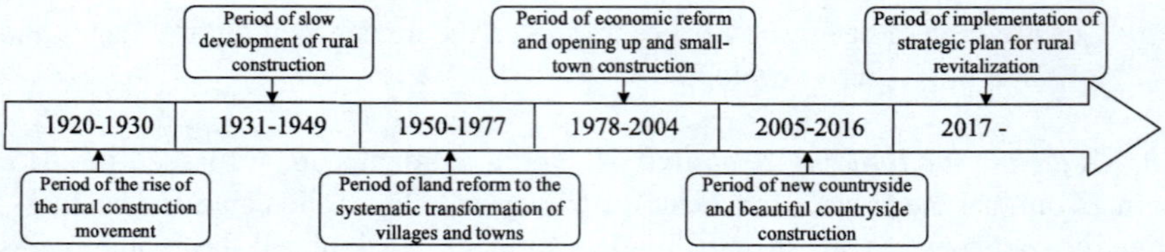

Fig. 2.2 Century of evolutionary and developmental processes of China's village and town construction

transformation period of village and town construction, reform and opening up, small-town construction, the construction of the new countryside and beautiful countryside, and rural revitalization strategy planning and implementation.

The century-old evolutionary and developmental processes of rural construction in China are shown in Fig. 2.2.

2.3 Change Influences of Rural Development on Green Village and Town Construction

The changes and development of rural society and the economy have had many influences on the development mode selection and planning content of green village and town construction. The main influencing factors are changes in the size and quantity of villages and towns, changes in the rural population, changes in construction land areas, changes in construction methods in villages and towns, and changes in willingness to build capacity.

2.3.1 Influences of Changes in the Size and Quantity of Villages and Towns

The construction and development of green villages and towns has been carried out under the current establishment and management mode of village and town organizations, which need to adapt to changes in village and town organizational scales and the influence of development.

1. Influences of changes in the number of towns

According to incomplete statistics, when the People's Republic of China was founded in 1949, there were approximately 2,500 towns in the country, and this number increased to 5,402 in 1953. By 1956, it had been reduced to 3,672 and by 1966 to 2,800. In 1978, the number of towns in China decreased to 2,176; afterward, it increased from year to year.

In 1983, there were 2,968 towns and 35,514 townships, and the ratio of the number of towns to the number of townships (referred to as the ratio of towns to townships) was 1:11.97, indicating that the development of urbanization at that time was at a low level. After the reform and opening up and the construction and development of small towns from 1978 to 2004, the number of townships increased to 19,883 in 2004. Additionally, the number of townships then gradually decreased to 17,451 due to revoking townships and setting up towns. The ratio of towns and townships changed to 1:0.88, and the number of towns exceeded the number of townships.

After the construction and development of new rural areas from 2005 to 2017, by 2017, the number of towns in China was 21,116, the number of townships was reduced to 10,529, and the ratio of towns and townships was adjusted to 1:0.50, indicating that the scale of villages and towns was constantly being optimized, which also provided opportunities for high-quality green village and town construction.

The number of towns and townships in China from 1978 to 2017 is shown in Table 2.1, and the changing trend analysis of the number of towns and townships in China is shown in Fig. 2.3.

2. Influences of changes in the number of villages

The focus of green village and town construction is the village. The number of villages set up in China from 1990 to 2017 is shown in Table 2.2.

As shown in Table 2.2, the number of villages in China has been decreasing from year to year, from 3.773 million in 1990 to 2.449 million in 2017. Over the past 28 years, the number of villages has decreased by 1.324 million, an average annual decrease of 49,000, with an average annual decrease rate of 1.30%, in line with the trend of actual rural population decline during the same period. Long-term planning for the construction and development of green villages should adapt to the changing trend in the number of villages.

3. Influences of changes in the number of village committees and rural households

Green village and town construction is organized and participated in mainly by village committees. The village committee is the most basic autonomous organization in rural areas under the jurisdiction of townships and towns. The number of village committees set up and the number of rural households in China from 1978 to 2017 can be obtained from the national data network, *China Social Statistics Yearbook* [16] and *China Agricultural Yearbook* [17].

The establishment of village committees and the number of rural households in China from 1978 to 2017 are shown in Table 2.3.

As shown in Table 2.3, with the advancement of urbanization and a large number of farmers entering cities to become urban residents, the number of village committees decreased from 690,888 in 1978 to 554,218 in 2017, a decrease of 19.78%, while the number of rural households increased from 173.47 million in 1978 to 275.0533 million in 2014, an increase of 55.95%. On the one hand, this trend reflects the decrease in the average number of rural households; on the other hand, it reflects the increasing emphasis of farmers on retaining rural household registration and

Table 2.1 Number of towns and townships in China from 1978 to 2017 (Unit: pcs)

Year	Number of towns set up	Number of townships set up	Year	Number of towns set up	Number of townships set up
1978	2176	–	1998	19216	25712
1979	2361	–	1999	19756	24745
1980	–	–	2000	20312	23199
1981	2678	–	2001	20374	19341
1982	–	–	2002	20601	18639
1983	2968	35514	2003	20226	18064
1984	7186	85290	2004	19883	17451
1985	9140	82450	2005	19522	15951
1986	10718	61417	2006	19369	15306
1987	11103	58739	2007	19249	15120
1988	11481	45195	2008	19234	15067
1989	11873	44624	2009	19322	14848
1990	12084	44397	2010	19410	14571
1991	12455	42654	2011	19683	13587
1992	14539	33827	2012	19881	13281
1993	15805	32445	2013	20117	12812
1994	16702	31463	2014	20401	12282
1995	17532	29502	2015	20515	11315
1996	18171	27056	2016	20883	10872
1997	18925	25966	2017	21116	10529

Source National Bureau of Statistics of People's Republic of China (PRC), http://data.stats.gov.cn/easyquery.htm?cn=C01&zb=A0101&sj=2018
Note The "–" in the table indicates that these data are missing from the statistics

safeguarding their land rights and interests. Additionally, it provides usable resources for using idle lands and farmhouses to develop green rural industries in green villages and towns.

2.3.2 *Influences of Changes in Rural Population*

The planning and green village and town construction should be in line with the changing trend of the rural population and its demand for residential use. Since the reform and opening up in 1978, China's rural population and the rural proportion of the total population have been decreasing. Compared with urban construction and development and the rapid and continuous increase in population in the same period, the construction and development level of villages and towns in terms of planning

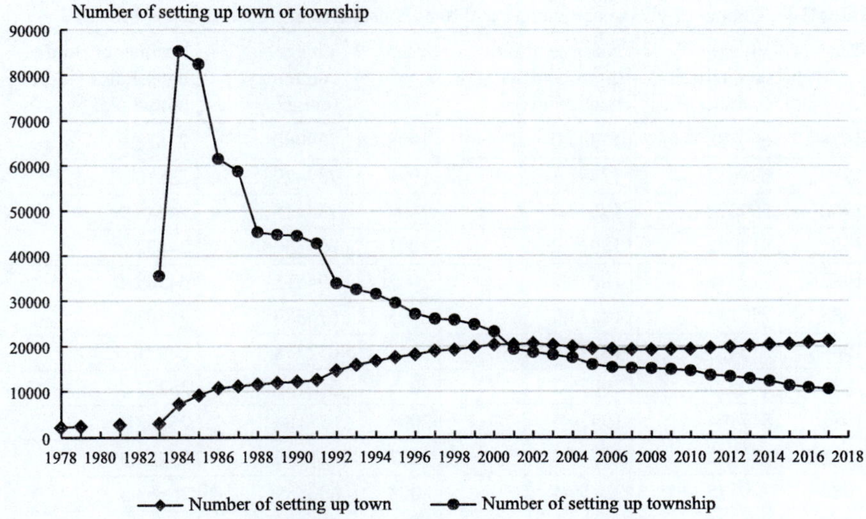

Fig. 2.3 Changing trends in town and township numbers in China from 1978 to 2017

Table 2.2 Number of villages in China from 1990 to 2017 (Unit: 10 thousand)

Year	Number of villages	Year	Number of villages	Year	Number of villages
1990	377.3	2000	353.7	2010	273.0
1991	376.2	2001	345.9	2011	266.9
1992	375.5	2002	339.6	2012	267.0
1993	372.1	2003	–	2013	265.0
1994	371.3	2004	320.7	2014	270.2
1995	369.5	2005	313.7	2015	264.5
1996	367.6	2006	270.9	2016	261.7
1997	365.9	2007	264.7	2017	244.9
1998	355.8	2008	266.6		
1999	359.0	2009	271.4		

Source Ministry of Housing and Urban-Rural Development of the PRC. China Urban-rural Construction Statistical Yearbook 2017
Note The "–" in the table indicates that these data are missing from the statistics

method, construction mode, investment channels, quality standards, operation and management, and many other aspects resulted in a widening gap between urban and rural areas.

From 1978 to 2017, the changes in the rural population and the total population of China are shown in Table 2.4. The trend analysis of the proportion of the rural population in the total population is shown in Fig. 2.4.

Table 2.3 Number of village committees and rural households in China from 1978 to 2017

Year	Village committees (number)	Number of rural households (10,000 households)	Year	Village committees (number)	Number of rural households (10,000 households)
1978	690388	17347.0	1998	739980	23678.0
1979	698613	17491.1	1999	737429	23810.5
1980	709820	17672.7	2000	734715	24148.7
1981	718022	18016.2	2001	709257	24432.2
1982	719438	18278.6	2002	694515	24569.4
1983	750141	18523.2	2003	678589	24793.1
1984	933485	18792.6	2004	652718	24971.4
1985	940617	19076.5	2005	640139	25222.6
1986	847894	19574.5	2006	631184	25268.4
1987	830302	20168.3	2007	621046	25434.9
1988	740375	20859.4	2008	603589	25663.5
1989	746432	21504.0	2009	599078	25975.7
1990	743278	22237.2	2010	594658	26384.6
1991	804153	22566.2	2011	589874	26607.0
1992	806032	22849.0	2012	588407	26802.3
1993	802352	22983.8	2013	588547	26948.5
1994	802052	23165.3	2014	585451	27053.3
1995	740150	23281.7	2015	580856	–
1996	740128	23437.7	2016	559186	–
1997	739447	23406.2	2017	554218	–

Source National Bureau of Statistics of the PRC, http://data.stats.gov.cn/easyquery.htm?cn=C01&zb=A0D01&sj=2018
Note The "–" in the table indicates that these data are missing from the statistics

Table 2.4 shows that in 1978, the total population of China was 962.59 million, and the rural population was 790.14 million, accounting for 82.08% of the total population. In 2017, the total population of China was 1390,080 million, while the rural population decreased to 576.61 million, accounting for 41.48% of the total population. From 1978 to 2017, the proportion of the rural population in the total population decreased by 1.02% annually.

How to adapt to the decreasing rural population and the changing trend of small family size while fully meeting the increasing demand of the majority of farmers for quality improvements in the new rural production and living ecosystem, thus continuously improving the living environment of farmers, are important factors affecting the planning and green village and town construction.

Table 2.4 Changes in rural population in China from 1978 to 2017

Year	National population (10 thousand)	Rural population (10 thousand)	Ratio of rural population to national population (%)	Year	National population (10 thousand)	Rural population (10 thousand)	Ratio of rural population to national population (%)
1978	96259	79014	82.08	1998	124761	83153	66.65
1979	97542	79047	81.04	1999	125786	82038	65.22
1980	98705	79565	80.61	2000	126743	80837	63.78
1981	100072	79901	79.84	2001	127627	79563	62.34
1982	101654	80174	78.87	2002	128453	78241	60.91
1983	103008	80734	78.38	2003	129227	76851	59.47
1984	104357	80340	76.99	2004	129988	75705	58.24
1985	105851	80757	76.29	2005	130756	74544	57.01
1986	107507	81141	75.48	2006	131448	73160	55.66
1987	109300	81626	74.68	2007	132129	71496	54.11
1988	111026	82365	74.19	2008	132802	70399	53.01
1989	112704	83164	73.79	2009	133450	68938	51.66
1990	114333	84138	73.59	2010	134091	67113	50.05
1991	115823	84620	73.06	2011	134735	65656	48.73
1992	117171	84996	72.54	2012	135404	64222	47.43
1993	118517	85344	72.01	2013	136072	62961	46.27
1994	119850	85681	71.49	2014	136782	61866	45.23
1995	121121	85947	70.96	2015	137462	60346	43.90
1996	122389	85085	69.52	2016	138271	58973	42.65
1997	123626	84177	68.09	2017	139008	57661	41.48

Source The National Bureau of Statistics of the People's Republic of China (PRC), http://data.stats.gov.cn/easyquery.htm?cn=C01&zb=A0301&sj=2018

2.3.3 Influences of Changes in Construction Land Area

The primary task of constructing green villages and towns is to save construction land resources and strictly protect arable land. The main factors that affect the land-use plan of green villages and towns are per capita cultivated land occupied area, built-up occupied area, homestead area of villager houses, public green areas, and road construction land areas.

1. Changes in cultivated land per capita

The analysis of the quantity and per capita possession of cultivated land in rural areas is an important basis for compiling and implementing the construction land planning

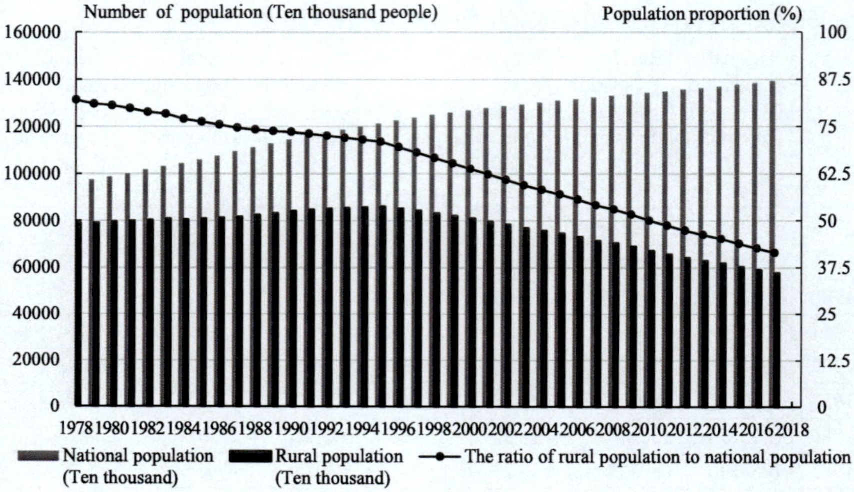

Fig. 2.4 Changing trend of rural population proportion of total population in China from 1978 to 2017

of green villages and towns. China is one of the countries with the largest number of farmers worldwide, but the per capita cultivated land area is far lower than the world average. Therefore, it has also implemented the strictest farmland protection system and construction land approval system.

Table 2.5 compares the per capita cultivated land area in China and in the world from 1978 to 2016, and Fig. 2.5 shows the trend analysis of the comparison between China and the world in from 1978 to 2016.

Table 2.5 and Fig. 2.5 show that the per capita cultivated land area of China fluctuated by approximately 0.1 ha over the past 39 years. After 2014, the total cultivated land area of China gradually increased, strictly maintaining the red line of cultivated land protection. In 2016, China's per capita cultivated land area was 0.086 hm^2, which was only 44.8% of the world's per capita cultivated land area of 0.192 hm^2 in the same period. Therefore, the basic principle of green village planning and construction is to strictly protect cultivated land and save construction land.

2. Changes in built-up areas of villages and towns

As a core area integrating rural life, non-agricultural production and public services, rural built-up areas are the main components of rural construction land. The scale of land use and the level of intensive land use in built-up areas have an important impact on the construction and development of green villages and towns.

See Table 2.6 for changes in the built-up areas of villages and towns in China from 1990 to 2017 and see Fig. 2.6 for the changing trend in the built-up areas of villages and towns in China from 1990 to 2017.

Table 2.6 and Fig. 2.6 show that the construction land of China's administrative villages and towns shows an overall increasing trend, indicating that farmers'

Table 2.5 Comparison of farmland and per capita possession of farmland between China and the world from 1978 to 2016

Year	National farmland (billion hm^2)	Chinese per capita farmland (hm^2/people)	World per capita farmland (hm^2/people)
1978	0.964	0.101	0.276
1979	0.965	0.100	0.272
1980	0.962	0.098	0.267
1981	0.968	0.097	0.262
1982	1.017	0.101	0.260
1983	1.073	0.105	0.256
1984	1.140	0.110	0.255
1985	1.201	0.114	0.254
1986	1.205	0.113	0.251
1987	1.202	0.111	0.247
1988	1.209	0.110	0.243
1989	1.222	0.109	0.239
1990	1.238	0.109	0.236
1991	1.249	0.109	0.232
1992	1.229	0.105	0.258
1993	1.213	0.103	0.253
1994	1.207	0.101	0.249
1995	1.196	0.099	0.245
1996	1.193	0.098	0.240
1997	1.195	0.097	0.237
1998	1.199	0.097	0.234
1999	1.193	0.095	0.231
2000	1.182	0.094	0.227
2001	1.161	0.091	0.224
2002	1.145	0.089	0.220
2003	1.117	0.087	0.218
2004	1.116	0.086	0.216
2005	1.123	0.086	0.214
2006	1.087	0.083	0.210
2007	1.085	0.082	0.207
2008	1.081	0.082	0.205
2009	1.077	0.081	0.202
2010	1.072	0.080	0.199
2011	1.065	0.079	0.200
2012	1.059	0.078	0.199

(continued)

Table 2.5 (continued)

Year	National farmland (billion hm^2)	Chinese per capita farmland (hm^2/people)	World per capita farmland (hm^2/people)
2013	1.057	0.078	0.197
2014	1.057	0.077	0.195
2015	1.190	0.087	0.194
2016	1.189	0.086	0.192

Source World Bank, https://data.worldbank.org.cn/indicator/AG.LND.ARBL.HA.PC?view=chart

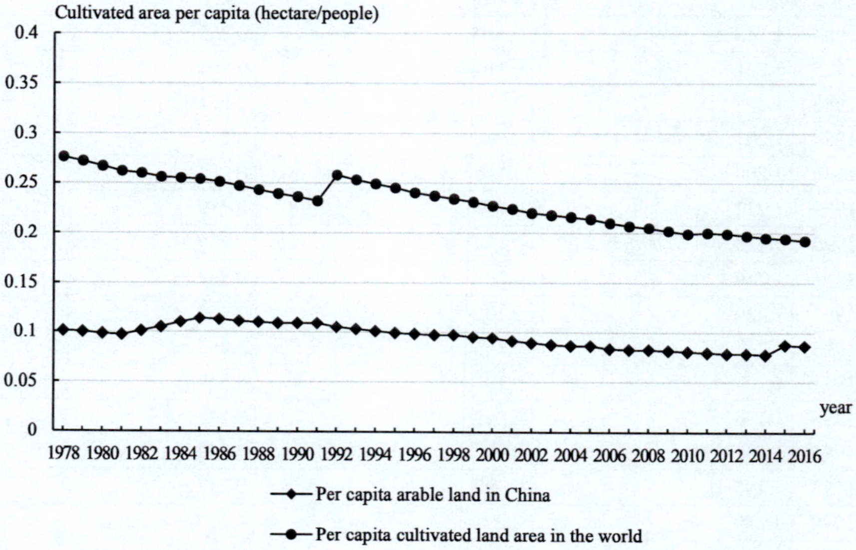

Fig. 2.5 Comparison of per capita arable land area between China and the world from 1978 to 2016

production and living construction activities are increasingly active and expanding in scale. The main reason for the decline in built-up rural areas is the gradual decrease in the number of townships caused by revoking townships and setting up towns. This also reflects that farmers have a more concentrated lifestyle and a more intensive distribution of rural resources, which is conducive to improving the agglomeration effect and comprehensive benefits of green rural environmental infrastructure and agricultural housing construction.

3. Changes in house-site areas in the countryside

It is an important task for green village and town construction to strictly control the standard of homestead construction. As the main part of the completed land of villages and towns, changes in the area also reflect changes in the scale of building houses in

Table 2.6 Changes in built-up areas of villages and towns in China from 1990 to 2017 (Unit: 10 thousand hm^2)

Year	Built-up area of established town	Built-up area of township	Area of village land
1990	82.51	110.06	1140.10
1991	87.05	109.35	1127.20
1992	97.49	98.12	1187.74
1993	111.87	99.86	1202.66
1994	118.81	101.19	1243.82
1995	138.63	103.71	1277.13
1996	143.73	95.20	1336.11
1997	155.32	95.71	1366.43
1998	162.97	93.70	1372.58
1999	167.53	92.56	1346.28
2000	181.98	90.66	1355.34
2001	197.15	79.73	1396.11
2002	203.24	79.14	1388.85
2003	–	–	–
2004	223.60	78.06	1362.71
2005	236.85	77.80	1404.20
2006	311.99	92.83	–
2007	284.33	75.89	1389.90
2008	301.60	81.15	1311.70
2009	313.10	75.76	1362.80
2010	317.90	75.12	1399.20
2011	338.60	74.19	1373.80
2012	371.40	79.55	1409.00
2013	369.00	73.69	1394.30
2014	379.50	72.23	1394.10
2015	390.80	70.00	1401.30
2016	397.00	67.30	1392.20
2017	392.60	63.38	–

Source Ministry of Housing and Urban-Rural Development of the PRC, China Urban-rural Construction Statistical Yearbook 2017
Note The "–" in the table indicates that these data are missing from the statistics

rural areas and the living standard of farmers. Song Wei et al. [18] calculated changes in rural homesteads in China from 1996 to 2005 and pointed out that the total area of rural homesteads in China increased from 9.0515 million hm^2 in 1996 to 9.1161 million hm^2 in 2005, an increase of approximately 0.71%. The homestead area per household decreased from 386.19 m^2 per household to 361.43 m^2 per household, a decrease of approximately 6.41%. The main reasons for this phenomenon are the

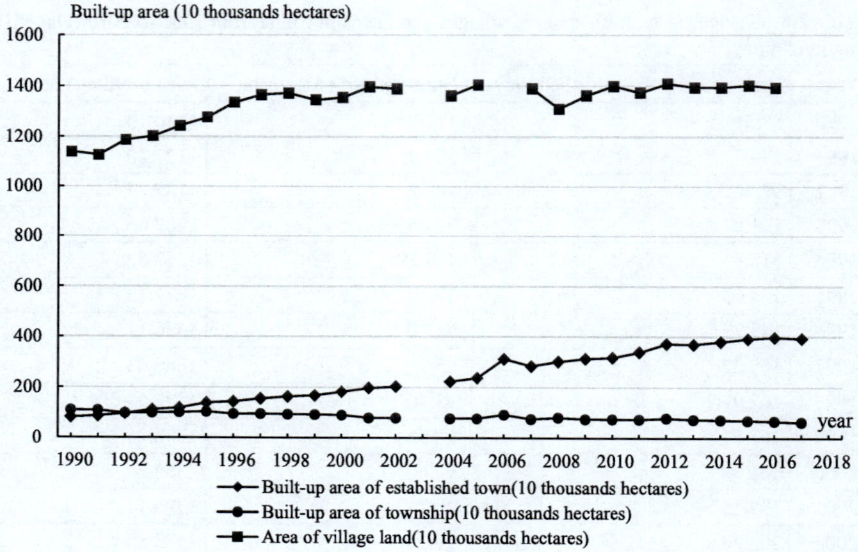

Fig. 2.6 Changing trend in built-up areas of villages and towns in China from 1990 to 2017

miniaturization of the scale of farmers and the increasingly built-up form of rural housing.

To properly manage the relationship between property rights and compensation for farmland, idle homesteads that farmers voluntarily withdraw from in accordance with the law are used for rural construction as a whole, which is also the development trend for saving construction land in green villages and towns.

4. Changes in public green areas in villages and towns

Green area construction plays an important role in protecting the rural ecological environment and improving the quality of life of farmers. It is an important part of the construction of green villages.

From 1990 to 2017, changes in green spaces in township parks (public) and per capita parks (public) in China are shown in Table 2.7.

As Table 2.7 shows, the area of public green space and per capita public green space in villages and towns decreased in 2006, which was caused by changes in statistical caliber. The public green space and per capita public green space in towns and townships in the two stages before and after 2006 showed an increasing trend, indicating that the rural greening effect was obvious and the ecological environment was improving steadily.

5. Changes in rural road construction land

Road construction is an important component of rural construction land and plays a leading and supporting role in the economic and social development of green villages

Table 2.7 Changes in the green area of parks (public) and per capita public green area of parks in townships in China from 1990 to 2017

Year	Green area of parks (public) of established town (10 thousand hm^2)	Green area per capita of parks (public) of established town (m^2/people)	Green area of parks (public) of township (10 thousand hm^2)	Green area per capita of parks (public) of township (m^2/people)
1990	0.8537	1.40	0.6353	0.88
1991	1.0634	1.62	0.8334	1.19
1992	1.2157	1.68	0.8704	1.32
1993	1.3700	1.74	0.9142	1.40
1994	1.7403	2.21	1.1087	1.76
1995	1.9957	2.15	1.0935	1.73
1996	2.2691	2.30	1.0783	1.79
1997	2.6078	2.50	1.1451	1.91
1998	2.9911	2.74	1.3099	2.22
1999	3.3236	2.86	1.3230	2.23
2000	3.7053	3.02	1.3531	2.33
2001	4.3895	3.38	1.3552	2.56
2002	4.8409	3.54	1.3135	2.54
2003	–	–	–	–
2004	6.0080	4.17	1.4137	2.57
2005	6.8100	4.60	1.3707	2.65
2006	3.30	2.40	0.29	0.85
2007	2.72	1.80	0.24	0.66
2008	3.09	1.90	0.26	0.72
2009	3.14	1.92	0.30	0.84
2010	3.36	2.00	0.31	0.88
2011	3.45	2.00	0.30	0.90
2012	3.73	2.10	0.32	0.95
2013	4.33	2.40	0.35	1.08
2014	4.48	2.40	0.34	1.07
2015	4.69	2.50	0.34	1.10
2016	4.79	2.50	0.33	1.11
2017	5.24	3.10	0.40	1.65

Source Ministry of Housing and Urban-Rural Development of the PRC, China Urban-rural Construction Statistical Yearbook 2017
Note The "–" in the table indicates that these data are missing from the statistics

and towns. Changes in rural road construction land can be analysed by changes in township road mileage. In 2017, the total length of rural roads in China reached 3.2540 million km, and the quality of road construction was improving.

Table 2.8 shows changes in road mileage in Chinese villages and towns from 1990 to 2017.

Table 2.8 Changes in the mileage of rural roads in China from 1990 to 2017 (Unit: 10 thousand km)

Year	Road length of established towns	Road length of townships	Road length of villages	Total length of rural roads
1990	7.72	15.20	262.08	285.00
1991	8.39	14.90	240.04	263.33
1992	9.65	14.22	262.86	286.73
1993	10.89	14.21	268.69	293.79
1994	12.55	14.03	263.18	289.76
1995	13.40	14.43	274.98	302.81
1996	15.48	14.36	279.32	309.16
1997	17.79	14.48	283.20	315.47
1998	18.68	14.27	290.33	323.28
1999	19.42	14.37	287.33	321.12
2000	21.00	13.73	287.02	321.75
2001	22.82	12.09	283.64	318.55
2002	24.32	12.10	287.34	323.76
2003	–	–	–	–
2004	27.55	12.63	285.14	325.32
2005	30.06	12.41	304.00	346.47
2006	26.00	7.00	221.90	254.90
2007	21.60	6.20	–	–
2008	23.40	6.40	–	–
2009	24.50	6.30	–	–
2010	25.80	6.60	–	–
2011	27.40	6.50	–	–
2012	29.10	6.70	–	–
2013	31.00	6.80	228.00	265.80
2014	32.70	7.00	234.10	273.80
2015	34.50	7.10	239.30	280.90
2016	35.90	7.30	246.30	289.50
2017	33.50	6.60	285.30	325.40

Source Ministry of Housing and Urban-Rural Development of PRC, China Urban-rural Construction Statistical Yearbook 2017

Note The "–" in the table indicates that these data are missing from the statistics

2.3.4 Change Influences in Construction Scale and Quality

As the scale of rural housing construction in China continues to expand, the construction quality and rural infrastructure gradually improve, the focus of rural housing construction in villages and towns is gradually shifting from the subsistence construction mode, which solves mainly the housing difficulties of farmers, to the construction mode of green villages and towns and beautiful liveable countryside.

1. Scale of rural housing construction is expanding

In 1957, China's rural housing construction area was 6.13 billion m^2, and the per capita housing area of farmers was approximately 11.3 m^2 [19]. For historical reasons, during the construction period from 1958 to 1977, the rural population continued to increase, but the housing construction speed was slow. In 1978, the per capita housing area of rural residents was only 8.1 m^2.

Since the reform and opening up in 1978, with the recovery and development of the rural economy, village and town construction in China has entered a period of large-scale and rapid development. By 2017, there was 29.282 billion m^2 of housing in rural areas, including 25.585 billion m^2 of residential buildings, 1.562 billion m^2 of public buildings, and 2.135 billion m^2 of production facility construction, accounting for 87.38%, 5.33%, and 7.29% of the total areas, respectively. The per capita floor space of rural residents increased to 46.7 m^2, 5.77 times that in 1978. In the same period, the per capita housing area of urban residents was only 36.9 m^2, lower than that of rural residents.

With the continuous increase of the existing rural housing construction stock, it has become an important content of green village and construction to carry out the green renovation of dilapidated rural housing and to implement the century-old rural housing construction system.

See Table 2.9 for the change in the per capita living area of rural residents in China from 1978 to 2017, and see Fig. 2.7 for the analysis of the change trend of the per capita living area of rural residents in China from 1978 to 2017.

2. Quality of rural housing construction has improved step by step

With the development of the rural economy and society and the improvement in farmers' living standards, the quality standards of farmers' houses have undergone a transformation from maintenance type to comfortable type to prosperous type, and the types of rural houses have also undergone a development and evolution from single-story houses with a civil structure to two-story buildings with a brick-and-wood structure to multi-story buildings with a reinforced concrete structure. By 2011, the per capita area of rural houses with reinforced concrete structures had surpassed that of brick-and-wood structures [18]. The construction of green and energy-saving buildings in line with rural characteristics has become the main content of green village and town construction.

From 1981 to 2012, the per capita area of houses with different structures in rural China is shown in Table 2.10.

Table 2.9 Changes in the per capita housing area of rural residents in China from 1978 to 2016

Year	Per capita housing area of rural residents (m^2/person)	Year	Per capita housing area of rural residents (m^2/person)	Year	Per capita housing area of rural residents (m^2/person)
1978	8.1	1992	18.9	2006	30.7
1979	8.4	1993	20.7	2007	31.6
1980	9.4	1994	20.2	2008	32.4
1981	10.2	1995	21.0	2009	33.6
1982	10.7	1996	21.7	2010	34.1
1983	11.6	1997	22.5	2011	36.2
1984	13.6	1998	23.3	2012	37.1
1985	14.7	1999	24.2	2013	–
1986	15.3	2000	24.8	2014	–
1987	16.0	2001	25.7	2015	43.9
1988	16.6	2002	26.5	2016	45.8
1989	17.2	2003	27.2	2017	46.7
1990	17.8	2004	27.9		
1991	18.5	2005	29.7		

Source National Bureau of Statistics, People's Republic of China, http://data.stats.gov.cn/easyquery.htm?cn=C01&zb=A0A0L02&sj=2018
Note "–" in the table indicates that these data are missing from the statistics

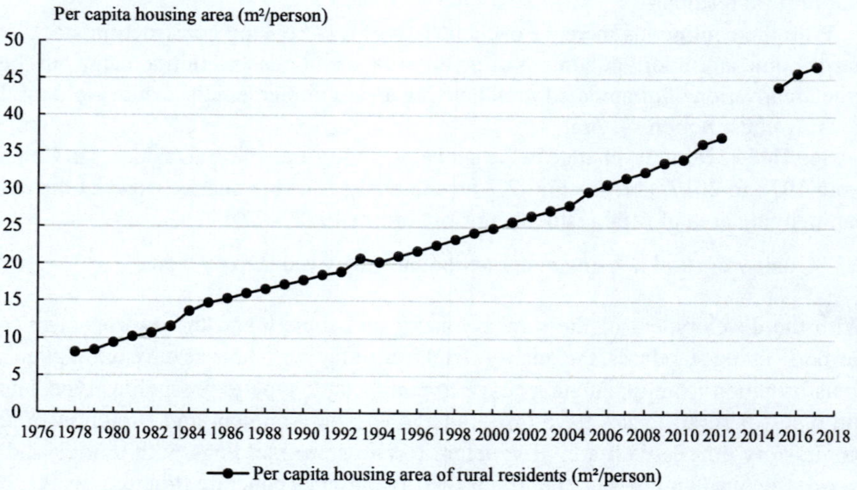

Fig. 2.7 Analysis of the trend of per capita housing area of rural residents in China from 1978 to 2017

Table 2.10 Per capita area of houses with different structures in rural China from 1981 to 2012 (Unit: m^2/person)

Year	Reinforced concrete structure of rural housing	Brick-and-wood structure of rural housing	Year	Reinforced concrete structure of rural housing	Brick-and-wood structure of rural housing
1981	0.1	4.9	1997	5.1	11.9
1982	0.1	5.2	1998	5.7	12.2
1983	0.3	6.2	1999	6.4	12.3
1984	0.2	6.8	2000	6.2	13.6
1985	0.3	7.5	2001	6.9	13.8
1986	0.4	8.2	2002	7.7	13.9
1987	0.6	8.6	2003	8.5	14.1
1988	0.7	9.0	2004	9.2	14.1
1989	0.9	–	2005	11.2	14.1
1990	1.2	9.8	2006	11.8	14.6
1991	1.6	10.4	2007	12.5	14.8
1992	1.8	10.7	2008	13.4	14.9
1993	2.3	11.8	2009	14.5	15.1
1994	2.7	11.5	2010	15.1	15.2
1995	3.1	11.9	2011	16.5	15.9
1996	4.4	11.7	2012	17.1	16.3

Source National Bureau of Statistics of the People's Republic of China, http://data.stats.gov.cn/easyquery.htm?cn=C01
Note "–" in the table indicates that these data are missing from the statistics

3. Rural infrastructures have gradually improved

Environmental infrastructure construction is the key content of green village construction. The National Bureau of Statistics conducted surveys and developed statistics on infrastructure construction and basic social services in 31,925 townships and 596,450 villages in six areas: transportation, energy and communications, environmental health, culture and education, medical and social welfare institutions, and market construction.

The statistical results of rural infrastructure construction and the basic social services census in China in 2016 are shown in Table 2.11.

Table 2.11 shows that compared with the past, village and town infrastructure construction in China has made remarkable progress, but there is still a large gap between rural and urban areas in terms of the functional quality and supporting capacity of environmental infrastructure construction, which is the key content for the planning and construction of green villages and towns.

Table 2.11 Statistics of rural infrastructure construction and basic social services in China in 2016

Number	Category	Area	Statistics and results
1	Traffic	Township	Townships with railway stations, wharves and expressways account for 8.6%, 7.7% and 21.5% of the total, respectively
		Village	61.9% of main roads in villages have street lamps. The distance from the village committee to the farthest natural village and residential settlement is mainly within 5 km
2	Energy and communications	Village	99.7% of villages have electricity, 11.9% have natural gas, 99.5% have telephones, 82.8% have cable TV, 89.9% have broadband internet, and 25.1% have e-commerce distribution sites
3	Environmental health	Township	91.3% of townships have a centralized or partially centralized water supply, and 90.8% of township household garbage collection is centralized or partially centralized
		Village	73.9% of village household garbage collection is centralized or partially centralized, 17.4% of village sewage is treated or partially treated centrally, and 53.5% of villages have completed or partially completed toilet improvements
4	Cultural education	Township	96.5% of townships have kindergartens and nurseries, 98.0% have primary schools, 96.8% have libraries and cultural centres, 11.9% have movie theatres, 16.6% have stadiums, and 70.6% have parks and recreation fitness plazas
		Village	32.3% of villages have kindergartens and nurseries, 59.2% have sports venues, and 41.3% have peasant amateur cultural organizations
5	Medical and social welfare institutions	Township	99.9% of townships have medical and health facilities, 98.4% have licenced (assistant) physicians, 66.8% have social welfare units, and 56.4% have homes for the elderly established by their governments

(continued)

Table 2.11 (continued)

Number	Category	Area	Statistics and results
		Village	81.9% of villages have clinics, and 54.9% have licenced (assistant) doctors
6	Market construction	Township	68.1% of townships have commodity markets, 39.4% have specialized markets for grain, oil, vegetables and fruits, 10.8% have specialized markets for livestock and poultry, and 4.3% have specialized markets based on aquatic products
		Village	47.5% of villages have general stores or supermarkets larger than 50 m^2, 4.9% have tourist reception services, and 30.0% have licenced restaurants

Source National Bureau of Statistics of the People's Republic of China, http://www.stats.gov.cn/tjsj/tjgb/nypcgb/qgnypcgb/201712/t20171215_1563589.html

2.3.5 Change Influences on Willingness to Build Capacity

Rural economic development provides economic support to promote the construction and development of green villages and towns. Farmers' income growth and willingness to build houses are the main sources of investment in green village and town construction as well as the important index basis for formulating construction and development plans for green villages and towns, carrying out construction projects and evaluating green village and town construction according to local conditions.

1. Changes in rural economic development level

Agriculture refers to planting, forestry, animal husbandry, fishery and other industries, including services directly related to pre-production, production and post-production. The main indicator of rural economic development is the total output value of agriculture. China's total agricultural output value grew continuously from 1978 to 2017, providing a reliable economic guarantee for the sustainable development of rural construction.

The growth of the total output value of agriculture, forestry, animal husbandry and fishery in China from 1978 to 2017 is shown in Table 2.12, and the growth trend is shown in Fig. 2.8.

2. Changes in investment and structure of village and town construction

With the continuous development of the rural economy, the construction and investment scales of villages and towns in China have also continued to increase. According to statistics, from the founding of the People's Republic of China to the end of 2016,

Table 2.12 Growth of total output value of agriculture, forestry, animal husbandry and fishery in China from 1978 to 2017 (Unit: 100 million yuan)

Year	Total output value (100 million yuan)	Increase over the previous year (%)	Year	Total output value (100 million yuan)	Increase over the previous year (%)	Year	Total output value (100 million yuan)	Increase over the previous year (%)
1978	1397.00	11.49	1992	9084.71	11.37	2006	40810.83	3.45
1979	1697.60	21.52	1993	10995.53	21.03	2007	48651.77	19.21
1980	1922.60	13.25	1994	15750.47	43.24	2008	57420.77	18.02
1981	2180.62	13.42	1995	20340.90	29.14	2009	59311.32	3.29
1982	2483.26	13.88	1996	22353.70	9.90	2010	67763.13	14.25
1983	2750.00	10.74	1997	23788.40	6.42	2011	78836.98	16.34
1984	3214.13	16.88	1998	24541.90	3.17	2012	86342.15	9.52
1985	3619.50	12.61	1999	24519.10	−0.09	2013	93173.70	7.91
1986	4013.01	10.87	2000	24915.77	1.62	2014	97822.51	4.99
1987	4675.70	16.51	2001	26179.65	5.07	2015	101893.52	4.16
1988	5865.27	25.44	2002	27390.80	4.63	2016	106478.73	4.50
1989	6534.73	11.41	2003	29691.80	8.40	2017	109331.72	2.68
1990	7662.10	17.25	2004	36238.99	22.05			
1991	8157.03	6.46	2005	39450.89	8.86			

Source National Bureau of Statistics of the People's Republic of China, http://data.stats.gov.cn/easyquery.htm?cn=C01&zb=A0D04&sj=2018

the cumulative total investment in village construction exceeded 10 trillion yuan [20]. In 2016, China's total investment in rural construction reached 1.5908 trillion yuan, with housing construction and municipal public facilities accounting for 74.7% and 25.3% of the total, respectively. Of this, 908.3 billion yuan was invested in rural development, accounting for 57.1% of the total [21].

See Fig. 2.9 for the analysis of the investment structure of rural construction in China in 2016.

3. Changes in township enterprises and professional cooperatives

Township enterprises are important investors and participants in green village and town construction and are also an important component of green village and town construction. According to incomplete statistics [20], in 1978, there were 1.524 million township enterprises in China, with an operating revenue of 43.15 billion yuan. By 2013, the number of township enterprises had increased to 32.044 million, with an operating revenue of 64202.14 billion yuan, an average annual increase of 1833.11 billion yuan over the past 35 years.

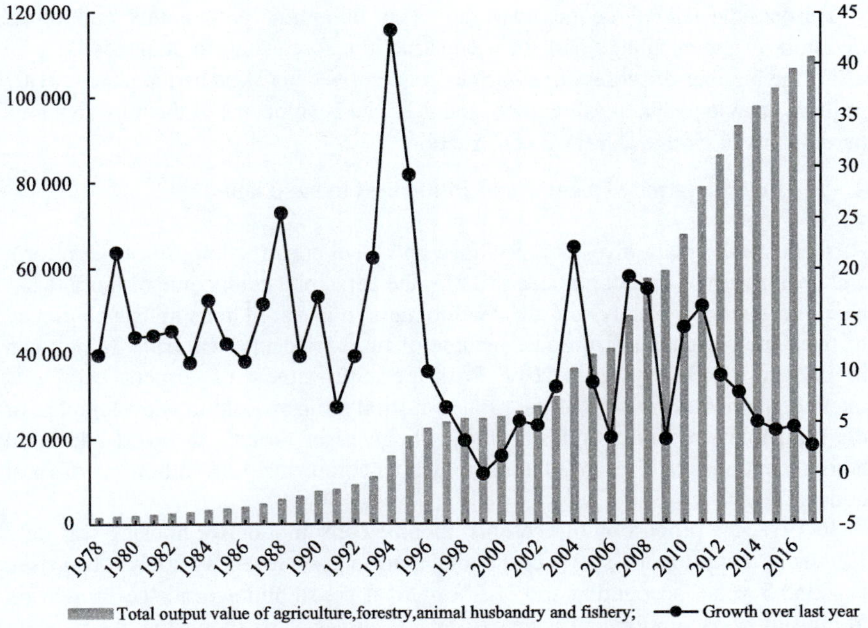

Fig. 2.8 Growth trend of total output value of agriculture, forestry, animal husbandry and fishery in China from 1978 to 2017

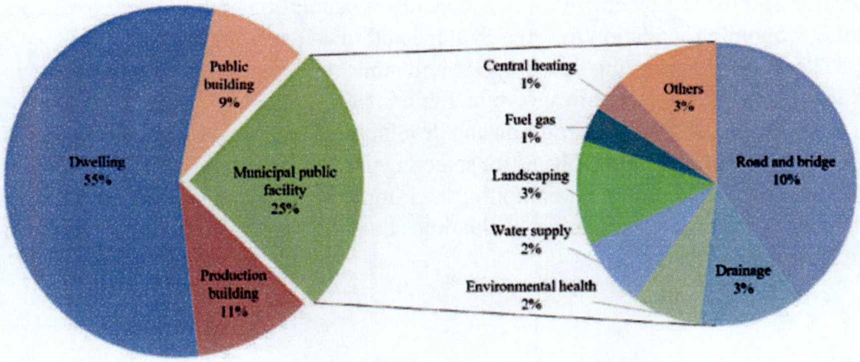

Fig. 2.9 Investment structure of China's village and town construction in 2016

With the rapid development of township enterprises, they have become an important leading party and participant in the green development of rural areas and green village and town construction.

Farmer specialized cooperatives are also important participants and direct investors in green village and town construction. According to statistics [22], in 2017, the number of professional farmer cooperatives in China had reached 201.07 million, showing an increasing trend, and they had become one of the most dynamic participants in village and town construction.

4. Changes in farmers' income and willingness to build houses

Farmers are the main investors in village and town construction. Since the reform and opening up of the countryside in 1978, the per capita net income of farmers has been increasing gradually, and their willingness to invest in housing has continued to rise. The per capita disposable income of rural residents rose from 133.6 yuan in 1978 to 13,432.4 yuan in 2017. With the improvement of farmers' economic income, the transformation of the rural industrial structure and the development of the rural tourism service industry, the willingness of farmers to invest and build houses continues to rise. In 2016 alone, China completed 1.06 billion m^2 of rural housing.

In 1978, the proportion of peasants' income consumption for housing was only 12 yuan, accounting for 10.3% of the total consumption level. By 2017, it had risen to 2,353.5 yuan, accounting for 21.5% of total consumption, and the proportion of housing consumption had increased by 2.09 times. This shows that the housing consumption of farmers is sustainable, and their willingness and ability to consume housing are increasing. During the same period, the proportion of farmers' income consumption for food and clothing decreased significantly. The increase in farmers' income and the enhancement of their housing consumption ability have provided a solid economic foundation for green village and town development.

The changes in housing consumption and other consumption of rural residents in China from 1978 to 2017 are shown in Table 2.13.

This chapter studies the evolution and development of green villages and towns in China, which is an important basis for selecting the construction model, constructing the construction community, compiling and implementing construction and development plans, and formulating evaluation standards for green village and town construction.

Table 2.13 Changes in housing consumption and other consumption of rural residents in China from 1978 to 2017

Year	Housing (yuan/person)	Percentage (%)	Food (yuan/person)	Percentage (%)	Clothing (yuan/person)	Percentage (%)	Other (yuan/person)	Percentage (%)
1978	12.0	10.3	78.6	67.7	14.7	12.7	10.8	9.3
1979	16.0	11.9	86.0	64.0	17.6	13.1	14.8	11.0
1980	22.5	13.8	100.2	61.8	20.0	12.3	19.6	12.1
1981	31.6	16.6	114.1	59.8	23.8	12.5	21.3	11.2
1982	35.6	16.1	133.5	60.6	25.0	11.4	26.2	11.9
1983	42.0	16.9	147.6	59.4	28.0	11.3	30.7	12.4
1984	48.4	17.7	162.3	59.3	28.9	10.6	34.2	12.5
1985	57.9	18.2	183.4	57.8	30.8	9.7	45.4	14.3
1986	70.3	19.7	201.5	56.5	33.0	9.2	52.2	14.6
1987	79.8	20.0	222.1	55.8	34.2	8.6	62.3	15.6
1988	96.3	20.2	257.4	54.0	41.1	8.6	81.8	17.2
1989	105.2	19.7	293.4	54.8	44.5	8.3	92.2	17.2
1990	101.4	17.3	343.8	58.8	45.4	7.8	94.0	16.1
1991	102.3	16.5	357.1	57.6	51.1	8.2	109.4	17.6
1992	104.9	15.9	379.3	57.6	52.5	8.0	122.3	18.6
1993	106.8	13.9	446.8	58.1	55.3	7.2	160.7	20.9
1994	142.3	14.0	598.5	58.9	70.3	6.9	205.7	20.2
1995	182.2	13.9	768.2	58.6	89.8	6.9	270.2	20.6
1996	219.1	13.9	885.5	56.3	113.8	7.2	353.8	22.5
1997	233.2	14.4	890.3	55.1	109.4	6.8	384.2	23.8
1998	239.6	15.1	849.6	53.4	98.1	6.2	403.0	25.3

(continued)

Table 2.13 (continued)

Year	Housing (yuan/person)	Percentage (%)	Food (yuan/person)	Percentage (%)	Clothing (yuan/person)	Percentage (%)	Other (yuan/person)	Percentage (%)
1999	232.7	14.8	829.0	52.6	92.0	5.8	423.7	26.9
2000	258.3	15.5	820.5	49.1	96.0	5.7	495.3	29.7
2001	279.1	16.0	830.7	47.7	98.7	5.7	532.6	30.6
2002	300.2	16.4	848.4	46.2	105.0	5.7	580.8	31.7
2003	308.4	15.9	886.0	45.6	110.3	5.7	638.6	32.9
2004	324.3	14.8	1031.9	47.2	120.2	5.5	708.3	32.4
2005	370.2	14.5	1162.2	45.1	148.6	5.8	884.4	34.6
2006	469.0	16.6	1217.0	43.0	168.0	5.9	975.0	34.5
2007	573.8	17.8	1389.0	43.1	193.4	6.0	1067.7	33.1
2008	678.8	18.5	1598.7	43.7	211.8	5.8	1171.4	32.0
2009	805.0	20.2	1636.0	41.0	232.5	5.8	1320.0	33.1
2010	835.2	19.1	1800.7	41.1	264.0	6.0	1481.9	33.8
2011	961.5	18.4	2107.3	40.4	341.3	6.5	1811.0	34.7
2012	1086.4	18.4	2323.9	39.3	396.4	6.7	2101.3	35.6
2013	1233.6	18.6	2495.5	37.7	438.3	6.6	2458.1	37.1
2014	1762.7	21.0	2814.0	33.6	510.4	6.1	3295.5	39.3
2015	1926.2	20.9	3048.0	33.0	550.5	6.0	3697.9	40.1
2016	2147.1	21.2	3266.1	32.2	575.4	5.7	4141.2	40.9
2017	2353.5	21.5	3415.4	31.2	611.6	5.6	4574.0	41.8

Source Department of Rural Socio-economic Survey, State Statistics Bureau of the People's Republic of China. China Rural Statistical Yearbook 2018

References

1. Li L (2007) Rural settlements: form, type, and evolution. Southeast University Press, Nanjing, p 32
2. Zhao XS, Gao CS, Xu C et al (2015) The history of the Republic of China on local politics and rural changes. Nanjing University Press, Nanjing, p 238
3. Shen YL (1981) Preparing for the establishment of the constitution in the late Qing dynasty, vol 1. Wenhai Press, Taipei, pp 727–741
4. Zheng DH (2000) The rural construction movement of the Republic of China. Social Sciences Academic Press, Beijing, pp 82–83
5. Yan YC (1937) Ten years of rural construction in China. The Commercial Press, Shanghai, p 97
6. Zhang SG (2013) Zuofu Lu and Beibei experiment. https://epaper.gmw.cn/gmrb/html/2013-06/01/nw.D110000gmrb_20130601_2-11.htm. Accessed 1 June 2013
7. Zhou LH (2010) The official administration of the county autonomy law of the national government of Nanjing. Soc Sci Res (04):172–175
8. Fei XT (1939) Peasant life in China. Routledge, London
9. Fei XT (2019) Peasants life in China. The Commercial Press, Beijing, pp 33–36, 93–103
10. Execution of the Republic of China (1946) County organizations at all levels. Zhengzhong Press, Nanjing, pp 10–11
11. Zheng CL (2016) Compilation of national statistics during the Republic of China period. National Library Press, Beijing, p 380
12. China's Land Reform Editorial Board (1988) Compilation of historical data on China's land reform. National Defense University Press, Beijing, pp 422–423
13. Zhang YQ, Zhao QJ (1985) History of land reform in China. Wuhan University Press, Wuhan, p 317
14. Research Group on 70 years' Achievements in China's Rural Construction (2019) The 70th anniversary of the founding of the People's Republic of China village development history review. Constr Small Towns 37(9):1–12
15. Editorial Board of China's Science and Technology Development Strategy for Urbanization and Village Construction (2010) China's science and technology development strategy for urbanization and village construction. Science Press, Beijing
16. Statistics Department of Social Science, Technology and Cultural Industries, National Bureau of Statistics (2018) China social statistics yearbook. China Statistics Press, Beijing
17. China Agricultural Yearbook Editorial Committee (2010–2016) China agricultural yearbook. China Agricultural Press, Beijing
18. Song W, Chen BM, Yang H et al (2008) Analysis of current situation of rural homestead resources in China. Agric Resour Reg China (03):1–5
19. Ministry of Housing and Urban-Rural Development of the People's Republic of China (2009) Looking back at the 60 years of rural construction achievements. http://www.mohurd.gov.cn/zxydt/200910/t20091021_196067.html. Accessed 9 Oct 2009
20. Ministry of Housing and Urban-Rural Development of the People's Republic of China (2017) Statistical yearbook of China's urban and rural construction 2017. China Statistics Press, Beijing
21. Wang X (2017) Study on the social network of construction project community of rural settlement. Dissertation, Huazhong University of Science and Technology
22. Comprehensive Planning Department of the State Administration for Market Regulation (2019) The basic development of market entities nationwide in 2018. http://www.samr.gov.cn/zhghs/tjsj/201902/t20190228_291539.html. Accessed 28 Feb 2019

Chapter 3
Patterns and Mechanisms of Green Village and Town Construction

Abstract Green village and town construction in China should be based on local conditions and choose a construction pattern suitable for local rural characteristics and development needs. This chapter analyses the concepts and characteristics, studies the constituent elements, proposes the main types and application cases, and establishes the promotional mechanism of the green village and town construction pattern.

China has a vast territory, and the geographical climate distribution, natural resource endowment, economic development mode and social development level of villages and towns in different regions vary. Based on the analysis of the various parties involved in green village and town construction and the constituent elements of the construction pattern, green village and town construction should be optimized according to local conditions. The pattern should be determined, and the promotional mechanism for implementing the green village and town construction pattern should be formulated.

3.1 Concepts and Characteristics of the Green Village and Town Construction Pattern

3.1.1 Concept of Green Village and Town Construction Pattern

Pattern refers to the standard style of things, such as the scientific experimental mode, economic development mode, engineering construction mode, and is an intermediary link between theory and practice [1]. Pattern also refers to the formation, design, creative style and typical successful examples of things. It has the characteristics of generality, simplicity, repeatability, structure, stability, and operability.

Pattern and way are different [2]. Pattern refers to the methodology for solving a certain type of problem and it can summarize the methods for solving a certain type

X. Wang and H. Hua, *Green Village and Town Construction in China*,
https://doi.org/10.1007/978-981-16-2098-0_3

of problem to a theoretical level. Way refers to the method and form of words and deeds, which are based on a certain level of productivity development. It expresses the level of human natural science and technology and social development, and more embodies the connotation of social and economic development.

The green village and town construction pattern refers to the general pattern of green village and town construction activities.

Green village and town construction activities involve local government administrative departments at all levels, town governments or village committees, construction engineering enterprises, various social organizations, and village and town residents. Green village and town construction involves system bidding, project investment and financing, engineering design, engineering construction, project management, and completion acceptance. It is also closely related to construction land management, ecological agricultural development, and environmental and ecological protection.

According to different geographical areas, different economic levels and different types of development of villages and towns, establishing a scientific green village and town pattern and making reasonable choices are important prerequisites for the construction development of green villages and towns.

3.1.2 *Characteristics of the Green Village and Town Construction Pattern*

The characteristics of the green village and town construction pattern are reflected in the following three main aspects.

1. Diversity of construction subjects

1) By participants

The construction subjects classified by participants refers to the participation of units as well as people in green village construction projects, including the township government, rural residents, village committee, farmers organization, construction unit, planning architect, enterprise financial institution and other social organizations, with prominent diversified characteristics.

2) By construction projects

The subject of construction classified by construction project refers to the construction projects of green villages and towns, including the planning of construction development, environmental infrastructure construction projects, and construction and transformation projects of green buildings in villages and towns.

2. Diversity of building environmental conditions

1) Diversity of construction patterns in different regional environments

China's rural areas are vast, and the landforms, resources, climatic conditions, and production approaches vary greatly. According to the regional characteristics of resource conditions, green villages and towns can be divided into five basic types: ecological development in the eastern region, cold winters and hot summers in the central region, multiethnic ecology in the western region, a tropical island climate in the southern region and extreme cold conditions in the northern region.

2) Diversity of construction standards at different economic levels

The collective economy of villages and towns and the income level of farmers' families directly affect the content and quality standards of village and town construction. According to two indicators—the collective annual average income of villages and towns and the per capita net income of farmers [3]—the quality of villages and towns can be divided into four types: wealthy, well-off, moderately comfortable, and poor. Different types of development goals impose different contents and quality standards on green village and town construction that constitute the diversity of green village and town construction standards.

3) Diversity of construction methods for different technique conditions

Green village and town construction has outstanding local diversity in terms of construction technology. For example, wooden and stone structures are commonly used for rural houses in mountain villages and towns, raw-soil buildings and cave dwellings are used in villages and towns in loess regions, and bamboo structures and hanging-foot buildings are used in minority areas. Therefore, in the planning and construction technology selection of green villages and towns, architectural design, building materials, construction techniques, architectural functions, landscape construction, etc. all reflect the diversity of local characteristics, design styles and construction methods.

3. Diversity of construction investment method

1) Government investment-led construction methods

Government investment is an important source of investment in environmental infrastructure and public service facilities commonly used in the early stages of green village and town construction. For example, national competent authorities or provincial, municipal, and county governments may provide direct investment or subsidy funds to carry out infrastructure construction projects such as rural roads, water supply and drainage facilities, sewage treatment plants, garbage treatment, clean energy, government investment or subsidies for construction projects for medical treatment or public service facilities such as health, culture, education and sports facilities in villages and towns.

2) Corporate investment-led construction methods

Enterprise investment is usually suitable for industry-driven green village and town construction projects. For example, with the development of the rural economy, an increasing number of township enterprises and foreign enterprises have established raw material bases, production and processing bases and tourism and leisure bases in villages and towns and have directly invested in infrastructure such as village and town roads, water supply and power supply, and pollutant treatment equipment. As another example, professional companies directly invest in green ecological tourism projects and simultaneously build and transform green villages and towns.

3) Financial sector-supported construction methods

Inspired by the national rural revitalization strategy and preferential policies, commercial banks and financial companies have continuously increased their support for rural revitalization and agriculture-related finance. They have also actively supported agricultural and rural infrastructure construction, digital rural construction, rural environmental improvement, agricultural industry, leisure agriculture and rural tourism projects. The financial sector can provide a stable and reliable source of construction funds, and local enterprises and farmers can also apply for various loans to invest in green village and town construction.

4) Social capital investment-supported construction methods

As China's overall economic strength continues to increase, social capital investment has played an increasingly important role in green village and town construction. For example, in the process of village and town construction, social capital has been invested in village remediation and overall land circulation operations in a "village-enterprise cooperation" pattern and has used construction land saved through village remediation and land consolidation to prioritize support for the development of new industries and new formats in villages and towns. Encouraging and attracting social capital investment is an important way to accelerate green village and town construction.

3.1.3 Components of the Green Village and Town Construction Pattern

The constituent elements of the green village and town construction pattern can be divided into three categories: subject, object, and content.

1. The subjects of green village and town construction

The subject, in a general sense, is the main part of a thing. In civil law, it refers to citizens or legal persons who enjoy rights and bear obligations. In contract law, it refers to the natural person or legal person who signs a contract and other organizations. In philosophy, it refers to people who have the knowledge and practical ability to manage the object.

The subjects of green village and town construction includes the construction party, the investing party, the planning and design party, the contractor and other relevant parties.

1) Construction party

The construction party is the parties or consortiums directly carrying out green village and town construction projects and is the contractor for the project construction contract, also known as party A. It includes mainly township governments, county government departments, villager committees, rural economic cooperation organizations, villagers, and enterprises that have been awarded contracting rights for construction projects and are the core responsible bodies for green village and town construction.

2) Investing party

The investing party is the enterprises and units that provide construction funds and financing channels for green village and town construction, including the project construction party (self-raised funds), various types of banks, financial investment institutions, enterprises and institutions, social organizations and other investors that are the main providers of funds for green village and town construction.

3) Planning and design party

The planning and design party is the departments and units that provide project planning, plan compilation, survey and design, and technical consulting services for green village and town construction, including planning and design institutes, architectural design and survey units, professional consulting units, professional technical personnel and other planning and design parties that are the main technical support for green village and town construction.

4) Contractor

The contractor is the party that undertakes the actual construction and includes construction enterprises with construction qualifications, various professional subcontracting enterprises, material and equipment suppliers, local construction enterprises or farmer construction teams, villagers with experience in building houses. And the contractor is the subject of construction for green village and town construction.

5) Other relevant parties

In addition to the above four parties, other relevant parties directly involved in green village and town construction mainly include social public welfare groups, volunteer organizations, experts and scholars, farmers' legal consultants and other relevant parties that provide the main social support for green village and town construction.

2. The Objects of green village and town construction

The object is anything that can be perceived or imagined. It is relative to the subject and is the subject's recognition and activity object. It includes both things that exist objectively and can be perceived subjectively and things that open up a thinking process.

The objects of green village and town construction are objects targeted by the construction project contract, including three categories: "objects", "behaviour", and "intelligence achievements".

1) Object

"Object" is generally economically valuable means of production and consumption that can be used by people. In green village and town construction, they are all kinds of material facilities for construction projects that exist within the village and town system. These include mainly environmental infrastructure construction, green farmhouse construction and renovation, ecological landscape construction and environmental restoration. In the broad sense, "things" are also "property", which can be divided into capital and physical property. For example, village and township builders need to find investors to sign investment and financing agreements or loan contracts for construction projects, and village and township collective common physical assets and farmers' private physical property can also be used to finance construction projects through various methods, such as mortgages and valuations.

2) Behaviour

"Behaviour" generally refers to human behaviours, that is, activities that people manifest under the control of thought. In green village and town construction, behaviour refers mainly to the activities undertaken by the contractor in order to complete the tasks of various construction projects. The main types of green village and town construction activities are contracting environmental infrastructure construction projects, undertaking green farmhouse construction and renovation projects, and undertaking ecological environmental protection and restoration projects.

3) Intellectual achievement

"Intellectual achievement" refers to the spiritual wealth or spiritual products created by people through intellectual labour, and the right generated by intellectual achievement is called intellectual property. Intellectual achievements have the characteristics of creativity, immateriality, and openness. The intellectual achievements of green village and town construction are various innovative technical products provided by relevant construction entities in order to complete green village and town construction projects, including construction consulting planning results, construction planning and architectural design results, new construction techniques and technologies, construction achievements such as brand-name registration, green product trademarks, geographical product mark certification, identification of historical and famous villages and towns, excellent historical buildings, and intangible cultural heritage inheritance.

3. Content elements of green village and town construction

Content refers to the substantive aspects of things or the sum of the internal factors of things. Content exists relative to form and determines form. Form depends on content and changes with the development of content. Form can react to content and promote the development of content.

The content of green village and town construction refers to the rights and obligations stipulated in the construction project contract and the law and includes construction development of green villages and towns and construction plan design, village and town construction project financing and cooperation, village and town environmental infrastructure and rural housing construction project contracting and outsourcing, village and town tourism scenic spot development and environmental protection restoration construction.

3.2 Main Types of the Green Village and Town Construction Pattern

Green village and town construction patterns can be divided into three basic types: construction-led type, construction-finance-driven type, and development-oriented type. According to different construction entities, financing channels and development methods, each type of construction pattern can be subdivided into several specific construction patterns. These specific construction patterns can also be combined and adjusted to form a new construction pattern. Decision makers (town governments, village committees, etc.) and participants (investors, enterprises, village and town residents, etc.) in green village and town construction should choose the construction pattern that best meets their actual needs in terms of the benefits and special conditions of village and town construction.

3.2.1 Dominant Pattern of Green Village and Town Construction

In green village and town construction, it is very important to choose the main organizer to play a leading role in the whole construction process. The main dominant pattern of green village and town construction is divided into the following six types [4].

1. Construction pattern led by village committees

In this construction pattern, the villagers are led by the village committee to jointly carry out green village construction and actively organize other participants.

2. Construction pattern led by county and town governments

In this construction pattern, county or town governments guide and encourage the unified planning and implementation of green villages and towns by means of policy support, capital investment in infrastructure construction and ecological environmental management.

3. Construction pattern led by peasants' organizations

In this green village and town construction pattern, the peasant organizations take the lead, organizing villagers to develop together and become prosperous through the development of green economy industries.

4. Construction pattern led by local enterprises

This construction pattern promotes the development of green economy industries in villages and towns through local township and village enterprises and promotes the simultaneous development of a green township economy and construction.

5. Construction pattern led by external enterprises

This construction pattern involves encouraging foreign enterprises to rent and operate property, adopting different forms such as enterprises, joint households and individuals to jointly invest in green development projects and promoting foreign investment to develop green villages and towns.

6. Construction pattern led by social organizations

Social organizations choose villages and towns for cooperation, and the town government, village committee and rural residents combine to jointly carry out innovative development, demonstration and promotion of green villages and towns through publicity on the green development concept and technical support for green construction.

3.2.2 *Financing-Driven Pattern of Green Village and Town Construction*

Obtaining construction investment is key to the success of green villages and towns. The construction pattern driven by project financing is the future development trend of green villages and towns. The financing-driven pattern of green village and town construction can be divided into two types: the basic PPP construction financing pattern and common financing pattern.

1. Basic PPP construction financing pattern

The public-private-partnership (PPP) pattern refers to partnership between the government and private organizations to provide certain public goods and services based on concession agreements.

Green village and town construction provides outstanding public goods and social services, and public welfare is strongly associated with such construction projects. It The basic PPP construction financing pattern should be preferentially adopted to raise much-needed long-term, high-investment, large-scale construction project funds. For example, partners should be entrusted to invest in the construction and operation of village and town sewage treatment plants, garbage collection and transportation, clean energy facilities, cooperation in building green villages and towns, joint development of eco-tourism scenic spots, and cooperative implementation of village and town ecological environment restoration projects.

In green village and town construction, the PPP pattern transfers the responsibilities of some county governments and township governments to social entities (such as tourism development companies, environmental protection companies, and production enterprises) in the form of franchise rights. The project construction community relationship of "sharing, risk sharing, and full-course cooperation" has the advantages of reducing not only the government's financial burden but also the investment risk of social entities.

2. BOT financing pattern of construction project

The common financing pattern of specific construction projects in green villages and towns is the build-operate-transfer (BOT) pattern. The BOT pattern can evolve into BT, BLT, TOT, BOOT, BOOST, BOO and other modes to match the choice of investment financing methods for different types of construction projects.

1) BOT pattern

The BOT pattern is a common method of infrastructure investment, construction and operation and is applicable mainly to rural infrastructure construction projects.

The operational feature of the BOT pattern is that a project to be constructed by the government or a company authorized by the government can be financed, invested, constructed, operated and maintained by another investment enterprise, as agreed upon in the contract. The investment enterprise can obtain profits through operation and bear corresponding risks within the period specified in the agreement. The government or a company authorized by the government reserves the right to supervise and control the project during this period. Upon expiration of the agreement, the project is transferred by the authorized investment enterprise to the government or the company authorized by the government.

2) BT pattern

The build-transfer (BT) pattern is a financing pattern in which the government uses non-governmental funds to carry out non-operational infrastructure construction projects. It is applicable mainly to the situation of a project that cannot provide products or charge fees to the public after being completed, so the government can only pay the project funds directly to the sponsor and allow the project to serve the public. For example, public schools or roads and bridges are built in villages and towns without charge.

The operational characteristics of the BT pattern are that the operation of a project is transferred to the owner through the project company's general contracting, financing and construction acceptance. The owner pays the total investment of the project plus a reasonable return to the investor. The investor authorized by the agreement is responsible only for the investment and for the financing and construction of the project. Upon the completion and acceptance of the project, the government or the authorized project owner purchases the project according to the contract.

3) BLT pattern

In the build-lease-transfer (BLT) pattern, the county, town government or township enterprises entrust investors with project construction. During the project operational period, the government or township enterprises have the obligation to become the lessee of the project. After the end of the lease period, all assets are transferred to the government or township enterprises. This pattern is often used for the development of green village and town tourist attractions and environmental infrastructure construction projects.

4) TOT pattern

The transfer-operate-transfer (TOT) pattern refers to the transfer of property rights or management rights for a certain period of time by government departments or township enterprises to investors for the purpose of operation and management. The investor recovers the entire investment through operation within the agreed-upon time limit and obtains a reasonable return. After the expiration of the contract between the parties, the investor returns the project to the government department or the original enterprise.

This pattern is often used for traffic facilities or tourist attractions in villages and towns, for example, infrastructure projects, government departments or township enterprises through franchise agreements. The project is put into production and turned over to the investors to reap the earnings over the next several years, with the facility at the same time raising a sum of money for the construction of new transport infrastructure or other tourist attraction infrastructure projects. After the expiration of the franchise, the investor transfers the facility back to government departments or township enterprises free of charge.

5) BOOT pattern

The build-own-operate-transfer (BOOT) pattern is applicable mainly to basic village and township industry projects financed by private partnerships or international consortiums. After the project is completed, the partnership or consortium will own and operate the property for a specified period of time. After the expiration of the period, the project will be transferred to the government.

6) BOOST pattern

The build-own-operate-subsidize-transfer (BOOST) pattern is a construction pattern that adds "subsidize" to the BOOT pattern. For example, the county and township governments subsidize certain tolls per ton of sewage or garbage for companies that undertake village and town sewage treatment plants or garbage disposal operations.

7) BOO pattern

The build-own-operate (BOO) pattern refers to a social investor who constructs and operates an infrastructure project in accordance with a concession granted by the government but does not transfer this infrastructure project to the government or other authorized owners. That is, a contractor constructs and operates a construction project according to a concession granted by the government but does not transfer the project to the public sector in the future.

3.2.3 Green Village and Town Guided Development Pattern

Drawing on the successful experience of other villages and towns in construction and development, giving full play to the advantages of local villages and towns is an important basis for choosing and determining local green village and town construction patterns and development methods. The dominant green village and town development mode is divided into the following two categories, each of which can be subdivided into several specific patterns.

1. Beautiful rural construction guide development pattern

Around 2010, Zhejiang Province began to explore the beautiful rural construction pattern that was subsequently promoted to the whole country, which is an important innovation and successful practice in green village and town construction. In 2014, based on the resource endowment and industrial development characteristics of rural areas in different regions, the former Ministry of Agriculture proposed ten guided development patterns for the construction of beautiful villages in China [5]; see Table 3.1.

Table 3.1 Development pattern of beautiful village construction

Type	Pattern	Suitable area	Construction feature	Typical village or town
1	Industrial development type	Relatively developed areas such as the eastern coast	It has obvious industrial advantages and characteristics. Farmers' specialized cooperatives and leading enterprises have a good foundation for development and a high level of industrialization. It initially formed the concept of "one village, one product" and "one township, one industry"	Yonglian Village, Nanfeng Town, Zhangjiagang City, Jiangsu Province
2	Ecological protection type	Areas with beautiful ecology and low environmental pollution	It has superior natural conditions, abundant water resources and forest resources, traditional pastoral scenery and rural characteristics, and obvious ecological environmental advantages. It has great potential to turn ecological environmental advantages into economic advantages and is suitable for the development of eco-tourism	Gaojiatang Village, Shanchuan Township, Anji County, Zhejiang Province

(continued)

2. Production-Living-Ecosystem (P-L-E) integration construction guides development pattern

Green village and town construction requires the simultaneous P-L-E coordination. For example, when constructing a green farmhouse, it is necessary to consider not only the comfort and safety of the villagers but also the convenience of developing the farmer's courtyard economy and the convenience of the farmhouse as well as the harmony between the individual building and the village's overall ecological environment.

Table 3.1 (continued)

Type	Pattern	Suitable area	Construction feature	Typical village or town
3	Suburban intensive type	Suburbs of large and medium cities	It has good economic conditions, high-quality public facilities and infrastructure, convenient transportation, intensive agriculture, a high-functioning operational level, high land yield, relatively high income level of farmers, and is an important "vegetable basket" base for large and medium-sized cities	Maogang Town, Songjiang District, Shanghai
4	Comprehensive governance type	Villages and towns with large populations, large scales, and relatively concentrated living	Its location conditions are good, economic foundation is strong, driving effect is strong and infrastructure is relatively perfect	Guangfa Village, Gongpengzi Town, Fuyu City, Songyuan City, Jilin Province
5	Cultural heritage type	Areas with special cultural landscapes, including ancient villages, buildings, and houses and traditional culture	It has rich rural cultural resources, strong folk culture and immaterial culture, and great potential for cultural display and inheritance	Pingle Village, Pingle Town, Mengjin County, Luoyang City, Henan Province
6	Fishery development type	Traditional fishing areas in coastal and water areas	Its industry is mainly fishery, which promotes employment, increases fishers' income and enables the rural economy to prosper through fishery development. Fishery plays a leading role in the agricultural industry	Fengmasan Village, Hengli Town, Nansha District, Guangzhou City, Guangdong Province

(continued)

Table 3.1 (continued)

Type	Pattern	Suitable area	Construction feature	Typical village or town
7	Grassland type	China's pastoral areas, semi-pastoral areas, and counties, accounting for more than 40% of the country's land area	Grassland animal husbandry is the basic industry of pastoral economic development and the main source of herdsmen's income	Brainstem Hada Gacha, Haoltugaole Town, Xilingol League, Inner Mongolia
8	Environmental remediation type	Rural dirty and messy areas	Its rural environmental infrastructure construction lags, environmental pollution problems are serious, and local farmers strongly desire environmental improvement	Hongyan Village, Lianhua Town, Gongcheng Yao Autonomous County, Guangxi
9	Leisure travel type	Areas suitable for the development of rural tourism	It is rich in tourism resources, accommodation facilities, and catering; leisure and entertainment are complete; transportation is convenient; and it is close to the city, suitable for leisure vacations, and has great potential for developing rural tourism	Jiangwan Town, Wuyuan County, Jiangxi Province

(continued)

In 2016, the *National Agricultural Modernization Plan (2016–2020)* issued by the State Council clearly put forward the basic principle of adhering to the coordinated promotion of production, living and the ecosystem for the purpose of properly managing the relationship between agricultural production, farmers' increased income, environmental management and ecological restoration.

The P-L-E integrated construction development guidance pattern refers to a new green village and town construction pattern aimed at achieving the coordinated promotion of production, life and ecology in rural areas. It also emphasizes "production and life integration", "life and ecological integration" and "production". Multiple construction features, such as "integration with ecology", can be divided into the following three patterns.

Table 3.1 (continued)

Type	Pattern	Suitable area	Construction feature	Typical village or town
10	Efficient agricultural type	China's main agricultural production areas	It mainly develops agricultural crop production, farmland water conservancy and other agricultural infrastructure; the commercialization rate of agricultural products and agricultural mechanization level is high; the per capita cultivated land resources are rich; and the crop straw yield is large	Sanping Village, Pinghe County, Zhangzhou City, Fujian Province

1) Green liveability and ecological agriculture integrated construction pattern

This pattern combines green village and town construction with local characteristic ecological agricultural developmental planning and projects. On the one hand, it can obtain higher economic and ecological benefits through the development of characteristic ecological agriculture and provide financial support for green village and town construction. On the other hand, through green village and town construction to promote the development of new formats such as leisure and tourism agriculture and experiential agriculture, it provides sustainable infrastructure guarantees for the development of characteristic ecological agriculture and attracts and retains young and high-quality modern agricultural producers and operators.

2) Green liveability and eco-tourism integrated construction pattern

This pattern combines green village and town construction with local rural eco-tourism development plans and projects. On the one hand, the development of rural eco-tourism attracts investment in projects needed for green village and town construction, increases farmers' income and building capacity, and enhances the collective income of villages and towns and the ability to invest in public service projects. On the other hand, it protects and improves the quality of the ecological environment through the green village and town construction, encourages farmers to operate rural tourism projects such as Agricola, and provides sustainable infrastructure and social participation platforms for the development of rural eco-tourism.

3) Rural complex construction pattern suitable for living, industry and tourism

This pattern promotes the integrated development of modern agriculture, leisure tourism, and pastoral communities, with farmers' cooperatives as the main construction carrier, on the premise of the full participation and benefit of farmers, and features the comprehensive development of circular agriculture, creative agriculture, and agricultural experience. This new pattern of green village construction and development aims to promote the coordinated development of facilities suitable for living, industry and tourism in villages and towns.

Based on the above analysis, a green village and town construction pattern system is shown in Fig. 3.1.

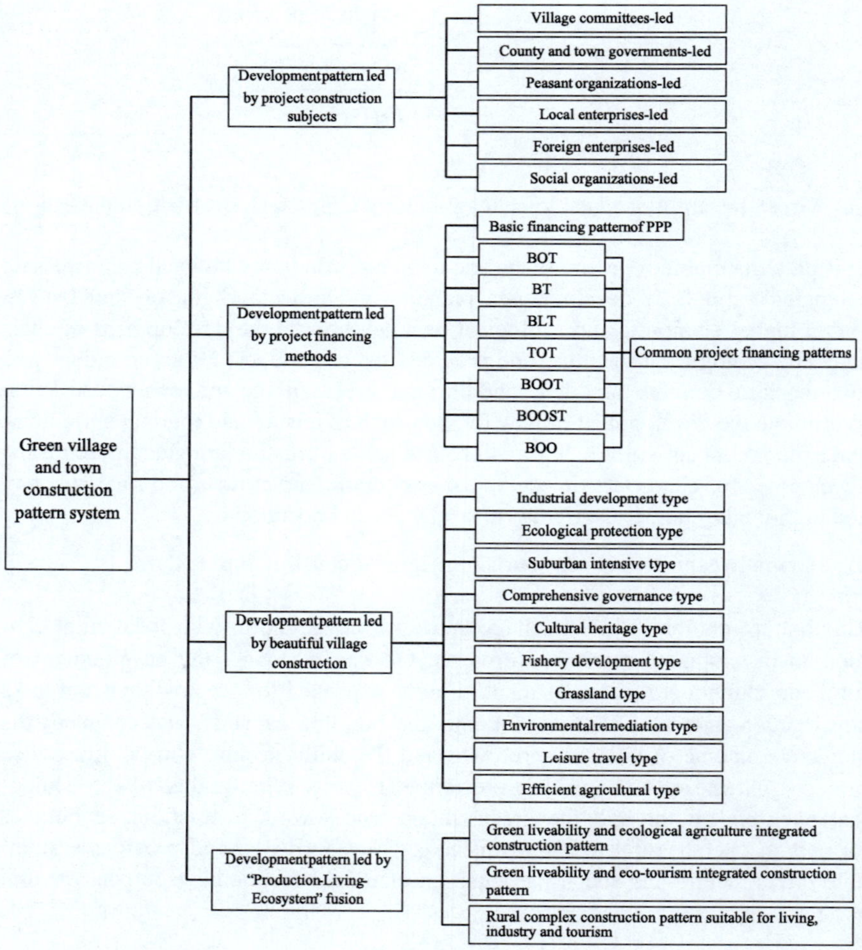

Fig. 3.1 Green village and town construction pattern system

3.2.4 Application Cases of Green Village and Town Construction Patterns

This case takes the construction of a green ecological village in Yanhe Village, Gucheng County, Hubei Province, as an example. The Yanhe Village construction pattern is mainly the "main construction-led" pattern. The Yanhe Village construction project was led by the village committee, and all villagers participated. With the support of the county and town government and incentive policies, under the guidance of experts and scholars in green village planning and design and construction technology training and demonstration projects, and with the participation of village tourism cooperatives and many external enterprises, after efforts over nearly two decades and through the completion of a series of construction projects, Yanhe Village was transformed from a poor mountain village into a national ecological village, a national green well-off village, a national forest village, an AAA national tourist scenic spot, a national village governance demonstration village and a beautiful and liveable country village.

1. Main construction subjects of Yanhe Village

The subject of the construction of Yanhe Village was composed of the village committee, villagers, experts and scholars participating in the construction activities. The village committee was the core organizer of the construction. The specific components of the construction of Yanhe Village were as follows:

(1) Construction. This was led by the village committee and formed together with the villagers and Yanhe Ecotourism Cooperative.
(2) Investors. The Yanhe Village Ecotourism Cooperative, composed of villagers' self-raised construction funds, scenic tourism development and investment enterprises, tea production and processing enterprises, special government subsidies at all levels, construction project lenders, and donations from social organizations, was the core investor.
(3) Planning and design. Planning and design were undertaken by Huazhong University of Science and Technology, Wuhan University of Technology, Huazhong Agricultural University, Beijing Green Cross Ecological Cultural Communication Center, Gucheng County Urban and Rural Construction Plan and Design Institute and other units.
(4) Project contractor. The contractor was a combination of the Wushan Town Construction and Construction Team, Wuhan Chuanmu Doors and Windows Co., Ltd., Wuhan Xinjincheng Environmental Protection Equipment Co., Ltd., Wuhan Leclerian Technology Co., Ltd., and the villagers' housing construction cooperation team.

2. Typical Yanhe Village construction patterns

(1) Self-preparation to build village-level road. In 2000, combined with the construction of a green well-off village, the Yanhe Village committee agreed upon the development concept of "building roads before getting rich and not

relying on construction", and the first construction project was to mobilize all villagers to raise funds and labour. Through loans and applications for state subsidies, the village committee took the lead in 2001 to build a village-level road leading to the town, thus introducing enterprises and funds for Yanhe Village to develop organic tea plantations and economic forest planting and processing. In addition, a "Agricola" was established to provide basic conditions for the development of rural eco-tourism.

(2) Implementing garbage classification. In 2005, Yanhe Village was established as an ecological civilization village. With the support of the Beijing Green Cross Ecological Cultural Communication Center and other social groups, Hubei Province took the lead in conducting rural household waste classification and collection and established the province's first village-level garbage transfer station and resource recovery centre. The construction achievements of the rural household garbage classification collection and treatment demonstration project became a bright business advantage for the construction of Yanhe as a green ecological village.

(3) Establish a construction demonstration base for a green ecological village. From 2006 to 2012, Yanhe Village seized the development opportunities offered by the national-level green ecological village demonstration base. The village committee and the expert team from Huazhong University of Science and Technology signed a cooperation and co-construction agreement. The team of experts compiled the development plans called *Economic and Social Development Plan for Yanhe Ecological Village in Hubei Province (2008–2020)* and *Construction Plan for Yanhe Village in Wushan Town, Gucheng County (2008–2020)* and determined the key construction projects and technical requirements of the ecological village, including green farm building construction, infrastructure construction, public facility construction, environmental improvement construction, and industrial development and construction for a total of 106 construction projects. These projects were all completed on schedule, and a national-level construction demonstration base of a green ecological village was formed in Yanhe Village.

3. Yanhe Village financing patterns

For the Yanhe New Village construction demonstration project, the financing method adopted was the BOO pattern. In 2008, to improve the living and living conditions of farmers in Yanhe Village and develop the rural tourism industry, the village committee decided to add the "Agricola" construction project and use the tidal flats on the banks of the Yanhe River to construct Yanhe New Village.

To this end, the Yanhe village committee established the Yanhe Village Ecotourism Cooperative to attract the villagers to invest in shares. At the same time, according to the voluntary principle, the farmers themselves contributed funds, the village committee identified the land for building, and experts were organized to provide drawings and techniques for designing and building the houses. The construction team built in a unified way. After completion, the ownership of the new village farmhouse belonged to the farmers who invested in the construction. The "Agritainment" project is managed by the villagers themselves. The operation and maintenance of the new village are managed by the ecological tourism cooperative.

Yanhe New Village covers an area of 4700 m^2, with a total construction area of 3732 m^2 and a total investment of 2.63 million yuan, including 13 peasant houses and 1 reception centre. The construction area of the village reception centre is 936 m^2, and the construction cost for the village collective was 560,000 yuan; the farmers' housing construction funds were self-raised by the farmers. The average farmer contributed approximately 159,000 yuan, and the average cost per m^2 was 749 yuan. The design of Yanhe New Village was provided by the Beijing Green Cross Ecological Cultural Communication Center. Huazhong University of Science and Technology was responsible for the design and construction technical guidance for the structures and for building energy conservation and sewage treatment facilities. The project construction started in February 2008 and ended in December 2008.

The completed Yanhe New Village has various architectural forms, beautiful architectural styles, complete infrastructures, small bridges, flowering trees and shade trees. It has become a famous ecological village tourism site in Yanhe Village and even in Xiangyang City and surrounding areas. In 2011, the Yanhe Ecological New Village Group was selected for the National "Eleventh Five-Year" Major Scientific and Technological Achievements Exhibition. In 2017, the annual income of tourism operation households in Yanhe New Village was above 100,000 yuan.

Yanhe Village adopted the "construction subject-led" construction pattern, which greatly promoted the development of the village's eco-tourism industry. In 2017, the eco-tourism cooperatives alone received more than 300,000 tourists, with an operating income of more than 80 million yuan, and the per capita income of farmers in the village increased by more than 5,000 yuan; the village basically achieved the "four ones" common prosperity plan development goal for the farmers, namely, one poverty alleviation project for each household, one small western-style building per household, one car per household, and one deposit of more than 100,000 yuan per household.

3.3 Promotional Mechanism of the Green Village and Town Construction Pattern

3.3.1 Promotional Mechanism for Construction Land Management

Saving land resources is the main goal of green village and town construction. Regardless of the construction mode adopted, a strict green village construction land security promotional mechanism must first be established.

1. Strictly enforce regulations for the construction land of villages and towns

Land for construction green villages and towns must strictly implement the provisions of the "*Land Administration Law of the People's Republic of China*" and effectively guarantee the landownership and use rights of village collectives and farmers. Land in rural areas and suburbs of cities, except for that owned by the state as stipulated by law, belongs to farmers' collectives; homesteads, self-retained land and self-retained mountains also belong to farmers' collectives.

Land for green village and town construction should coordinate the productive land, living land and ecological land use to meet the reasonable demand for land use for industry and infrastructure in villages and towns, protect and improve the ecological environment, and promote improved land conservation and utilization.

2. Strictly comply with land management requirements for villages and towns

In green village and town construction, if the conversion of agricultural land into construction land is involved, the approval procedures for the conversion of agricultural land should be followed; township enterprises, township (or town) or village public facilities, public welfare projects, rural villagers' houses and other construction projects should follow the village plan and the market plan and should have a rational layout and comprehensive development approach; supporting construction is also needed. These actions should go through approval procedures; if rural collective economic organizations use the construction land selected by the township (or town) land and confirmed in the overall plan for starting an enterprise or cooperating with other units or individuals, the selection should go through the examination and approval procedures according to law. The improvement of the management level of construction land should be promoted.

3. Improve security management methods for the use of farmers' homesteads

The rural housing base refers to members of the village collective economic organization who meet the requirements and build on their own rural land in accordance with the standards stipulated by laws and regulations.

In 2018, the No. 1 Document of the Central Committee of the CPC clearly proposed improving the policy for idle homesteads and idle farmhouses; exploring the separation of the three rights of homestead ownership, qualification rights, and use rights; implementing collective ownership of homesteads; protecting the rights of homestead farmers and farmers to house property, appropriately releasing homesteads and farmers' right to use houses [6]. The "three rights separation" reform policy for peasants' homesteads provided important resources for using a large number of idle homesteads and farmhouses in rural areas as construction sites for green towns.

In green village and town construction, the use of peasant house sites must be strictly supervised and managed in accordance with the *Land Administration Law of the People's Republic of China*. A rural villager can own only one house site, and the area of the house site must not exceed the standards stipulated by provinces, autonomous regions, and municipalities; the construction of a house by a rural villager must conform to the overall township (town) land-use plan and village plan, and the house must not occupy a permanent basic farmland. The original homestead and free land in the village should be used as far as possible; rural villagers who sell, rent, or donate their homes and then apply for homesteads will not be approved; and the country allows rural villagers who settle in cities to voluntarily withdraw from their homesteads in accordance with law. Rural collective economic organizations and their members are encouraged to make active use of idle homesteads and homes. The improvement of farmers' homestead monitoring and management capabilities is promoted.

4. Improve the land rights protection system for village collectives and peasants

Green village and town construction is a long-term dynamic construction and operational management process. The construction method and landownership may be affected by different construction patterns and partners. The long-term effective protection of the rights of the village collective and farmers' land should be established in accordance with the law. The system should be maintained.

(1) Effectively protect the wishes of the villagers. The transfer, lease, etc. of collectively owned construction land should be approved by more than two-thirds of the members of the collective economic organization or two-thirds of the villagers' representatives.
(2) Maintain interest rights when land usage changes. The right to use collectively owned construction land acquired through transfer or other means can be transferred, exchanged, contributed, donated, or mortgaged. The rural collective economic organization can report those who do not use the land for the approved purpose to the people's government to obtain approval to recover the original approved land. Then, they can take back the right to use the land and protect the rights and interests of the village collectives and peasants after land circulation and use through methods such as land shares, industrial contracting, benefit distribution, employment and social security.

3.3.2 Promotional Mechanism for Construction Investment

The difficulty in green village and town construction is how to obtain sufficient and sustainable investment in construction projects. The key promotional targets for green village and town construction investment methods are environmental infrastructure, eco-tourism facilities, and green farm building construction projects.

1. Strive for governmental funding policy support

For example, in green village and town environmental infrastructure construction projects, full use should be made of special funds for rural infrastructure construction and support subsidies and incentives provided by the state and local governments at all levels to obtain safe drinking water and sewage treatment facilities, garbage recycling treatment facilities, and ecological landscape restoration. Additionally, comprehensive environmental management, village and village appearance improvement, road facility construction and other project funds should be sought, and the project promoters should actively strive for infrastructure construction funds and projects supporting the development of ecological agriculture; with the support of the government and enterprises, they can choose to use the BOT financing pattern, etc. Solutions to construction funding shortages should be promoted.

2. Apply for a financial institution construction loan

For example, in the development of rural tourism industry and scenic area construction projects, the project promoters should jointly create rural tourism scenic areas and green leisure villages and towns with brand-name tourism companies or development companies and jointly strive for construction project loans from agricultural banks, construction banks, agricultural cooperatives and social financial investment institutions. Additional financial support can be attained through special loans for the construction of beautiful villages, the development and use of clean energy, and the transformation of environmental facilities. The expansion of investment channels for green village and town construction should be promoted.

3. Attract corporate investment and win-win cooperation

For example, in the process of implementing construction and development plans for green villages and towns, project promoters should make full use of the value of the economic developmental resources, ecological landscape resources and historical human resources of the village or town; take the initiative to attract investment; and attract foreign enterprises and local enterprises to invest in rural tourism and special ecological agricultural construction projects. They should induce leading enterprises to carry out integrated investment cooperation in green ecological agriculture and green village and town construction, forming a long-term mechanism of "win-win cooperation". The formation of village and town construction investment communities should be promoted.

4. Guide village residents to invest in building houses

For example, in green farmhouse construction and renovation projects, farmers' cooperatives should be established to mobilize farmers' enthusiasm for building houses and use farmers' own capital input, and preferential policies and subsidies should be adopted for developing green farms. Construction and eco-tourism projects should focus on the green renovation of old and dilapidated rural houses in villages and towns. Farmers' sense of construction satisfaction should be improved.

3.3.3 Promotional Mechanism for Construction Community

1. Promote the construction community development of villages and towns

A green village and town project is a social construction activity that can be completed only with the participation and close cooperation of multiple participants. Combined with the selected green village and town construction pattern, a green residential plan, green economy, and green environment should be promoted as common development goals, and a green village and town construction community should be composed of all participants in the construction project to ensure the smooth implementation of village planning and construction projects.

2. Formulate guiding policies and incentive mechanisms

For example, in the construction-led pattern, it is necessary to formulate and implement a variety of guiding policies and incentive mechanisms and continuously optimize the cooperative relationship among county governments, village committees, villagers, farmers' organizations, local and foreign enterprises, and social organizations to form a construction Heli. As another example, in the P-L-E integrated construction guided development pattern, it is necessary to vigorously promote the close cooperation and simultaneous implementation of green liveable construction, eco-agricultural construction, eco-tourism construction and other participants to achieve the sharing of resources and optimal green village and town construction results.

The specific content and construction method of the green village and town construction community will be studied in Chap. 4 of this book.

3.3.4 Promotional Mechanism for Construction Project Management

Green village and town construction project management includes construction and development plan management, environmental infrastructure construction management, and green building construction and renovation management.

1. Dynamic management of village and town construction development plan

The implementation of a village and town construction and development plan is a dynamic adjustment and optimization process. The dynamic promotion methods for the implementation of construction and development plans are as follows:

(1) Require that the plan be operable, adjustable, reproducible and evaluable and can stimulate the internal power of villagers and farmers and cultivate their "haematopoietic" function [7];

(2) Activate the internal and external development factors of village and town construction through planning and implementation [8]. Internal motivation factors refer to the self-development ability of villages and towns, and external promotion factors include the driving force of regional development and urbanization foreign aid for villages and towns; and
(3) According to the development stages and characteristics, dynamically manage the whole process of the preparation and implementation of the green village and town construction plan.

2. Overall management of environmental infrastructure construction

The key points of the whole process management of environmental infrastructure construction in villages and towns are as follows:

(1) At the early stage of construction, attract market investment and social capital and implement the conditions of environmental infrastructure construction;
(2) During the construction implementation stage, supply safe water and gas, sewage and black and odorous water treatment, and garbage treatment as the focus for ensuring the quality of environmental infrastructure construction; and
(3) In the stage of completion and use, promote the green operation of environmental infrastructure and an innovative village and town environmental infrastructure operational management system.

3. Demonstration and promotion of green farmhouse construction projects

The construction of green farmhouses is key to green village and town construction and requires pattern innovation and demonstration guidance. Attention should be paid to the experiential summary and demonstration project promotion of the green farmhouse construction mode, construction technology and management. Demonstration, promotion and promotion methods include the following:

(1) Through the key construction of green farmhouse demonstration projects, provide a pattern for rural residents to build houses and promote the construction expansion of green farmhouses;
(2) Formulate incentive policies and carry out technical training to mobilize rural residents to participate in the construction of green farmhouses; and
(3) Give full play to the role of village and town enterprises, and promote the development of the green building materials industry and green village and town construction enterprises.

References

1. Chen SQ (2010) Symmetrical economics. China Times Economic Press, Beijing
2. Lu MF, Zhang L (2010) Strategic thinking on constructing a diversified SME financing pattern. Enterp Res (07):50–53

3. Wang JX (2008) Study and reflection on village economic transformation. J Guangxi Univ Natl (Philos Soc Sci Ed) (03):2–6
4. Wang XM, Wang X, Wu JX et al (2017) Modern rural settlement construction engineering community and network analysis pattern. J Civ Eng Manag (1):1–9
5. Bao GX (2018) Top 10 creating patterns of "beautiful countryside" in China. http://www.jzsbs.com/index.php/Home/Index/detail?id=9633. Accessed 7 May 2008
6. Lan HY (2019) The legal system of rural collective land under the vision of rural revitalization has been improved. Hebei Law Sci (04):119–129
7. Zou DC, Wang K, Tan J et al (2019) The development planning strategy of village and town in China under the background of new urbanization. Chin Eng Sci (02):1–5
8. Zhang FG, Liu YS (2008) The dynamic mechanism and development pattern of regional rural development in China. J Geogr (02):115–122

Chapter 4
Green Village and Town Construction Community

Abstract Green village and town construction in China is a sustainable construction that requires the cooperation of multiple parties. It is necessary to build a green village and town construction community. This chapter studies the formation and development of the construction community, proposes the establishment method and operational mode of the construction community, proposes the social network composition method of the construction community, constructs the social network analysis model of the construction community, and conducts a case study of social network analysis and the application of the green village and town construction community.

Green village and town construction is a kind of cooperative construction activity that has diverse development patterns, multiple participants, and strong social influences. Establishing a construction community suitable for rural characteristics is the developmental motivation and organizational guarantee of realizing the green village and town planning and construction target, and it plays an irreplaceable role in green village and town construction.

4.1 Formation and Development of the Green Village and Town Construction Community

4.1.1 Formation and Characteristics of the Construction Community

1. Formation of the construction community

The construction community refers to the collection of all subjects involved in construction project activities. Under the same goal, all members of the community,

X. Wang and H. Hua, *Green Village and Town Construction in China*,
https://doi.org/10.1007/978-981-16-2098-0_4

according to different levels and roles of distribution and through mutual communication, coordination and cooperation, form a construction body system with a close functional connection, mutual influence and diversified interests [1].

Construction activity is a social activity that can be completed only with the joint participation and close cooperation of multiple participants, and green village construction activity is no exception. The construction community has become a rapidly developing research and innovation application field in engineering since the beginning of this century [2]. Green village and town construction is no exception.

The concept of community was first put forward by Aristotle [3]: "Every polis (city) we see is a kind of community". In 1887, Tennis proposed connotations of "community", believing that a community is a closely connected entirety formed on the basis of blood, geography or spirit, such as a family, neighbourhood, or village, whose core is "default agreement" [4].

Since the beginning of humankind's engagement in construction activities, these activities have been associated with an embryonic form of the construction community. In ancient China, with the construction of large public buildings such as palaces, a relatively simple construction community was formed, dominated by the king, organized and managed by craftsmen, provided by material suppliers and participated in by numerous construction workers.

In modern times, with the emergence of architects, the formation of urban planning, and the emergence of the industrial operational system of construction factories, owners, architects, construction factories, material and equipment suppliers, banks and financial institutions, and governments at all levels jointly constitute the modern construction community.

Currently, with the development of construction consulting services, engineering insurance, supervision, consulting and other industries have developed rapidly, and the construction community has expanded and become more complicated. Owners, investors, designers, engineering contractors, suppliers of materials and equipment, supervisors, consultants, insurance companies, banks, governments, social organizations, community organizations and other stakeholders constitute the modern construction community [5].

China's modern construction community concept is based on the development of engineering sociology in the early 21st century [6, 7]. *Introduction to Engineering Sociology: A Study of Engineering Community* [8] and other works published in 2010 mark the initial research on China's modern construction community.

The formation and evolution of the construction community in China can be seen in Fig. 4.1.

2. Characteristics of the construction community

In the process of evolution and development, the modern construction community has gradually developed the following characteristics:

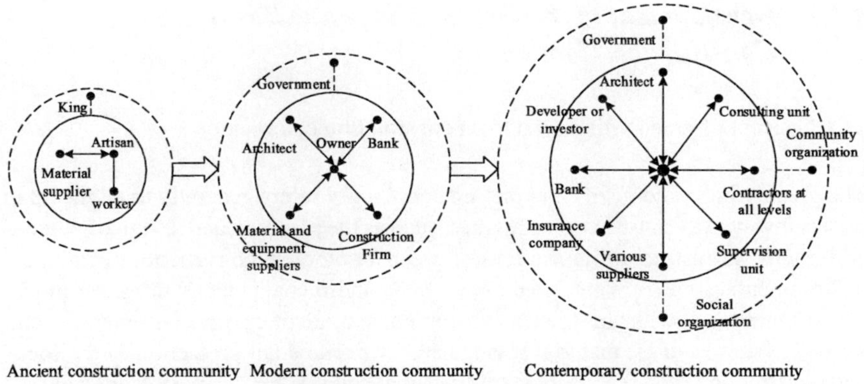

Fig. 4.1 Formation and evolution of the construction community in China

1) A plurality of subjects

The construction community consists of two major types of stakeholders involved in construction projects. The first category consists of direct participants, including owners, architects, development investors, contractors at all levels, various suppliers, supervision units, consulting units, banks and insurance companies. The second category consists of indirect participants, including government departments, community organizations and social organizations. Through the implementation of the construction project, all the participants not only realize common interests but also pursue their own demands.

2) A complex social network

There are complex and changeable social network relationships in the construction community: first, the internal relationship between direct participants; second, the external relationship between direct participants and indirect participants; and third, the interactions between the entire community and the external society.

3) Member flow and recombination

As the construction community is generally concentrated on a specific engineering activity, when that activity is completed, the project-associated engineering community is generally also dissolved. The members of the community will reorganize and develop around new construction projects to form a new construction community. However, green village and town construction is a long-term and sustainable construction activity, and the construction community can be divided into a long-term development community and a short-term project community.

4.1.2 Development of the Green Village and Town Construction Community

1. Concept of green village and town construction community

The green village and town construction community refers generally to all the main bodies involved in construction activities, such as the planning and design of villages and towns, infrastructure construction, green farmhouse construction, ecological environmental improvement, green operations and management, and green industrial development, with comprehensive implementation of energy conservation, land savings, water savings, material savings and environmental protection as the focus and construction of safe, liveable buildings that suitable for business as the goal.

2. Evolution of the green village and town construction community

The green village and town construction community is guided by the theory of the modern construction community and the theory of green development and gradually develops and forms on the basis of inheriting the construction community of rural settlements. Basically, it undergoes two stages of evolution.

1) From traditional village community to modern village and town community

Rural settlements, which correspond to urban settlements, refer to places where the rural population resides, produces and lives their lives. These settlements have their own dependencies and local characteristics. Rural settlement construction is a complex social system with multiple levels and multiple subjects, and the people involved are the core of rural settlement construction, which is the population settlement point of rural space [9, 10]. With the development of modern rural economic and social forms and the evolution of construction management, the traditional rural settlement construction mode has been gradually replaced by the village and town construction mode.

The traditional construction of rural houses and agricultural production facilities relies on the implementation of the "village community" mode [11]. Although this loosely organized community has a clear division of labour and independent construction ability, it cannot meet the new requirements for modern village and town construction and rural revitalization and development.

Traditional village and town construction is relatively closed to the outside world, a self-sufficient construction system of farmer communities, and an "acquaintance society". Closed living and production social relations are constituted by the consanguinity of geography and village culture. For example, traditional agricultural housing construction projects and agricultural production construction projects rely on the traditional "village community" [12, 13], which adopts the independent construction mode to complete projects jointly. For example, basic and public service facility construction projects in villages and towns are generally organized and implemented by ethnic groups, tribes and families, reflecting "collective construction".

Residences are constructed through the voluntary help of villagers, friendship, help exchanges, contractual help, and so on, which reflects a "co-construction" mode.

With the transformation of the social form and structure of villages and towns in China from a traditional, self-sufficient and closed rural society to a modern cooperative and open society, the objects and methods of village and town construction show a trend of diversified development, such as old village reconstruction, centralized settlement, villagers move out of the village and gather together, industrial park development, and commercial housing development. An increasing number of departments, organizations, individuals and groups are involved in village and town construction, and the structure of their social networks is becoming more complicated [14].

2) From modern community to green community in villages and towns

The connotations of the rural construction community are significantly different from those of the urban construction community. The living and working places of urban residents are generally constructed in different locations according to different planning zones. Therefore, the village and town construction community should not only undertake traditional construction tasks, such as the construction of agricultural houses, village and town infrastructure construction and environmental improvement but also undertake systematic engineering content, such as village and town industrial development, social service facility construction, the protection of historical features and the construction of rural civilization.

With the continuous development of green village and town construction, government departments, enterprises, financial institutions, social organizations, planning architects and other external organizations and professionals at all levels have become more involved and have become an important driving force for construction and development. At the same time, the green village and town construction mode is constantly innovating, and construction subjects show a trend of diversified development. Thus, the composition and development of the green village and town construction community also requires constant innovation.

The green village and town construction community is an "open society" focusing on land savings, energy savings, water savings, material savings and environmental protection. It is composed of two parts: the internal village and town community composed of blood relationships and geographical relationships, village culture, green construction and development demand, and the external village and town community participating in green construction activities.

4.2 Establishment and Operation of the Green Village and Town Construction Community

4.2.1 *Establishment Method of the Green Village and Town Construction Community*

The purpose of establishing the green village and town construction community is to achieve the construction tasks and expected goals of green villages and towns by organizing and optimizing the division of labour and cooperation among all participants in construction activities.

The internal part of the green village and town construction community includes the town government and village committee, village and town residents, farmers' organizations, and village enterprises. The external community includes county (district) governments, planning architects, construction units, financial institutions, and social organizations.

The analysis of the constituent elements of the green village and town construction community is shown in Fig. 4.2.

The basic types of green village and town construction community and the functions of each participant are shown in Table 4.1.

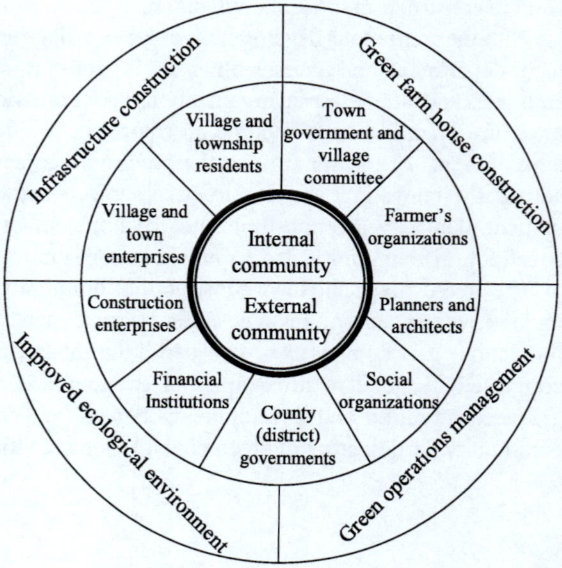

Fig. 4.2 Constituent element analysis of the green village and town construction community

Table 4.1 Basic types and participant functions of the green village and town construction community

Category	Key participants	Participant rules
Internal community	Town government and village committee	They are the main planners and managers of green village and town construction and perform construction supervision, project management, educational services and other functions
	Rural residents	They are the main investors, builders and users and the direct beneficiaries of green village and town construction
	Farmers' organizations	It is a rural economic cooperative organization, community service and cultural organization established or joined independently by rural farmers
	Village and town enterprises	They are established by villages and towns independently engaged in professional and large-scale production and management activities
External community	County governments	The county (district)-level government departments in charge of village and town construction are the guides, supervisors and promoters of green village and town construction
	Construction enterprises	They are investors, developers, construction enterprises, etc., that are external promoters and project partners of green village and town construction
	Planners and architects	They are planners of green village and town construction projects, designer of construction schemes and participants in the construction management process
	Social organizations	They are the social organizations and expert teams of scientific research institutions that participate in green village and town construction. They are the guiding communicators and professional partners
	Financial institutions	They are the rural credit cooperatives, agricultural banks, construction banks and other relevant financial organizations

4.2.2 Operational Method of the Green Village and Town Construction Community

The operational mode of the green village and town construction community requires that, on the basis of the scientific selection of the construction mode and according to the nature of the different construction projects, all participants should be optimized to build a construction community suitable for local characteristics, and the cooperative operational mechanism with the best construction benefits should be adopted to promote the realization of the green village and town plans and construction goals.

According to the analysis of the dominant mode of green village and town construction in Sect. 3.2.1, the operational mode of the construction community can be divided into six categories:

(1) Village committee-led type;
(2) County and town government-led type;
(3) Peasant organization-led type;
(4) Local enterprise-led type;
(5) Foreign enterprise-led type; and
(6) Social organization-led type.

The operational mode and main roles of the green village and town construction community are shown in Table 4.2.

Although different core participants dominate the operational mode of various green village and town construction communities, they all need close cooperation and mutual support from all participants.

The operational mode of the green village and town construction community is shown in Fig. 4.3.

The following introduces construction and operational application examples of six types of green village and town construction communities.

1. Construction community led by village committee

Yanhe Village is located in Wushan Town, Gucheng County, Xiangyang City, Hubei Province, and has an area of 12 km^2. It had 4 village groups, 303 households and 1050 people in 2017.

The green village construction community of Yanhe Village is led by the village committee and supported by policies of the Gucheng County and Wushan Town governments, with the villagers as the project subjects. The collective enterprise Yanhe Eco-tourism Economic Specialized Cooperatives at Yanhe Village provides financial support, and Huazhong University of Science and Technology, Beijing Green Cross and other social organizations provide technological support.

According to the requirements of the construction project launched by Huazhong University of Science and Technology, a green building team successively set up building wall insulation materials, energy-saving windows and doors, sewage disposal and an ecological environmental engineering equipment company, a solar equipment company, landscape greening engineering companies, and energy-saving

Table 4.2 Operational mode and main roles of the green village and town construction community

Number	Operational mode of construction community	Main roles of construction community
1	Village committee-led type	Leading villagers jointly carry out green village and town construction and organize other participants to join
2	County and town governments-led type	Through policy support and capital investment, government departments organize and guide the implementation of green village and town construction plans, environmental infrastructure construction, green farmhouse construction, ecological environment treatment and other projects
3	Peasant organizations-led type	The farmer organization acts as the leader lead the farmer to participate actively to develop the green economy industry to promote green village and town construction
4	Local enterprises-led type	Local enterprises promote the coordinated development of green economy industry and green village and town construction
5	Foreign enterprises-led type	Foreign enterprises perform renting and operations and manage enterprises, joint households, individuals and other modes of joint investment and development of green village and town construction
6	Social organizations-led type	Through publicizing and promoting green development concepts and green construction technology for villages and towns, social organizations guide other participants to cooperate in green village and town construction and demonstration

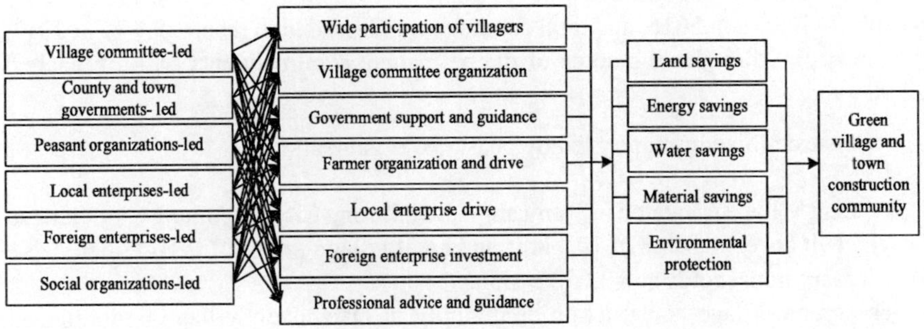

Fig. 4.3 Operational mechanism of the green village and town construction community

electrical equipment of more than 10 green construction-related companies, forming a village committee-led type of construction community and operational mode.

The Yanhe Village construction community has played a leading role in green village planning and construction, and Yanhe Village has been titled "The Most Beautiful Relaxing Village in China", a "National Civilized Village", a "National Green and Well-off Village", a "National Agricultural Tourism Demonstration Site", a "National AAA Scenic Spot" and other designations. The per capita net income of the villagers increased from 1,900 yuan in 2000 to 12,000 yuan in 2015, gradually improving their living standard.

2. Construction community led by county and town governments

Anji County is located in northwestern Zhejiang Province and covers an area of 1,882.45 km^2. It has jurisdiction over 8 towns, 3 townships, 4 streets and 187 villages, with a population of 313,300 in 2017.

The green village construction community of beautiful villages in Anji County is led by the Anji County government and its town governments. The county government is responsible for the overall planning, management organization, construction guidance and performance appraisal of the beautiful countryside. The town governments conduct overall coordination between higher- and lower-level departments, guide beautiful village construction, provide financial support and technical guidance, and coordinate and plan the connecting areas between villages. The established village is the core of beautiful village construction. Under the unified leadership of the county and town governments, the specific plan, construction implementation, and usage and maintenance of various construction projects are put into effect. The county and town governments have formed the county and town government-led construction community and operational mode.

Anji took the lead in suggesting that all established villages in the county achieve the beautiful village goal: "villages are perfect, households start businesses, everywhere is harmonious, and everyone is happy" [15], There were 164 high-quality beautiful villages in 2016, and total coverage was intended to reach 95.7%. In 2017, the per capita disposable income of the permanent rural residents of Anji reached 27,904 yuan.

3. Construction community led by peasants' organizations

Daizhuang Village is located in Tianwang Town, Jurong City, Zhenjiang City, Jiangsu Province. It covers an area of 10.4 km^2 and has 4 village groups. In 2017, there were 866 peasant households and 3,052 people.

The green village construction community in Daizhuang Village is dominated by local peasant organizations. Through the establishment of specialized organic agriculture cooperatives, Daizhuang Village fully respects the wishes of farmers, unitizes planning, production, sales and management, and establishes a cooperative mechanism of shared benefits and risks in which 90% of the farmers in the village participate. The cooperatives have strongly supported the construction of rural infrastructure, irrigation and water conservancy facilities, roads, bridges and greenhouses as well as irrigation facility renovation to promote the continuous development of village construction.

The Daizhuang Organic Agriculture Professional Cooperative has been named the "National Farmer Professional Cooperative Model Association". The village cooperatives are the core and the villagers' active participation is the basis of the peasant organization-led construction community and operational mode.

By establishing specialized organic agriculture cooperatives and giving full play to the guiding role of agricultural experts, Daizhuang Village has promoted the development of modern organic agriculture and increased farmers' income. The per capita net income increased from less than 3,000 yuan in 2003 to 27,000 yuan in 2018, laying an economic foundation for the construction of green ecological villages.

4. Construction community led by local enterprises

Located in Jiangnan Town, Fengman District, Jilin City, Jilin Province, Jianhua Village covers an area of 6.4 km^2 and has jurisdiction over 2 communities, 5 natural villages, 1098 households and 4,831 people.

The green village construction community of Jianhua Village in Jilin Province is dominated by local enterprises. Since 2001, Jianhua Village has successively established local enterprises such as the Longhua Logistics Company, Ecological Agriculture Park Company and Real Estate Development Company. The village collective has organized the concentrated residential area of Jianhua New Village and planned agricultural industrial parks, demonstration industrial parks and ecological leisure areas. It makes intensive use of land, and the construction of the rural settlements is carried out in an orderly manner. Jianhua Village adheres to the village level as the mainstay, takes the market as the guide, and implements the "village-enterprise unity" management model. It has formed a local enterprise-led construction community and operational mode with village-level enterprises as the core.

Driven by village-run enterprises, the service industry and ecological agriculture in Jianhua Village have developed rapidly, and village construction has also been synchronized. In 2015, the gross social production of the whole village was 260 million yuan, the village collective income was 28 million yuan and the per capita income of the villagers was 15,500 yuan. Jianhua Village has become a well-known "model village" and has won the honours of "National Civilized Village and Town", "National Model Village of Democracy and Rule of Law", "National Civilized Model Village" and "The Most Beautiful Village and Town in China".

5. Construction community led by external enterprises

Luoshuai Village is located in Yuanmen Town, Baisha Li Autonomous County, Hainan Province. It is a small village of Li nationality, with 56 households and 203 residents.

The green village construction community of Luoshuai Village is the foreign enterprise-led type. Since 2010, Tianya Station of the village has built 54 farmers' houses for all the villagers through the reconstruction of the old village. The company developed and built tourism service facilities such as accommodations, catering facilities, and leisure and sports facilities with the remaining rural collective construction land after the transformation, helped villagers independently develop personalized farmhouse business services through skill training, and recruited villagers to engage in tourism service work in the company on the basis of excellence. Based on the participation of all villagers, the foreign enterprise-led construction community and operational mode were formed.

The Hainan Tianya Station Tourism Project Development Company cooperated with local villagers in Luoshuai Village to develop the first green rural tourism demonstration base in Hainan, which promoted rural economic development and significantly improved the living conditions of the farmers. The development of rural tourism increased the average annual income of the villagers by 2500 yuan. Luoshuai Village has also won the honour of "National Minority Distinctive Village", "National Civilized Village and Town", "National Forest Village" and "Key Village of Rural Tourism in China".

6. Construction community led by social organizations

Located in the suburbs of southeastern Pingqiao District, Xinyang City, Henan Province, Haotang Village has an area of 20.7 km^2 and jurisdiction over 18 village groups and 2,385 people.

The green village construction community of Haotang Village is dominated by social organizations. In 2009, guided and supported by the China Rural Planning and Design Institute, Beijing Green Cross and other social organizations, Haotang Village began to comprehensively construct a sustainable experimental village development. The social organization team led Haotang Village to vigorously develop the rural economic cooperative organization and reform the rural financial system. The art, aesthetics, construction and environmental protection concepts of modern rural civilization were conveyed to farmers to assist village collective cooperatives establish the village, develop social organizations for mutual assistance of the elderly residents of the village and build village companies and ecological tourism as the leading industries of a green ecological village. Thus, the social organization-led type of construction community and operational mode has been applied.

In 2013, Haotang Village was the first "Beautiful and Liveable Demonstration Village" in the country, and the village collective assets increased to more than 20 million yuan [16]. It has also been named the "National Characteristic Landscape Tourism Village", "China's Most Beautiful Leisure Village", "China's Rural Tourism Model Village" and "China's Human Settlement Environment Award".

4.3 Social Network Analysis of the Green Village and Town Construction Community

4.3.1 Establishment Method of the Green Village and Town Construction Community Social Network

1. Concept and analytical method of social network

A social network refers to a collection of social actors and their relationships, consisting of multiple nodes and lines between nodes. The nodes in the social network represent social actors, and the lines represent the ties between actors. The actors can be individuals, business units, social organizations, communities, villages, or government departments [17]. The green village and town construction community is a typical social network, and the social network nodes of the construction community can be defined as all the participants in the construction activities, such as owners, architects, development investors, contractors at all levels, all kinds of suppliers, supervision units, banking institutions, consultants, insurance companies, government departments, social organizations, and community organizations. The social network can also be defined as the relationships between the participants, such as contractual relationships, management relationships, coordination relationships, communication relationships, and supervision relationships.

The purpose of social network analysis is to analyse the relationship between actors, construct a relationship model, analyse its structural characteristics, and study the effect and impact of the model structure on the overall network or individuals within it [18]. In recent years, scholars in China and abroad have gradually applied the social network analysis method to the study of rural construction, with a focus on constructing a social network model for various participants in rural construction [19, 20].

The method of social network analysis is to study the relationship between actors in the network, and the key objects are (1) to analyse the relationship structure among the social actors; (2) to analyse the basis of systematic empirical data; (3) to gain insight from drawing relationship graphs; and (4) to analyse the basis for the use of mathematics or computational models.

2. Social network structure of the green village and town construction community

Table 4.1 shows that the green village and town construction community can be divided into 2 categories: internal and external. Through mutual cooperation, communication and coordination, the participants jointly constitute the social network of the green village and town construction community.

In the social network of the green village and town construction community, the internal community consists of residents, town governments, village committees, peasant organizations and enterprises in villages and towns that aim to develop production and improve the living standard, forming the long-term and stable core partners of the green village and town construction community. The external community is composed of county (district) governments, construction units, planning architects, social organizations and financial institutions and is the cooperative and development partner of the green village and town construction community.

See Table 4.3 for the social network characteristics analysis of the green village and town construction community.

Table 4.3 Social network characteristics of the green village and town construction community

Number	Construction community operational mode	Construction community social network characteristics
1	Village committee-led type	With the village committee as the core, the cohesion of the village committee is strong, and the enthusiasm of the villagers is fully mobilized for green village construction
2	County and town government-led type	With the town government as the core, it invests more in village and town infrastructure and public service facilities, and the construction is completed quickly, which is suitable for the green town construction
3	Peasant organization-led type	With farmers' organizations, such as economic cooperatives, as the core, villagers' take the initiative in participating and benefit more, which is suitable for village and town construction with the strong organizational ability of farmers
4	Local enterprise-led type	Taking local township enterprises as the core can be a solution to the problem of construction funds, rural economic development occurs quickly, and there is strong stability, which is suitable for village and town construction with developed enterprises
5	Foreign enterprise-led type	With rural enterprises as the core, rural construction occurs quickly, but villagers' participation is relatively low, which is suitable for village and town construction with a developed export-oriented economy
6	Social organization-led type	With social organizations, such as universities, research and design institutes and other expert teams as the core, villagers, village committees and town governments are required to actively cooperate with and invest in the construction project, which is of high quality and strong demonstrability, which is suitable for the construction of pilot project and demonstration green villages and towns

See Fig. 4.4 for the relationship among the social network of the green village and town construction community. In the figure, the township government includes the town government and village committee, and the township enterprises include local enterprises and foreign enterprises.

3. Social network characteristics of the green village and town construction community

(1) Universality of network composition. The social network of the green village and town construction community consists of the internal community and external community, which has universality in the form and content of the network.
(2) Stages of network evolution. Generally, the evolution and development of the social network of the green village and town construction community will go

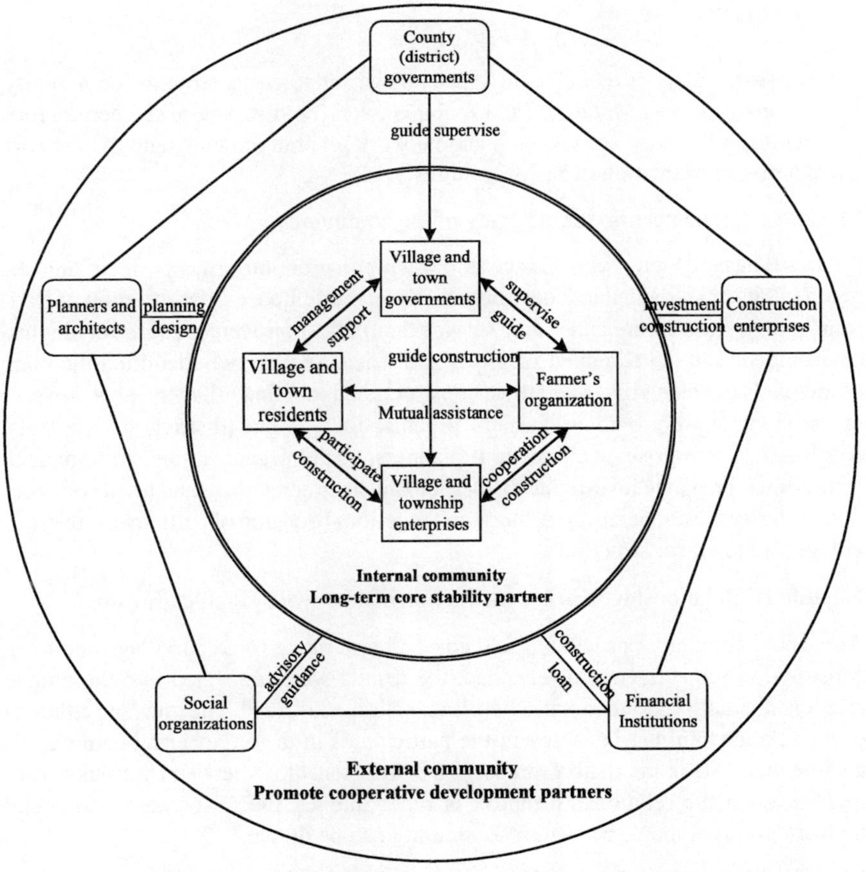

Fig. 4.4 Relationship of the social network of the green village and town construction community

through 4 stages: the embryonic stage, starting stage, growth stage and development stage. With improvements in the green village and town development and management levels, the network composition of the construction community will constantly evolve and develop.

(3) Optimization of network structure. The social network structure of the green village and town construction community should be constantly innovated and optimized in the construction process to meet the new needs of the green village and town development and construction model.

4.3.2 Network Analysis Model of the Green Village and Town Construction Community

1. Establishment method of the network analysis model of the construction community

In the network analysis model of the green village and town construction community, towns consist of 3 sub-modules: the social network field survey of the community, the relationship matrix construction and network graphic drawing, and the network characteristics evaluation of the community.

Module I: Social network field survey of the community.

First, we investigated green village and town construction projects, including the green village and town plan, construction of green farmhouses, construction of environmental infrastructure and public service facilities, improvement of the ecological environment and development of green industries. Second, we identified the main participants in green village and town construction, including villagers, town governments and village committees, farmers' organizations, county (district) governments, construction enterprises, architectural planners, social organizations, and financial institutions. Then, we investigate the participants in green village and town construction to clarify the cooperation methods and functional division of participants in green village and town construction.

Module II: Relationship matrix construction and network graphic drawing.

After the community participants and co-operation mode of green village and town construction projects were determined, the results were translated into the unique relationship matrix by the social network analysis method [21]. The rows and columns of the relationship matrix represent the participants in the engineering community, and the elements in the matrix represent the "relationships" between various participants. When the relationship matrix is input into UCINET software, the social network graph of the construction community can be drawn.

Module III: Network characteristics evaluation of the community.

Based on the network characteristics evaluation of the green village and town construction community, several key indexes in the social network analysis method, including degree centrality, betweenness centrality, closeness centrality, Katz index and others, were selected to characterize and optimally evaluate the construction community. Then, on the basis of the characteristics evaluation, optimization strategies were proposed for constructing village and town community networks.

2. Social network evaluation index of the construction community

The social network evaluation of the green village and town construction community towns consists of 4 core indexes, and the characteristics of each index are as follows:

Index 1: Degree centrality. The degree centrality of the network node refers to the number of other points directly connected to this point. It is expressed as

$$C_{Di} = \sum_{j=1}^{n} x_{ij},\ j \neq i \tag{4.1}$$

where x_{ij} is 0 or 1, which represents whether point j is related to point i.

In the social network of the green village and town construction community, if a participant has the highest degree centrality, then it is at the centre of the network, has the most power, and can be regarded as the leader of the green village and town construction project.

Index 2: Betweenness centrality. If a point is at the geodesic (the shortest path between 2 points) of many other point pairs, it has a higher betweenness centrality. Betweenness centrality measures the potential ability of a point to act as a mediator to control other points. Suppose that the number of geodesics between point j and point k is gjk; then, the probability that the third point i is on the geodesic between points j and k is $bjk(i)$, and the number of geodesics between points j and k passing through point i is $gjk(i)$; thus, $bjk(i) = gjk(i)/gjk$. The betweenness centrality of point i can be expressed as

$$CBi = \sum_{j}^{n} \sum_{k}^{n} bjk(i),\ j \neq k \neq i,\ j < k \tag{4.2}$$

In the social network of the green village and town construction community, if a participant persuades more individuals or organizations to join the village and town construction project, the participant is said to have higher betweenness centrality. This participant can be regarded as a bridge to other participants.

Index 3: Closeness centrality. This index measures the degree to which a point is not controlled by others. If a point is close to all other points in the network, it is said to have higher closeness centrality. The closeness centrality of a point is the reciprocal of the geodesic distance sum between it and all other points in the network. Its formula is expressed as

$$C_{Ci}^{-1} = \sum_{j=1}^{n} d_{ij},\ j \neq i \tag{4.3}$$

where d_{ij} is the geodesic distance between point i and point j.

In the social network of the green village and town construction community, if a participant is close to many other participants, the participant is said to have higher closeness centrality and can better communicate with other participants and play an important coordination role in village and town construction.

Index 4: Katz index. An influence index measures the degree to which an actor receives attention from or pays attention to others. The calculation of the influence index requires considering both the direct relationship and the indirect relationship of actors. Commonly used influence indexes include the Katz index, Huber index and Taylor index [22]. This evaluation method adopts the Katz index, and its calculation formula is

$$t = \left(\frac{1}{a}\text{I-C}'\right)^{-1} \tag{4.4}$$

where C is the relationship matrix between actors and a is the attenuation coefficient.

In the social network of the green village and town construction community, if a participant has a higher Katz index, it is said to have a greater or deeper impact on other participants, and it can play a strong role in guiding the development of village and town construction projects.

3. Social network optimization strategy of the construction community

By applying the above network analysis model and analysing the evaluation results of the network indexes, the function of the social network of the green village and town construction community can be optimized. The main methods are as follows:

1) Enhance the functions of participants with low indexes in the construction community

The participants with low indexes in the community are generally at the edge of the network. Participation methods should be optimized to enhance the links between these participants and the network core, to enhance the density and cohesion of the overall network, and to improve the overall coordination of construction projects.

2) Strengthen the network bridge and link impact of participants in the construction community

If the betweenness centrality of a participant in the network is much higher than that of other participants, it means that this participant has taken on more of an intermediary role. The burden of communication and coordination should be reduced by strengthening connections with other participants to make the distribution of network resources more uniform, to make information transmission smoother, and to improve the overall cooperation efficiency of the village and town construction community.

4.3.3 Case of Social Network Analysis of a Green Village and Town Construction Community

With the social network analysis of the construction community of Yanhe Village, Hubei Province, as an example, the analysis method is as follows.

1. Green village construction project analysis in Yanhe Village

Through long-term field investigation in Yanhe Village, 106 green village plans and construction projects in Yanhe Village over a 10-year period were analysed.

The classification and contents of the Yanhe Village green village construction project are shown in Table 4.4.

2. Social network construction of the Yanhe Village construction community

The construction project of Yanhe New Village introduced in Sect. 3.2.4 is taken as an example. The main participants included the Yanhe Village committee, 13 villagers, Yanhe Village Eco-tourism Cooperative, Wushan Town construction team, Huazhong University of Science and Technology, Wuhan University of Technology, Beijing Green Cross and 3 other social organizations, Wuhan Chuanmu Doors and Windows Company and 10 green construction enterprises. Through investigation and analysis, the relationship matrix between the participants in the construction project was constructed, and social network analysis software UCINET was used to draw a network diagram of the construction community [1].

The development of the Yanhe Village construction community experienced 4 stages: germination stage, starting stage, growth stage and development stage [18].

The network structural changes of the Yanhe Village construction community at different development periods are shown in Fig. 4.5.

3. Development stage analysis of the Yanhe Village construction community

1) Germination stage of construction

Since 1992, the village committee has led villagers in raising funds to build roads, vigorously carry out afforestation, and develop economic forests, tea gardens and tea-growing organizations. Some villagers took the lead in contracting tea farms for 3 village groups, enabling the tea industry to rapidly develop into the first green pillar industry in Yanhe Village.

In its germination stage, the Yanhe Village construction community consisted of the village committee, villagers, foreign enterprises and town government, and its network structure analysis is shown in Fig. 4.5a.

2) Start-up stage of construction

In 2003, the Beijing Green Cross, a non-governmental environmental organization, chose Yanhe Village as a pilot "ecological civilization village". The village committee

Table 4.4 Classification and contents of the Yanhe Village green village construction project

Number	Construction project classification	Construction project contents
1	Village construction plan	Yanhe Village social and economic development plan, Yanhe Village construction master plan, Yanhe Village ecological leisure tourism resort tourism development master plan, Yanhe Village development and management plan, Yanhe Village rural tourism development master plan, Bairishan scenic area construction plan, Luyuzhuang Park construction plan and 7 other items
2	Key rural house construction	Ordinary farmhouse, Tianyi farmhouse, strange stone museum, Yanhe New Village Group, Shengxian Farm, Ginkgo villa, Dushan house, reception centre, tea house and Tianyi tea house, for a total of 10
3	Infrastructure construction	Village power system capacity increase, fire-control system, circular road, village-level road, bus station, riverbank regulation, Luyu Manor Manshui Bridge, Yanwan Bridge, Tianyi Manshui Bridge, scenic suspension bridge, communication network, broadband network, and land consolidation, for a total of 13
4	Public facility construction	Yanhe Village primary schools, village service centre, Yanhe Village gate, mayor station, village clinic and scenic spot, weather stations, Yanhe Plaza, Jinyang ecological square, farmhouse, public toilets, scenic spot star toilet, village scene identification systems, landscape lamp, ecological corridor, Chaking Pavilion, Zhenwudian, Jiabandong road, palace, Changshenggong museum, Bairishan Mountain scenic area, reception centre, Bairishan Mountain gate, Bairishan Mountain parking lot, Luyu sculpture, Luyu estate, square LED display, etc., for a total of 25
5	Environmental improvement construction	Garbage classification, one construction and four renovation projects, sewage treatment stations, ecological roofing, ecological toilets, conversion of farmland to forest, village greening, environmental renovation, low-carbon solar garbage cans, etc., for a total of 9

(continued)

Table 4.4 (continued)

Number	Construction project classification	Construction project contents
6	Industrial development and construction	Yanhe ecological tea garden, seedling and flower planting base, *Eucomia ulmoides* planting base, mushroom and fungus planting base, pig and sheep breeding base, native chicken breeding base, fish-pond breeding, Yanhe reception centre, Ginkgo villa, 22 farmhouse construction projects, Zhenwu restaurant, Yanhe Incense Square snack shop, square tea shop, 7 supermarkets in Yanhe Village, mountain spring drinking water development, village tourism Internet development projects, etc., for a total 42

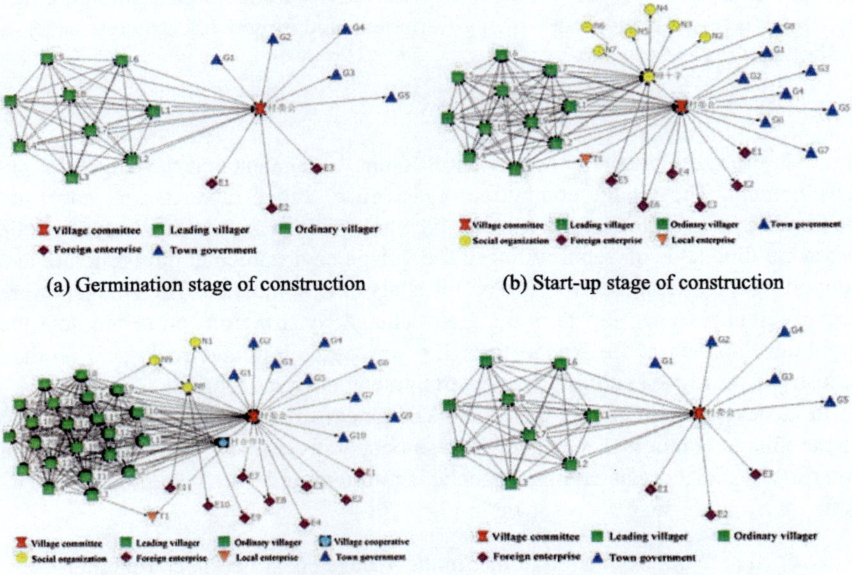

(a) Germination stage of construction

(b) Start-up stage of construction

(c) Growth stage of construction

(d) Development period of construction

Fig. 4.5 Structural changes in the social network of the Yanhe Village construction community at different developing stages

organized and worked out the economic development plan of the village and the development plan of the Yanhe scenic spot and guided villagers to invest in the construction of infrastructure such as roads and bridges and the renovation of old houses in the village. It organized villagers to set up farmhouse facilities so that ecological tourism quickly developed into the second green pillar industry in Yanhe Village.

In its start-up stage, the Yanhe Village construction community consisted of the village committee, villagers, local enterprises, social organizations, foreign enterprises and town government, and its network structure analysis is shown in Fig. 4.5b.

3) Growth stage of construction

In 2007, Yanhe Village set up an ecological tourism cooperative to attract villagers to become shareholders. Co-operatives develop eco-tourism through multi-channel financing, which promotes the rapid development of the village economy. During this period, Yanhe Village and Huazhong University of Science and Technology worked out the social and economic development plan and green village construction plan and jointly carried out the green and ecological well-off village construction demonstration. Thus, the village construction community grew rapidly.

In its growth period, the Yanhe Village construction community consisted of the village committee, villagers, village cooperatives, local enterprises, foreign enterprises, social organizations and town government, and its network structure analysis is shown in Fig. 4.5c.

4) Development period of construction

In 2012, the green village construction of Yanhe Village entered the stage of rapid development. The village committee organized a unified plan, design, construction and renovation for farmhouses of the whole village and gradually formed the green building style of Yanhe Village. The village environmental infrastructure and supporting facilities of eco-tourism scenic spots were completed. The villagers were actively guided to participate in the green village construction and to promote the rapid development of the Yanhe Village green agriculture and ecological tourism industry featuring tea culture, rural entertainment and rural tourism.

In its development period, the Yanhe Village construction community consisted of the village committee, villagers, village cooperatives, local enterprises, foreign enterprises, social organizations, financial institutions and town government, and its network structure analysis is shown in Fig. 4.5d.

4. Social network evaluation of the Yanhe Village construction community

1) Calculation of social network indexes of the construction community

UCINET was used to calculate the social network analysis indexes of the Yanhe New Village construction community, including degree centrality, betweenness centrality, closeness centrality and the Katz index. The calculation results reflect the network characteristics of each participant in the Yanhe Village construction project.

See Table 4.5 for the calculation results of the network indexes of the Yanhe Village construction community.

Table 4.5 Calculation results of the network indexes of the Yanhe Village construction community

Sequence	Degree centre		Intermediate centre		Closeness centrality		Katz index	
1	Village committee	25.000	Village committee	164.571	Village committee	0.879	Village committee	0.077
2	Villagers	17.000	Huazhong University of Science and Technology	88.000	Villagers	0.659	Villagers	0.053
3	Beijing Green Cross	16.000	Demonstration farmers	43.071	Beijing Green Cross	0.644	Beijing Green Cross	0.050
Last bit	Construction enterprises	1.000	All other participants	0.000	Construction enterprise	0.377	Enterprises	0.003

Table 4.6 Social network analysis and evaluation of the Yanhe Village construction community

Participants	Evaluation index analysis
Village committee	The village committee's degree centrality, intermediate centre degree, proximity centre degree and Katz index are all in the first place, indicating that the village committee was at the centre of the network, was an important bridge and link in the network and had a strong influence
Villagers	The degree centrality, proximity centre degree and Katz index of the villagers are second only to those of the village committee, which indicates that the villagers were the secondary centre of the network, were close to other parties in the network and had certain influence and attention
Huazhong University of Science and Technology	Huazhong University of Science and Technology is second only to the village committee in intermediate centrality. As an intermediary, it introduced a group of new participants to Yanhe Village and built a bridge of cooperation and development between internal and external participants
Beijing Green Cross	The Beijing Green Cross's degree of centrality, proximity to the centre and Katz index are in third place. Its position in the network followed those of the village committee and villagers, and it was an active external participant

2) Social network analysis and evaluation of the construction community

The social network analysis and evaluation of the Yanhe Village construction community are shown in Table 4.6.

3) Problem diagnosis of the construction community social network

(1) Enhance capacity of participants with low indexes in the community

Table 4.5 shows that the indexes of the participating enterprises are the lowest, and they are at the edges of the network. The reason is that enterprises are mostly introduced by external participants and have little contact with internal participants, especially villagers. Therefore, enterprises should focus on strengthening their understanding of rural construction needs and communication with villagers to promote the popularization of green construction technologies in rural areas.

(2) Enhance network bridge communication among community participants

Table 4.5 shows that the degree centrality and closeness centrality of the village committee are much higher than those of other participants, indicating that the village committee played a major leading role as a network bridge and link, which may easily have led to the functional failure of the overall network structure due to decision-making errors by the village committee. The village committee should focus on coordinating a close relationship between the internal community and external community, improving its scientific decision-making ability, and guiding the sustainable development of the green village.

4) Optimal path of construction community development

The construction and development process of Yanhe Village over a 20-year period was also a continuous innovation and optimization process of the construction community. The optimal development path of the Yanhe Village construction community is as follows:

(1) In the germination stage of construction, through joint exploration by the village committee and villagers, the structural optimization of the construction community led by the development of the tea industry and rural tourism was promoted with strong support from the county and township governments.
(2) In the start-up stage of construction, the village committee introduced social organizations such as colleges and universities to participate in the construction and introduced foreign enterprises and units to cooperate with and invest in the green village construction project, which promoted the structural optimization of the construction community led by the village committee.
(3) In the growth stage of construction, with the village committee and village farmer cooperatives as the core and with the help of Huazhong University of Science and Technology and other social organizations, the construction of a state-level well-off ecological model village promoted the optimization of the structure of a green village construction community with close cooperation between the internal community and external community.
(4) During the development stage of construction, Yanhe Village successively attracted many foreign enterprises to participate in the green village construction and built Yanhe Village into a national demonstration base for green village construction, which promoted the long-term cooperation mechanism of the construction community and the optimization of the benefits of green village construction.

See Fig. 4.6 for the development and optimization path of the Yanhe Village green village construction community.

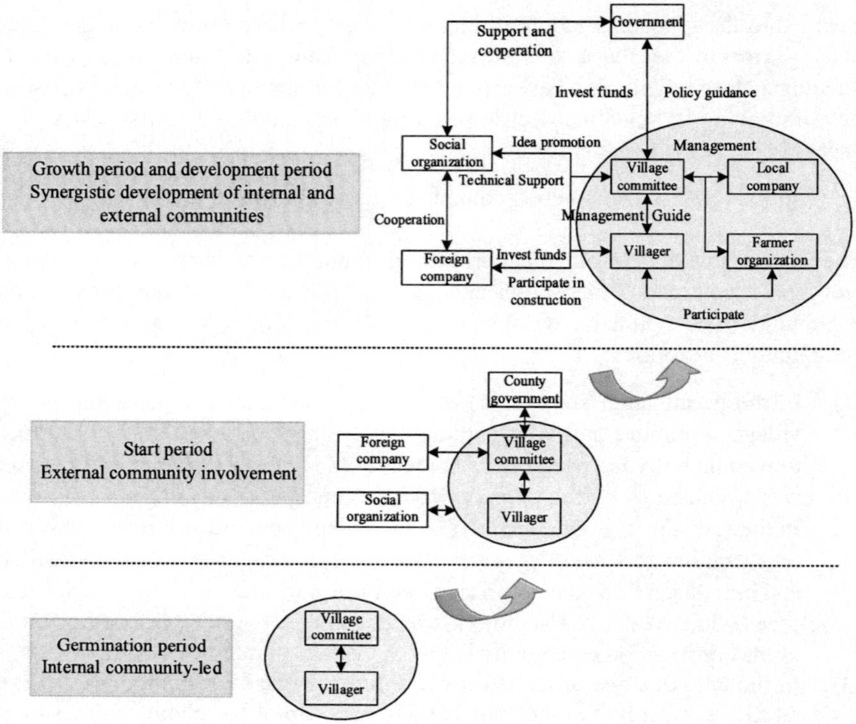

Fig. 4.6 Development and optimization path of the Yanhe Village green village construction community

References

1. Wang XM, Wang X, Wu JX, et al (2017) Modern rural settlement construction engineering community and network analysis model. J Civil Eng Manag 1:1–9
2. Zhang XH (2008) The nature of the engineering community. J Dialect Nat 6:43–47
3. Aristotle (1983) Political science. Commercial Press, Beijing, p 3
4. Tenis F (1999) Community and society: basic concepts of pure sociology. Commercial Press, Beijing, pp 65–71
5. Wang XM, Wang X, Wu JX, et al (2016) A preliminary study on the partnership model of green agricultural housing construction. Real Estate China (Academic Edition) 7:36–42
6. Tian P, Chen SJ (2015) "Internal engineering sociology" and "external engineering sociology"—an attempt to construct a system of engineering sociology. Eng Res Eng Interdiscip Persp 1:87–95
7. Tian P, Chen SJ (2016) From "science community" to "engineering community"—discussing with professor Li Bocong. Eng Res Eng Interdiscip Persp 1:55–62
8. Li BC (2010) Introduction to engineering sociology. Zhejiang University Press, Hangzhou, p 8
9. Han F, Cai JM (2011) Morphological evolution and reconstruction of rural settlements in semi-urbanized areas of China. Geog Res 7:1271–1284
10. Guo XD, Ma LB, Zhang QY (2012) Research on spatial evolution characteristics and driving mechanism of rural settlements in Qin' an county based on GIS. Econ Geog 7:56–62

11. Wang D (2007) Co-construction of rural settlements and integration of architects. Time Archit 4:16–21
12. Fei XT (2019) Peasants life in China. The Commercial Press, Beijing, pp 33–36, pp 93–103
13. Li GQ (2005) The controversy about Chinese village community—take Jieneng-Rongyuan as core. Sociol Res 5(6):194–213
14. Shan DQ, Yuan M (2008) 18 years since the reconstruction of Mulouzhai village in Rongshui—a re-reflection on the exploration of the reconstruction of traditional settlements in the western poverty-stricken areas. World Archit (7):21–29
15. Wu LC, Wu KF (2014) Four models of beautiful countryside construction and their comparison—based on the survey of Anji, Yongjia, Gaochun and Jiangning. J Huazhong Agricult Univ Social Science Edition) 1:15–22
16. Hao D (2015) Haotang village: make the countryside more like the countryside—the sustainable development experiment of Haotang village in Xinyang, Henan Province. Acad Rev 1:34–39
17. Wang X (2017) Study on the social network of construction project community of rural settlement. Dissertation, Huazhong University of Science and Technology
18. Liu J (2004) Introduction to social network analysis. Social Sciences Academic Press, Beijing, pp 24–26
19. Marquardt D, Möllers J, Buchenrieder G (2012) Social networks and rural development: LEADER in Romania. Sociol Ruralis 52(4):398–431
20. Nuñez-Espinoza JF, Figueroa Rodríguez OL, Jiménez-Sánchez L (2014) Elements available to analyze social networks for rural development in Mexico the RENDRUS case. Agricultura, Sociedad Y Desarrollo 11(1):1–24
21. Liu J (2009) Lecture notes of holistic network analysis. Gezhi Press, Shanghai
22. Liu J (2014) Holistic network analysis: a practical guide to UCINET software, 2nd edn. Gezhi Press, Shanghai

Part II
Green Village and Town Planning and Construction Methods

Chapter 5
Plan Compilation of Green Village and Town Construction

Abstract Green village and town construction and development should be guided by a construction plan. In this chapter, the nature and characteristics of the construction plan are analysed, methods of compiling the construction plan are proposed, the main contents and research methods of the construction plan are introduced, and construction plan compilation cases are provided.

The construction plan is the programmatic document of green village and town construction, and a supplementary special plan is also needed to guide green village and town construction development. The implementation of the construction plan underlies the whole process of green village and town construction and management. According to the actual development needs of different types of villages and towns, appropriate plan types and compilation methods should be flexibly selected.

5.1 Nature and Characteristics of the Green Village and Town Construction Plan

5.1.1 Nature of the Green Village and Town Construction Plan

A village and town plan refers to the comprehensive deployment, specific arrangement and implementation management for economic and social development, land use, spatial layout and various constructions in a village or town in a certain period of time. It is the comprehensive deployment plan formulated by the town government or the village committee to realize economic and social development goals; determine the nature, scale and direction of development; and coordinate the town or village layout and construction. It is the basis for village and town construction and management [1].

The green village and town construction plan refers to green development and the special plan for green construction in the current township planning system. It is a

X. Wang and H. Hua, *Green Village and Town Construction in China*,
https://doi.org/10.1007/978-981-16-2098-0_5

plan to guide a village or town in implementing green construction and sustainable development in a certain period of time. The green village and town construction plan is a special plan for green construction and development that should meet the requirements of village and town plans.

The compilation and implementation of the green village and town construction plan should guide the green development of construction and ensure the comprehensive implementation of green village and town construction measures focusing on land savings, energy savings, water savings, material savings and ecological environmental protection; realize the construction requirements for a liveable village or town; provide suitable jobs for villagers; promote resource conservation, a beautiful environment and safety; promote the coordination of the economy, society and ecological environment of the village or town; and realize harmonious co-existence between humans and nature.

The green village and town construction plan is a special plan to supplement the green construction and development content in the "town plan" or "village plan", and the plan content and scale should be flexibly adapted to local conditions. According to the purpose and contents of the plan, the scope of the green village and town construction plan can be selected from different areas, such as town region, village region or village built-up region.

5.1.2 Characteristics of the Green Village and Town Construction Plan

1. Adhering to green and sustainable village and town development

The essence of green village and town development is to achieve sustainable development. Therefore, the basic requirements for green village and town construction are the sustainable use of resources, the sustainable growth of economic industries, the sustainable improvement of the ecological environment, and the sustainable development of social life.

The process of green village and town construction is a green transformation and development process of the traditional village and town construction pattern. The principles of "people-oriented" and "green development" must be adhered to in the process of plan compilation and implementation to meet most rural residents' increasing demand for green living, green production and green ecological quality improvements.

2. Emphasizing harmonious co-existence between construction and the natural geographic situation

The green village and town construction plan should pay close attention to coordination with the natural geography of the area where the village or town is located and should preserve and excavate the green elements of local village and town construction. Three main aspects are reflected:

1) Adapt to local geographical and climatic conditions

In the vast rural areas of China, which cover 50° of latitude from south to north, the geographical and climatic conditions different regions vary greatly, leading to great diversity in traditional village plans and rural buildings with local characteristics compatible with the geographical and climatic conditions. For example, much attention is paid to winter protection, windproofing and warmth retention in cold northern regions; ventilation, heat dissipation, rain shelter and moisture-proofing are highlighted in hot and humid southern areas; typhoon and rainstorm resistance, ventilation improvement, heat dissipation and lighting conditions in building roadways and patios are emphasized in coastal areas; and in the northwest plateau area, earthen buildings such as cave dwellings are constructed to adjust temperature differences and achieve the proper temperatures at different times and in different seasons.

2) Coordinating with local topographic and geomorphic characteristics

Approximately two-thirds of China's land area is mountainous. Mountain residents generally use relatively flat land as farmland and usually construct buildings on slopes or mountainsides, forming a beautiful mountain village arrangement. In the southern Yangtze River regions with a developed water system, villages and towns are built along the river, forming a unique water-town landscape. Villages and towns in the plains are built on higher ground with dams and drainage ditches to prevent natural flooding, forming a village and town arrangement based on the natural ecotype and disaster prevention.

3) Meeting geological conditions and using local materials

Meeting local geological conditions and using local materials are principles to be followed in green village and town construction. For example, wood, bamboo and stone in mountainous areas are widely used as materials for buildings in different styles. The soil of the Loess Plateau in the northwest has good stability, and the dry climate and low groundwater level are conducive to the arched structure of earthen cave dwellings. The structures are not only warm in winter and cool in summer but also save precious wood and heating resources. The adobe buildings in rural areas of the Central Plains, the stone watchtowers and slab walls in Tibet, etc. are all typical examples of meeting geological conditions and using local materials. These building styles should be reasonably preserved and inherited in present-day green village and town the planning and construction.

3. Reflecting the characteristic features and diversity of villages and towns in different regions

Green village and town plans should fully reflect local characteristics. Village and town construction and development have generally undergone a long process of evolution and formed local characteristic features. In green village and town planning and construction, we should not simply or blindly apply a single "standard map" to build homogenized villages and towns. It is necessary to inherit and develop the plans and construction methods of rich and diverse traditional villages and towns and to build rural villages and towns that are more beautiful, greener and more ecological.

The typical characteristics of villages and towns in different regions of China should be effectively protected and inherited in the planning and construction of green village and towns [2]. There are five main characteristic types of village and town construction:

1) Flat type

This type is generally seen in the plains. For small-scale villages, the development of a main street is often taken as an axis, forming a group layout of heads, centre, surrounding areas, etc. The main streets are mostly east-west oriented; thus, buildings tend to be north-south oriented. For larger towns, cross streets are used as the development axis, and the cross street is the public activity centre and resident gathering place. As villages and towns develop, especially in aspects of business and the service industry, network-type streets and functional group layouts are adopted.

2) Mountain type

Most villages and towns of this type are built on hillsides. One type is arranged along the parallel of the contour line; the trend conforms to changes in the mountain potential, and the street is in the form of a curved strip. Another type is built on the mountain and laid perpendicular to the contour line. It sets the stage, forming layer changes. Regarding orientation, villages and towns in mountainous areas are located on the sunny slopes of the mountains and have good sunny exposure and shelter from the wind. Regarding elevation, most are built in the foothills. The terrain can resist flood erosion and facilitate external traffic. Regarding drainage, the backs of mountains are chosen and combined with drainage ditches to prevent damage from rainstorms and floods.

Cave-type villages are also common mountain type of traditional settlement found on the Loess Plateau in northwest China.

3) Back-mountain type

The first type is a village or town set against the mountainside facing the water. The mountainous terrain barrier can be used to resist cold winds, while the water body is helpful for convenient living and transportation. Generally, the sunny slope of the mountain close to the shore is chosen, and the construction method of masonry stone foundations or wooden lifted pillars is adopted so that the building is higher than the normal water level, preventing flood disasters. Thus, the mountain view, architecture and water form a beautiful landscape of "lucid waters and lush mountains".

The second type of village is set against the mountainside facing farmland. Generally, the sunny slope of the mountain close to the farmland is chosen, reflecting a blend of the natural environment and pastoral scenery.

4) Water type

Represented by the regions south of the Yangtze River and the large lakes and water towns in central China, villages and towns are built along one or both sides of the river in an interwoven river network system. The villagers engage mainly in agriculture, fishery and handicraft production. Small bridges are built to form 2 traffic networks of waterways to provide convenience for production and daily living. The towns are generally located at the confluences of rivers. Buildings generally face streets and back onto rivers to make full use of waterway transportation. Waterside villages and towns have the same common streets and lanes as ordinary villages and towns as well as waterfront streets and lanes and facilities such as bridges and docks, forming a waterside style of “tiny bridges, flowing brooks, and cottages”.

Fishing settlements are another typical waterfront type of village and town. China’s rivers, lakes and seas are rich in resources, and there are many islands. Fishing settlements are generally built on river forks or bays. The fishermen’s buildings are mostly built on the water’s edge, and some are supported directly on the water. The fishermen can board their boats directly from the buildings to facilitate water production. Large-scale fishing settlements have dedicated wharfs and aquatic product trading fairs, which have gradually evolved into market towns.

5) Scattered distribution type

This construction form is common in subtropical multi-ethnic regions. For example, the Dai nationality residents in Yunnan Province have a Buddhist temple as a public activity centre. They adopted a dry-type construction form and opening around bamboo buildings to gain high ventilation for problems such as local high temperatures, the damp and rainy environment, and low terrain features. The tower buildings in Guangdong Province and Fujian Province are also typical scattered villages and towns that have formed a colourful ethnic style and unique tourism cultural resources.

4. Fully embodying integration with local traditional culture and customs

Green village and town plans should fully inherit the historical context. China’s excellent rural traditional plans and construction concepts fully embody the characteristics of “green ecology”, such as the “co-existence of Yin and Yang” and “geomantic omen theory”, which have referential significance for modern green village and town planning and construction.

China’s traditional cultural concept of planning and construction is guided by the concepts “unity of man and nature” and “co-existence of Yin and Yang”. It advocates a people-oriented approach and coordination with the natural ecological environment; respects the religious and moral culture, family kinship and local living customs of different regions; builds production and trading places based on the natural ecology and harmonious “group living” community of villagers; and reflects multiple attributes of the relationships between humans and nature, humans and humans, and humans and culture. It reflects ancestral wisdom in terms of “liveability” planning and construction.

"Feng Shui" (geomantic omen), the traditional theory of Chinese planning, is an example. Feng Shui theory in Yi integrates Yin-Yang and five elements, four images and eight trigrams, and yuanqi theory into one to study the optimal architectural allocation of "qi, power, principle and form".

Chinese planning site selection for ancient towns, roads, villages, buildings, temples, and cemeteries paid close attention to the four directions and locations, was closely integrated with celestial phenomena, and pursued the "astronomical phenomena field" and "the help of heaven and man" to promote the mutually beneficial and prosperous development of humans and nature [3].

5.2 Compilation Method of the Green Village and Town Construction Plan

The green village and town construction plan is a special plan for green village and town construction. The plan compilation for green village and town construction should be carried out in accordance with laws and standards such as the *Urban and Rural Planning Law of the People's Republic of China, Standard for Planning of Town, Technique Code for Village Rehabilitation*, and *Guidelines for the Construction of Beautiful Village.*

5.2.1 *Main Content of Green Village and Town Construction Plan*

Generally, green village and town construction plans can be divided into three basic types: the economic and social development plan, the construction and development plan, and the comprehensive development plan.

1. Economic and social development plan

1) Contents of plan

The green village and town economic and social development plan is the comprehensive deployment, specific arrangement and management measures for green development goals and key implementation projects of village and town future economic and social development. The plan content mainly includes construction status and development resource analysis, an economic and social development goal plan, an economic industry development plan, an ecological tourism development plan, a social development plan, and an ecological environmental protection plan.

2) Time limit of plan

The time limit of the economic and social development plan is generally divided into 3 periods: the short term is 3–5 years, the medium term is 8–10 years, and the long term is 15–20 years. During the implementation of the plan, it can be revised according to the changing needs of the village and town economic and social development, the completion of construction projects and changes in environmental conditions. A new round of planning should be organized and compiled when the plan expires.

The outline of the green village and town economic and social development plan is shown in Fig. 5.1.

2. Construction and development plan

1) Main contents of plan

(1) The green village and town land-use and functional zoning optimization plan includes green village and town construction site selection optimization, with the key points being the site selection of the town residential area and village residential groups and buildings, and green village and town construction function zoning optimization, with key points being basic farmland protection areas, general agricultural land, rural construction land, independent industrial and mining areas, scenic and tourist areas, ecological and environmental safety control areas, natural and cultural heritage reserves, and forestry and animal husbandry reserves.

Outline of green village and town economic and social development plan

1 General Principles
1.1 Origin of task
1.2 Planning scope (town domain or town, village domain or village, etc.)
1.3 Planning purpose and significance
1.4 Basis of planning compilation
2 Basic Situation and Development Trend Analysis of Villages and Towns
2.1 Physical geography
2.2 Socio-economic profile
2.3 Overview of development resources
2.4 Overview of ecological environment
2.5 Analysis of social and economic development trend
2.6 Advantages and constraints of development
3 Guiding Ideology and Goals for Economic and Social Development of Green Villages and Towns
3.1 Planning guidelines (Green development, coordinated development, innovative development, sustainable development, etc.)
3.2 Basic principles of planning (people-Oriented, local conditions, overall planning, optimal comprehensive benefits)
3.3 Time limit of planning(divided into short-term, medium-term and long-term goals, with the previous year as the base year)
3.4 Economic and social development planning objectives (economic development, social development, ecological environment indicators)
4 Economic and Industrial Development Planning of Green Villages and Towns
4.1 Analysis of economic development environment (favorable conditions and constraints)
4.2 Industry status analysis and structural optimization
4.3 Economic and industrial development target planning
4.4 Leading industrial development planning (ecological agriculture, ecological tourism, etc.)
4.5 Key development areas and construction projects
5 Development Planning for Social Undertakings in Green Villages and Towns
5.1 Development advantages and bottlenecks
5.2 Development goal planning of social undertakings
5.3 Key development areas and construction projects
5.4 Mechanisms for promoting development of social undertakings
6 Ecological and Environmental Protection Planning for Green Villages and Towns
6.1 Status quo evaluation of ecological environment quality
6.2 Planning of ecological function zoning
6.3 Construction of natural resources and ecological environment protection system
6.4 Major areas and key tasks of ecological and environmental protection
7 Safeguard Measures for Planning and Implementation
7.1 Organizational guarantee
7.2 Fund guarantee
7.3 Social security
7.4 Promotion mechanism
8 Atlas and Accessories

Fig. 5.1 Outline of green village and town economic and social development plan

(2) The green village and town environmental infrastructure construction plan includes the construction of drinking-water source protection and water-saving facilities; construction of domestic sewage treatment facilities; construction of garbage treatment facilities; construction of clean energy utilization facilities; construction of greening and landscape facilities; and construction of disaster prevention and mitigation and early-warning facilities.
(3) The green village and town green building construction and renovation plan includes green building plans and design requirements; green renovation of existing rural buildings; and green operation and management of village and town communities.
(4) The green village and town ecological environmental restoration plan includes the ecological landscape construction pattern and the ecological environmental treatment and restoration plan.

2) Implementation period of plan

Generally, the green village and town construction development plan can be divided into three phases: 3 years for the short term, 5 years for the middle term, and 10–15 years for the long term.

3) Programming procedure

The compilation procedure of the green village and town construction development plan is as follows: collection of village plan data and field research → Optimization plan of village and town land use and functional zoning → Plan of environmental infrastructure construction → Plan of green building construction and old building reconstruction → Plan of ecological improvement and environmental restoration → Implementing safeguard measures in green village and town construction plan → Compiling atlas green village and town construction plan.

3. Comprehensive development plan for green village and town construction

The comprehensive development plan of green village and town construction, also referred to as the development plan of green village and town construction, is a comprehensive compilation of the main contents of the economic and social development plan and development plan. The content and planning period of the development plan of green village and town construction should be determined with reference to the above two types of plans.

5.2.2 *Investigation Methods of the Green Village and Town Construction Plan*

1. Investigation methods of the green village and town construction plan

The investigation methods of the green village and town construction plan are divided into data collection and field investigation.

1) Data collection

The materials needed to compile the green village and town economic and social development plan are mainly as follows:

(1) National, provincial and county current green development plans, policies and regulations for villages and towns;
(2) Current national, provincial, municipal, county and township economic and social development plans;
(3) Statistical data on economic and social development issued by statistical departments of governments at all levels;
(4) Natural resources and ecological environment of the village or town;
(5) Economic and industrial developmental conditions of the village or town;
(6) Developmental status of the social undertakings of the village or town;
(7) Various the village and town development plans in the early stage; and
(8) Research on the village or town and statistical data collected by the planning team.

The materials needed to compile the green village and town construction development plan are mainly as follows:

(1) National, provincial and county current construction plan of green village and town-related standards as well as technical provisions;
(2) Master construction development plan and special plan of villages and towns;
(3) Physical geography and developmental resources of the village or town;
(4) Economic and social development conditions of the village or town;
(5) Construction land and functional zoning of the village or town;
(6) Environmental infrastructure construction in the village or town;
(7) Rural building construction and old building reconstruction of the village or town;
(8) Ecological environment and improvements of the village or town; and
(9) Thematic research and statistical data of village and town conducted by the planning team.

2) Field investigation

The field investigation methods for compiling the green village and town construction plan are as follows:

(1) Data collection by relevant village and town planning and construction departments;
(2) On-the-spot reconnaissance of village and town construction;
(3) Environmental quality monitoring;
(4) Ecological landscape survey;
(5) Aerial photography investigation;
(6) Personnel interviews and symposiums;
(7) Questionnaire survey of village residents; and
(8) Other investigations of villages, towns and counties.

The specific content and methods of field investigation can be flexibly determined according to the needs of the plan and combined with the actual situation of the village or town.

2. Investigation case of the green village and town construction plan

The *Economic and Social Development Plan for Yanhe Ecological Village in Hubei Province (2008–2020)* and the *Construction Plan for Yanhe Village in Wushan Town, Gucheng County (2008–2020)* are taken as examples to briefly introduce the key points of the research on the compilation of the green village and town construction plan.

1) Physical geography

Yanhe Village is located 3 km northwest of Wushan Town, Gucheng County, Hubei Province. It is surrounded by mountains and rivers and has a beautiful environment. It is a typical hilly mountain village. The village covers an area of 1200 ha, with a basic farmland area of 64 ha and a tea garden area of approximately 66.67 ha. The village residential areas are mostly situated on the back of the mountain and facing the water and are integrated with the countryside and mountain forests. The rural people live simply, the ecological environment is beautiful, the climatic conditions are excellent, the landforms and landforms are diverse, and the natural resources are rich. The conditions for building a green village are good.

2) Population composition of the village

Yanhe Village is rich in labour resources, which are core resources for green village and town construction.

Basic information on the villager groups in Yanhe Village is shown in Table 5.1.

The demographic composition survey of Yanhe Village is shown in Table 5.2.

The age brackets in Yanhe Village in 2007 is shown in Fig. 5.2.

The proportion of population and labor force between groups in Yanhe Village in 2007 is shown in Fig. 5.3.

Table 5.1 Basic information of villager groups in Yanhe Village (2007)

Group category	Number of farmers	Population	Number of labourers	Cultivated land area (hm^2)
First group	69	250	135	15.39
Second group	67	227	115	16.59
Third group	83	296	165	22.29
Fourth group	23	64	31	9.90
Total	242	837	446	64.17

Table 5.2 Demographic composition survey of Yanhe Village (2008)

Total number of households at end of year	Total	Gender		Age			
		Male	Female	Under 18 years old	18–35 years old	35–60 years old	Above 60 years old
247	918	501	417	209	250	280	179

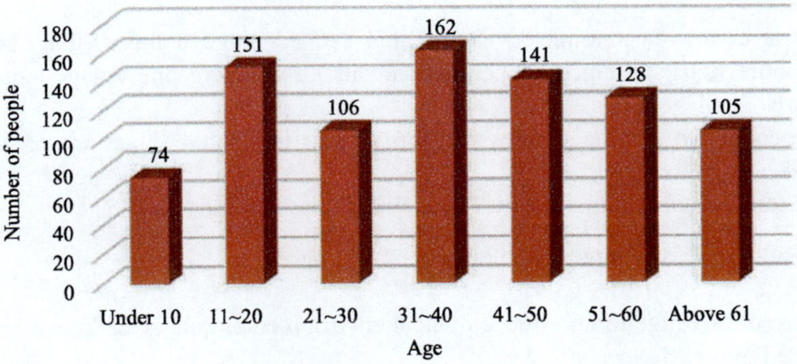

Fig. 5.2 Age brackets in Yanhe Village in 2007

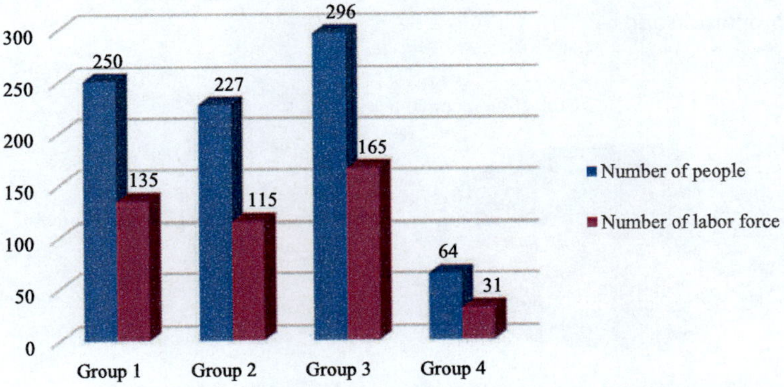

Fig. 5.3 Proportion of population and labor force between groups in Yanhe Village in 2007

3) Development of social undertakings

(1) Survey on the organizational structure of the village. Yanhe village-level organizations include the village CPC branch, the village committee, the villagers' meeting, the villagers' representative meeting and the villagers' group, and village affairs are managed by the village party branch and the village committee.
(2) Survey on the status quo of village construction, including village area location, village topographic climate and village construction projects. The key points are the construction quality of farm buildings, public service facilities, road and transportation facilities, water supply and drainage facilities, electric power facilities, telecommunications facilities, postal facilities, clean energy utilization facilities, environmental protection facilities, environmental sanitation facilities, and disaster prevention and reduction facilities.
(3) Honours won by the village. By the end of 2007, Yanhe Village had won more than 10 national, provincial and municipal advanced titles.

4) Foundation of economic development

The basic survey of economic development in Yanhe Village includes village collective income, household income, personal income of villagers, and village industrial structure.

The collective income sources and proportions in Yanhe Village are shown in Fig. 5.4.

The family monthly income in Yanhe Village is shown in Fig. 5.5.

5) Ecological environmental quality

The current investigation on the ecological environmental quality of Yanhe Village is divided into two aspects:

(1) Research on ecological environmental factors, including geographical environment, water environment, forest vegetation, and ecological agriculture proportion and

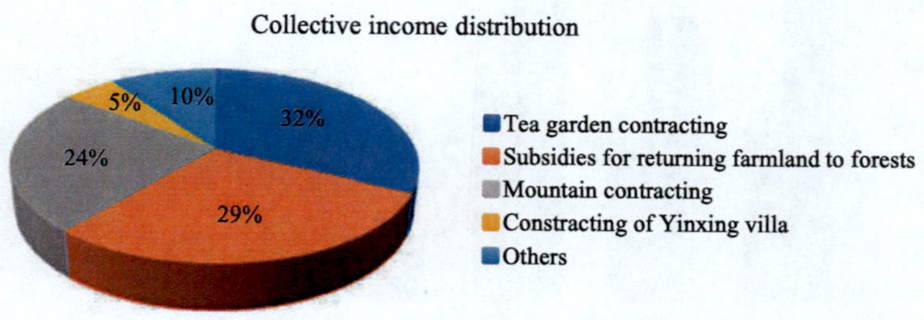

Fig. 5.4 Collective income sources and proportions in Yanhe Village in 2007

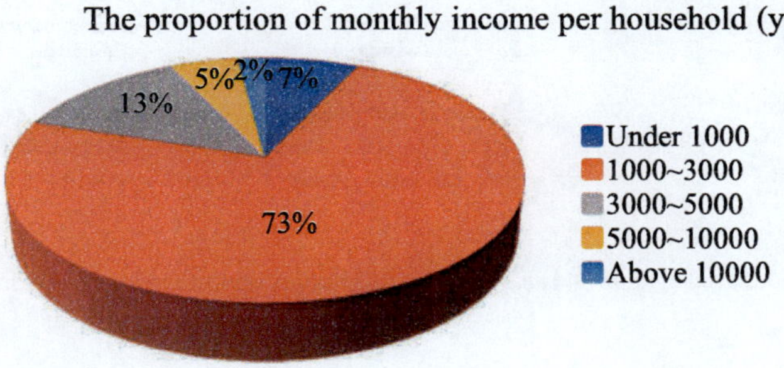

Fig. 5.5 Household income per month in Yanhe Village in 2007

(2) The current status of ecological environmental quality, including water environmental quality, air environmental quality, sound environmental quality, solid waste discharge, and industrial pollution source analysis.

The measured water environmental quality data in Yanhe Village are shown in Table 5.3.

The household investigation and analysis of domestic sewage treatment in Yanhe Village are shown in Fig. 5.6.

The household investigation and analysis of garbage collection and treatment modes in Yanhe Village are shown in Fig. 5.7.

6) Analysis of developmental resources

The developmental resources of Yanhe Village mainly include the following:

Table 5.3 Measured data of water environmental quality in Yanhe Village (2007)

Test items	Test sites			
	Yanhe	Macaogou	Gutterway in village	Mountain spring
COD_{cr} (mg/L)	38	135	417	22
BOD_5 (mg/L)	11.4	40.5	202.9	6.6

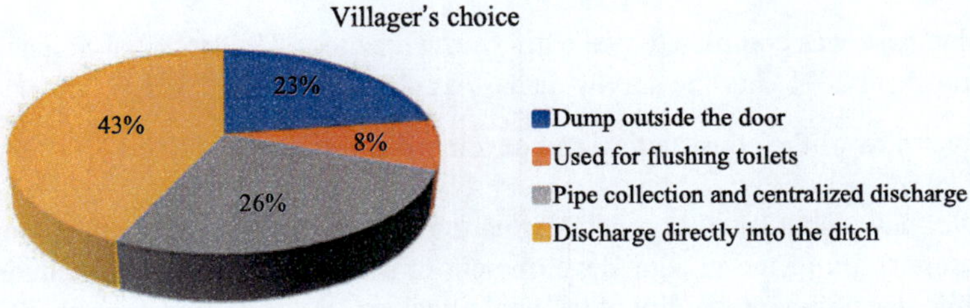

Fig. 5.6 Household investigation and analysis of domestic sewage treatment in Yanhe Village in 2007

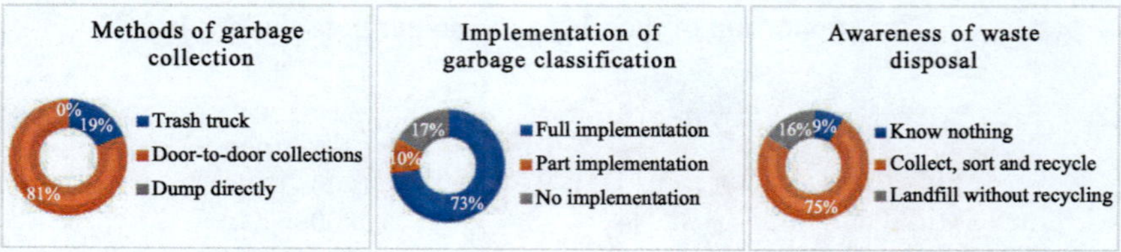

Fig. 5.7 Household investigation and analysis of garbage collection and treatment modes in Yanhe Village in 2007

(1) Land resources;
(2) Lake water resources;
(3) Ecological agricultural resources;
(4) Ecological tourism resources;
(5) Human resources; and
(6) Social network resources.

5.3 Compilation Cases of the Green Village and Town Construction Plan

The compilation of the green village and town construction and development plan should be guided by the green development of village and town construction and focus on the comprehensive implementation of land savings, energy savings, water savings, material savings and ecological environmental protection. In this section, the *Construction Plan for Yanhe Village in Wushan Town, Gucheng County (2008–2020)* and the *Construction Planning of Green Development Demonstration Town in Fushan Town, Chengmai County, Hainan Province (2016–2020)* compiled by the authors' team are taken as examples to introduce the key contents of the green village and green town construction and development plan.

5.3.1 Green Development Plan of Yanhe Village

This plan case was completed by a team from Huazhong University of Science and Technology in 2008. For the survey data on Yanhe Village, see Sect. 5.2.2.

1. Overview of the construction and development of Yanhe Village

The overview includes the general situation of the natural geography, population composition of the village, development of social undertakings, foundation of economic development, quality of the ecological environment, and current situation of village construction.

2. Green village construction target plan

(1) Overall planning objectives. Yanhe Village makes full use of its ecological environment and characteristic landscape resources to form a green and beautiful village with a superior living environment, good ecological benefits and reasonable industrial layout through the integration of the village's utilized space and building texture and environmental renovation.
(2) Key plan objectives. Taking the villagers as the subject, the objective is to meet the requirements of green housing construction in the village. Taking environmental improvement as the purpose, the objective is to upgrade the green environment of production and living in the village. Taking service provision as the purpose, the objective is to improve the village environmental infrastructure. Taking humanity as the context, the objective is to form a characteristic style of village construction. Taking guidance and control as tools, the objective is to establish the green village construction management mechanism.

3. Green village functional zoning plan

(1) Functional zoning. The construction and development of Yanhe Village are concentrated mainly in the residential area, which is divided into four residential groups and a comprehensive supporting block. According to the different planting products, industrial development is divided into agricultural sightseeing areas and tea sightseeing areas. The forest and green space of the whole village are classified as a natural ecological area to form the overall green development spatial layout of the village.
The master construction plan of Yanhe Village is shown in Fig. 5.8. The functional zoning plan of Yanhe Village is shown in Fig. 5.9.
(2) Village construction land. The landscape centre of Yanhe Village is the Yanhe New Village area, which is also the public service centre and tourist reception centre of Yanhe Village. The villagers' residential area is built in an orderly manner along the mountain in accordance with the original site, forming four groups of residents and enabling the construction to economize on land.
(3) Production land and forest and green space. Yanhe Village is divided into a basic farmland area and tea production area according to its industrial plan and land characteristics. To highlight the characteristics of planting, a fungus planting base was opened in the southern part of the village. Rural agricultural land is strictly protected.

The land-use plan of Yanhe Village is shown in Fig. 5.10.

4. Green building project plan

(1) Yanhe New Village green house construction demonstration. Yanhe New Village consists of 13 ecological farmers' residences and 1 village reception centre, among which 1 farmer's residence and the village reception centre were selected to be built as green, ecological and energy-saving demonstration buildings.
The green house demonstration construction plan of Yanhe New Village is shown in Fig. 5.11.

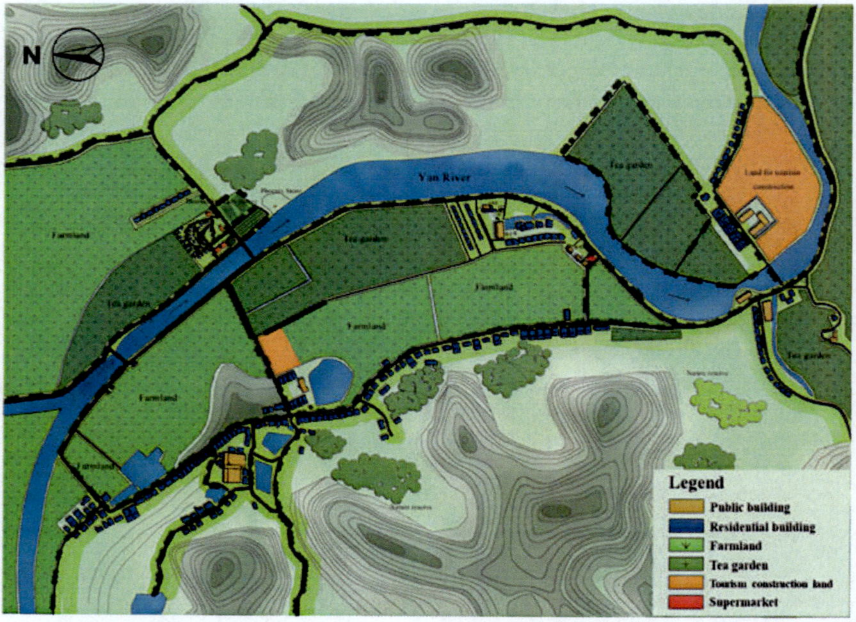

Fig. 5.8 Master construction plan of Yanhe Village

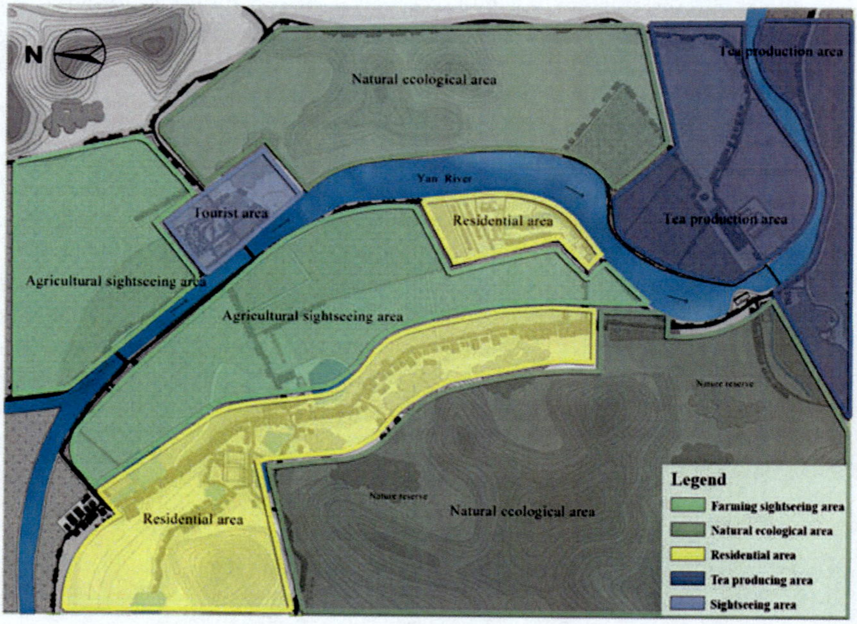

Fig. 5.9 Functional zoning plan of Yanhe Village

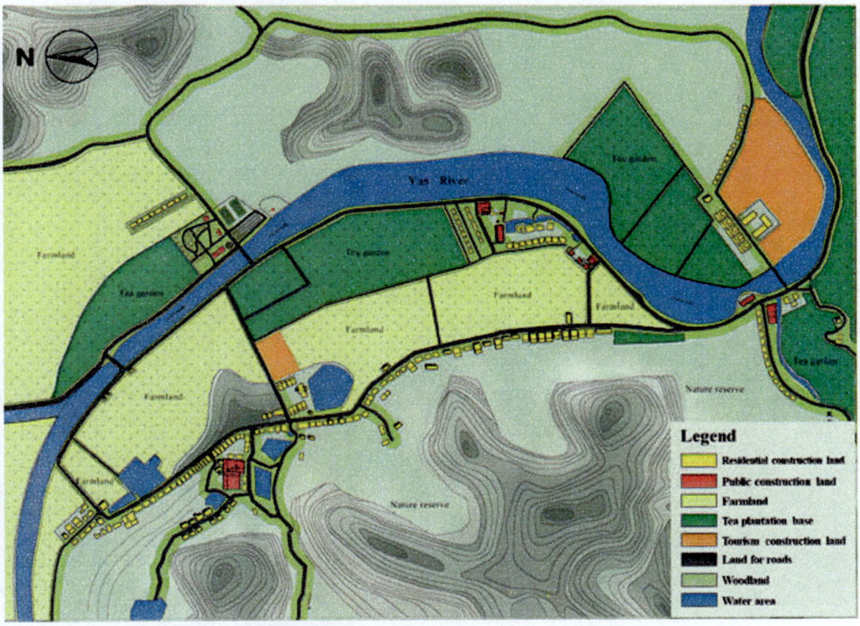

Fig. 5.10 Land-use plan of Yanhe Village

(2) Green transformation of old rural houses in the village. Key construction projects are as follows: the former collective warehouse of the village, built in the 1970s, was transformed into a farmhouse ecological tourism demonstration farm; the original tea-processing plant in the village was transformed into an ecological tea house and a leisure central square; the old village hall was transformed into a tourist reception centre; and the old village houses were transformed to beautify the appearance of the village. The green transformation inherited the historical context and architectural style of the village.

5. Special green village construction plan

(1) Road traffic plan. The width of roads outside the village was strictly limited to avoid encouraging large and passing vehicles, damaging the natural green ecological environment of the village and incurring noise pollution. Bike paths were built around the village to promote green village travel.

(2) Landscape system plan. This plan included afforestation planning, river regulation, landscape planning, and protection and restoration of village green landscape resources.

The ecological landscape system plan of Yanhe Village is shown in Fig. 5.12.

(3) Public service facility plan. This plan included village administration facilities; educational facilities; cultural, sports and health facilities; commercial and financial facilities; and other projects. There was co-operation between a green, moderately prosperous village construction target and the eco-tourism development project.

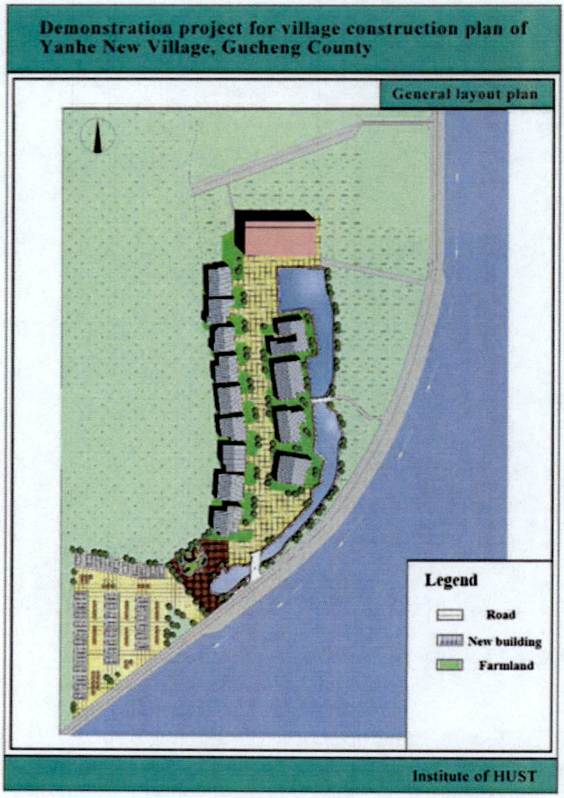

Fig. 5.11 Green house demonstration plan of Yanhe New Village

(4) Water supply and drainage plan. This plan included a water supply engineering plan, drainage engineering plan and sewage treatment plan. A green village water resources protection and utilization system was formed that focused on water source protection, water conservation and sewage treatment.
The sewage treatment plan of Yanhe Village is shown in Fig. 5.13.

(5) Power supply and communication plan. This plan included the power supply engineering plan, communication engineering plan, and information engineering plan.

(6) Clean energy plan. The main clean energy sources used in Yanhe Village are solar energy and biogas. Solar energy is used mainly for household water heaters and streetlamps in the village, and biogas is used for household energy. An energy-saving approach was developed for Yanhe Village.

(7) Sanitation facility plan. According to the principle of "classified collection, fixed point storage, regular transfer and centralized treatment", Yanhe Village set up a resource recycling station. Every 2–3 households set up a classified garbage collection; combined with rural tourism, 8 public toilets were set up,

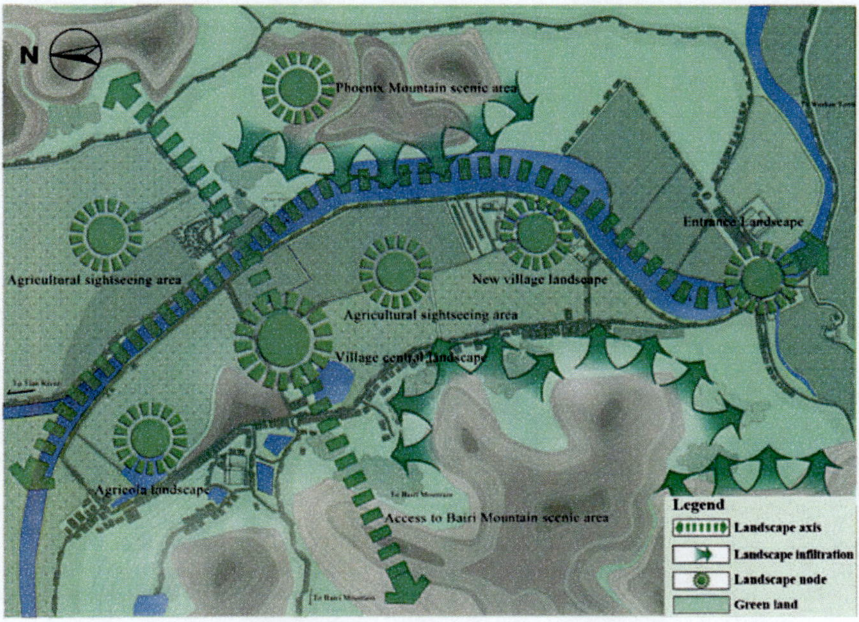

Fig. 5.12 Ecological landscape system plan of Yanhe Village

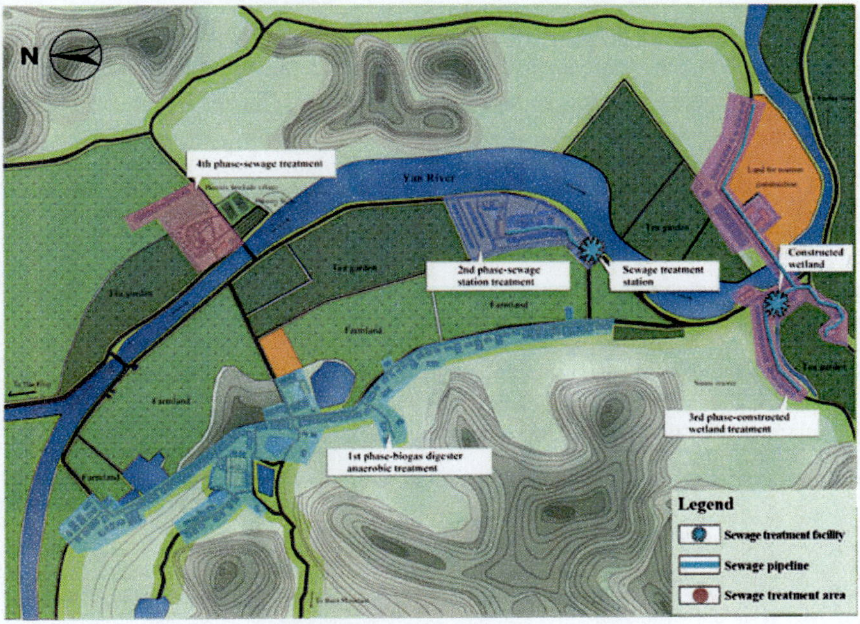

Fig. 5.13 Sewage treatment plan of Yanhe Village

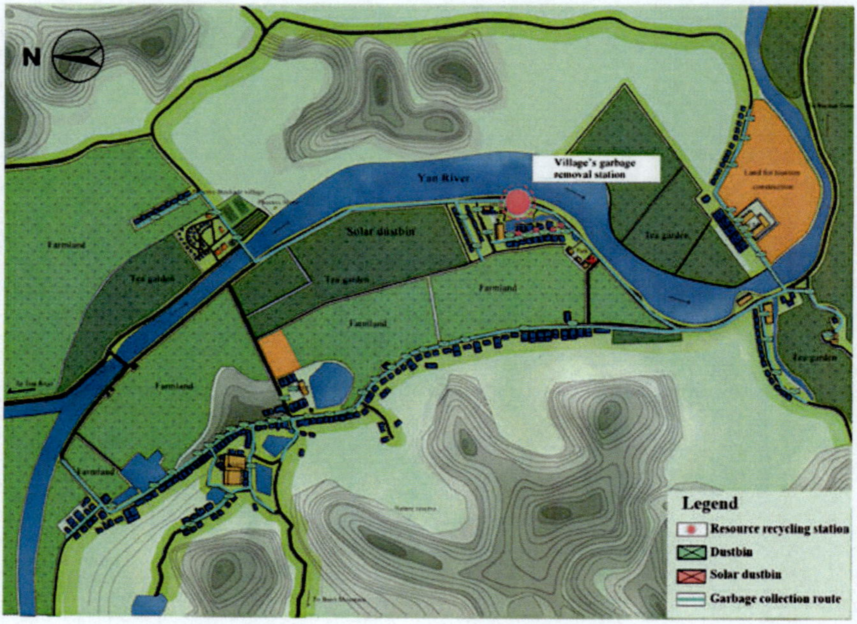

Fig. 5.14 Garbage collection and transportation plan of Yanhe Village

and the indoor toilets of the whole village were renovated to meet the standard. A clean and healthy Yanhe Village was built.

The garbage collection and transportation plan of Yanhe Village is shown in Fig. 5.14.

(8) Disaster prevention and mitigation plan. This plan included geological disaster prevention and control, fire control, seismic disaster prevention, flood control, and epidemic prevention. The goal was to build a safe Yanhe Village.

The disaster prevention and emergency system plan of Yanhe Village is shown in Fig. 5.15.

6. Green village industrial layout plan

The plan focused on the construction of five green ecological industrial zones:

(1) An agricultural sightseeing zone with the theme of agricultural and high-tech planting leisure and sightseeing;
(2) A tea tourism zone featuring tea production, tea processing and the tea country cultural experience;
(3) A green eco-tourism sightseeing zone covering the whole village;
(4) A farm leisure zone featuring the experience of farm life; and
(5) A natural ecological protection zone focusing on the economy of medicinal herbs, flowers, forest fruits and natural landscape tourism in mountainous areas.

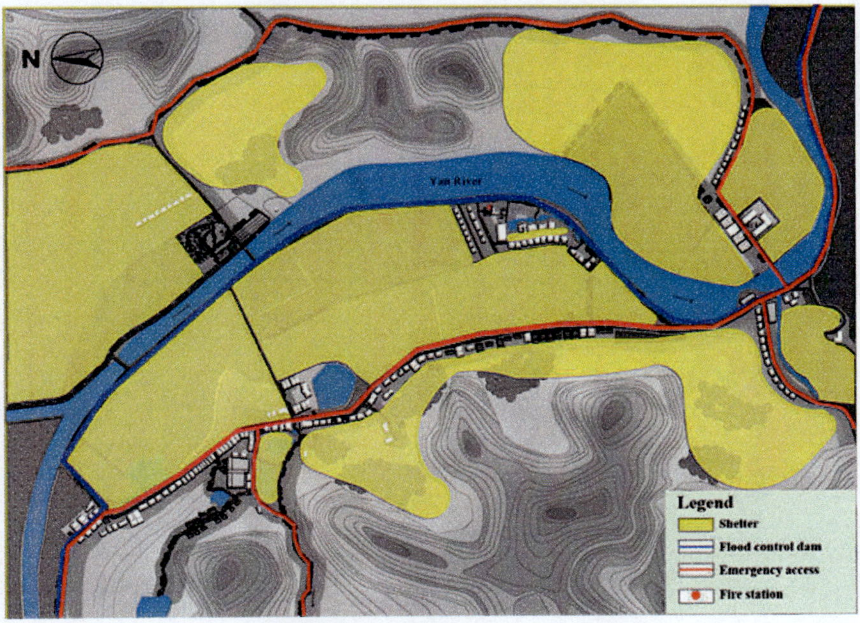

Fig. 5.15 Disaster prevention and emergency system plan of Yanhe Village

The eco-tourism plan of Yanhe Village is shown in Fig. 5.16.

The case of the Yanhe Village construction plan investigation and compilation introduced in this chapter is also the basis of the case analysis of the green village and town planning and construction evaluation in part III of this book.

5.3.2 Fushan Town Green Construction Development Plan

This plan case was the construction planning of a green development demonstration town in Fushan Town, Chengmai County, Hainan Province (2016–2020), completed by the team from Huazhong University of Science and Technology in 2016.

1. Overview of Fushan Town

Fushan Town is located in northern Chengmai County, Hainan Province, and covers an area of 79.30 km^2. Fushan Town is 18 km from Jinjiang Town, 49 km from the provincial capital, Haikou City, and 80 km from Meilan Airport. The west line expressway around the island, Yuehai Railway and Xihuan High-speed Railway all pass through the territory, so transportation is very convenient. Fushan Town has a tropical island monsoon climate, evergreens, warm temperatures, abundant rainfall, moist air, adequate sunlight, and rich soil.

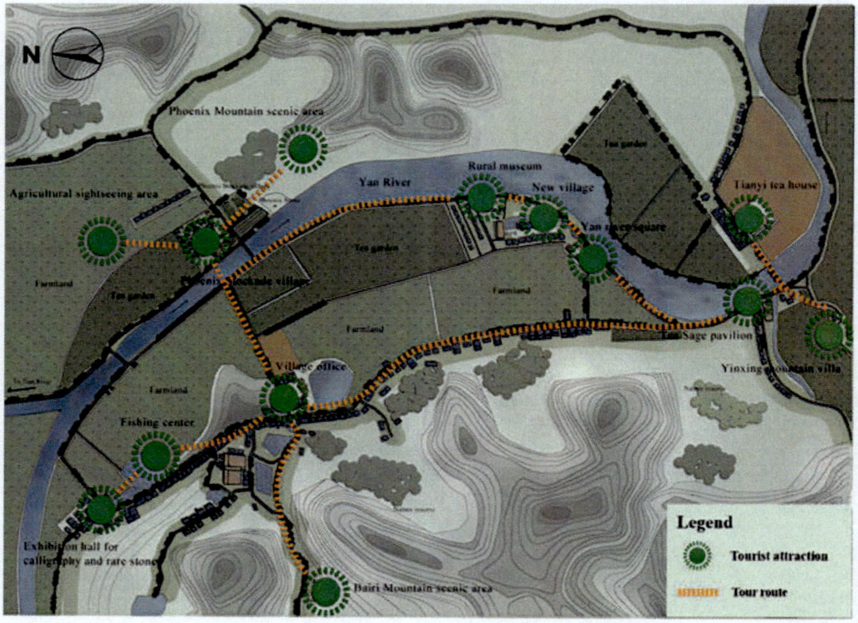

Fig. 5.16 Eco-tourism plan of Yanhe Village

Fushan Town has 7 village committees, 1 community neighbourhood committee, 44 natural villages, 69 village (residential) groups, and one state-owned farm. At present, the town has a permanent population of 19,000. Fushan Town's sustainable economic development is in good condition. In 2015, the town's GDP reached 927 million yuan, the per capita disposable income of urban residents reached 25,857 yuan, and the per capita disposable income of rural permanent residents reached 11,706 yuan.

Chengmai County is a famous green county in China. It was the first National Sustainable Development Community established in Hainan Province. Fushan Town is a demonstration base for Chengmai County to implement the green sustainable development strategy. Fushan Town has been designated a "national civilized town", "national beautiful liveable town", China's best cultural leisure tourism town", "China's ecological culture town (township)", "Hainan characteristic charming town", and "Hainan ecological civilization township" and has received many other honours, laying a good foundation for the establishment of a green development demonstration town.

2. Planning and construction objectives

(1) Overall planning objective. The objective was to build a green development demonstration base in the National Sustainable Development Community in Chengmai County featuring green industry, eco-tourism and green villages and towns by 2020. The green development demonstration town in Hainan

Province, a demonstration town of new urbanization green construction and an influential sustainable development town in China and the world, would promote sustainable economic and social development in Fushan.

(2) Planning guidelines. Guided by the overall strategy of "open development, coordinated development, green development, innovative development, and shared development" formulated by the state, the green development in Fushan Town was accelerated with a focus on promoting green industry, eco-tourism and green village and town development.

(3) Construction demonstration target. The objectives were to build a demonstration town of green industry development aiming at "appropriate employment for Fushan", to build a demonstration town of eco-tourism development with the goal of "suitable for tourism in Fushan"; and to build a "liveable Fushan" as the goal of a green development model town.

(4) Construction and development indexes. The planning and construction indexes of the green development demonstration of Fushan Town included overall development, green industry, eco-tourism and green village and town, with a total of 26 construction indexes, as shown in Table 5.4.

3. Field of construction demonstration

(1) Develop green industries vigorously. The green industry system of Fushan Town was formed with tropical island ecological agriculture and high-value-added agricultural product processing industry as the subject, ecological tourism service industry as the base, and low-carbon and environmentally friendly industry as supplements.

(2) Give priority to ecological tourism. By 2020, Fushan Town would be built into the most important ecological leisure tourism node on Hainan Island. Green scenic spots would be built, and green transportation and green consumption life patterns would be vigorously promoted.

(3) Demonstration of green village and town construction. Forty-four high-quality beautiful villages were constructed, and the first whole zone of a green model town of Chengmai County and Hainan International Tourism Island was built. It would become a demonstration base for the construction and development of green villages and towns in Hainan Province and Chengmai County.

4. Demonstration layout of green construction and development

The layout of the green development demonstration of Fushan Town is "one axis, three cores, four sections". One axis refers to the Fushan Town green industry development axis; three cores refers to the three green town core areas of Fushan Town, and four sections refers to the four development sections of the leading industry in Fushan Town.

The demonstration layout of green development and construction of Fushan Town, Chengmai County, is shown in Fig. 5.17.

Table 5.4 Planning and construction indexes of green development demonstration of Fushan Town

Index category	Number	Index	Unit	Index value (2020)
Overall development	1	Annual GDP of the town	100 million yuan	16.5
	2	Annual growth rate of GDP	%	10
	3	Per capita annual income of farmers	Yuan	≥18000
	4	The proportion of tertiary industry accounts to GDP	%	≥30
Green industry	5	Low-carbon economy proportion of GDP	%	50
	6	Per 10,000 yuan GDP energy consumption	Tons of standard coal	<0.492
	7	Per 10,000 yuan GDP water consumption	m^3	<110
	8	Coffee-planting area	hm^2	1333.34
	9	Characteristic fruit-planting area	hm^2	3333.35
	10	Proportion of organic, green and pollution-free product planting area	%	≥60
	11	Comprehensive utilization rate of manure in large-scale livestock and poultry farms	%	95
	12	Agricultural film recovery rate	%	85
Eco-tourism	13	Tourism proportion of GDP	%	25
	14	Eco-tourism output value	100 million Yuan	≥10
	15	Tourism specialty products	type	7
	16	Tourism public service facilities	Number	5
	17	Annual number of tourists	10,000 man-time	200
	18	Proportion of overseas tourists per year	%	10

(continued)

Table 5.4 (continued)

Index category	Number	Index	Unit	Index value (2020)
Green village and town	19	Per 10,000 yuan GDP construction land occupation area	hm^2	<60
	20	Number of beautiful rural demonstration villages	Number	12
	21	Concentrated water supply rate of drinking water	%	100
	22	Unconventional water source utilization	%	16
	23	Concentrated sewage treatment rate	%	80
	24	Garbage innocuous treatment rate	%	75
	25	Clean energy penetration rate	%	55
	26	Renewable energy utilization	%	10

5. Demonstration of green industry development

The green industry of Fushan Town is composed of ecological agriculture of a tropical island, a high-value-added agricultural product processing industry, an ecological tourism service industry and low-carbon environmentally friendly industry. Key demonstration projects of the construction and development of green industries include the following:

(1) The construction of a modern agricultural demonstration zone;
(2) The construction of a pollution-free fruit and vegetable base;
(3) The construction of the ecological agriculture sightseeing zone;
(4) The construction of the green agricultural product processing zone;
(5) The construction of the green agricultural products logistics centre; and
(6) Fushan green cultural industry construction.

6. Demonstration of ecological tourism development

The demonstration actions of the eco-tourism construction of the Fushan green development demonstration town include the following:

(1) Accelerating the upgrading of existing tourist attractions;
(2) Accelerate the construction of the ecological tourism culture;
(3) Accelerate the construction of themed leisure resorts; and
(4) Accelerate the construction of ecological tourism service facilities.

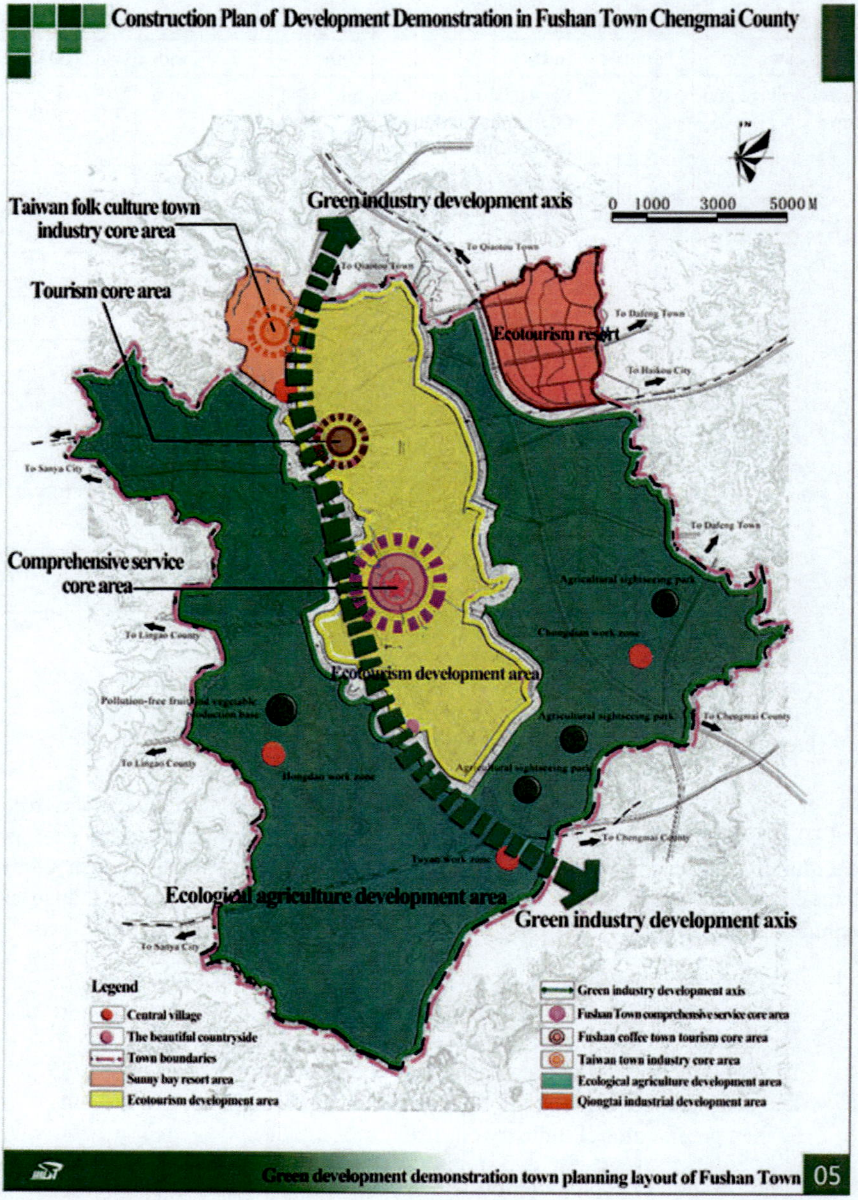

Fig. 5.17 Demonstration layout of green development and construction of Fushan Town, Chengmai County

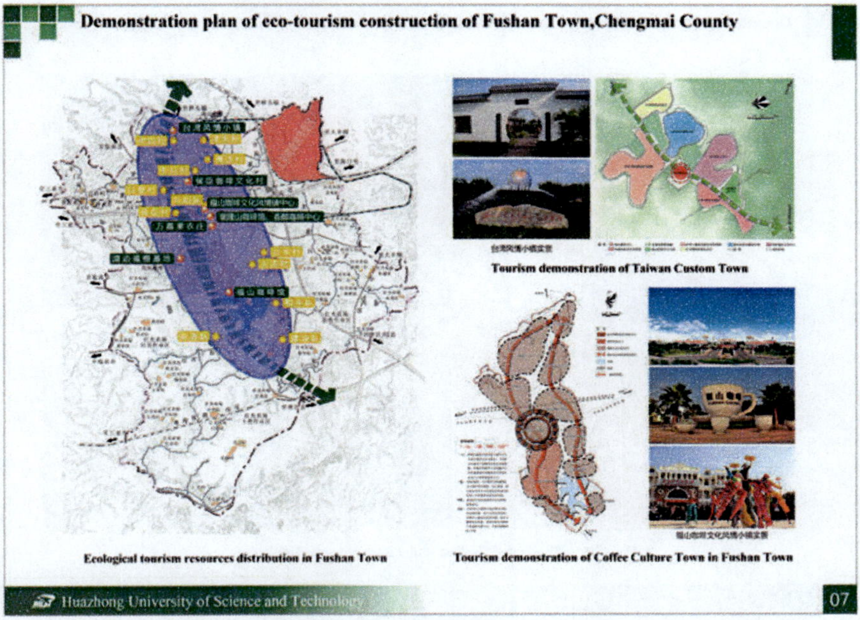

Fig. 5.18 Demonstration plan of eco-tourism construction of Fushan Town, Chengmai County

The demonstration plan of the eco-tourism construction of Fushan Town, Chengmai County, is shown in Fig. 5.18.

7. Demonstration of beautiful village construction

The plan goal was to build Fushan Town into a demonstration base for the construction and development of green villages and towns in Hainan Province. Key demonstration projects include the following:

(1) Beautiful rural belt construction;
(2) Environmental infrastructure construction in villages and towns;
(3) Improvement of people's livelihoods in towns and villages; and
(4) Internet + village construction.

The demonstration plan of beautiful village construction of Fushan Town, Chengmai County, is shown in Fig. 5.19.

8. Promotional mechanism of construction plan implementation

The implementation and promotional mechanism of the construction plan demonstration of green development of Fushan Town, Chengmai, includes the following:

(1) The innovation-driven mechanism for green development is the driving force guarantee for the implementation of the plan;

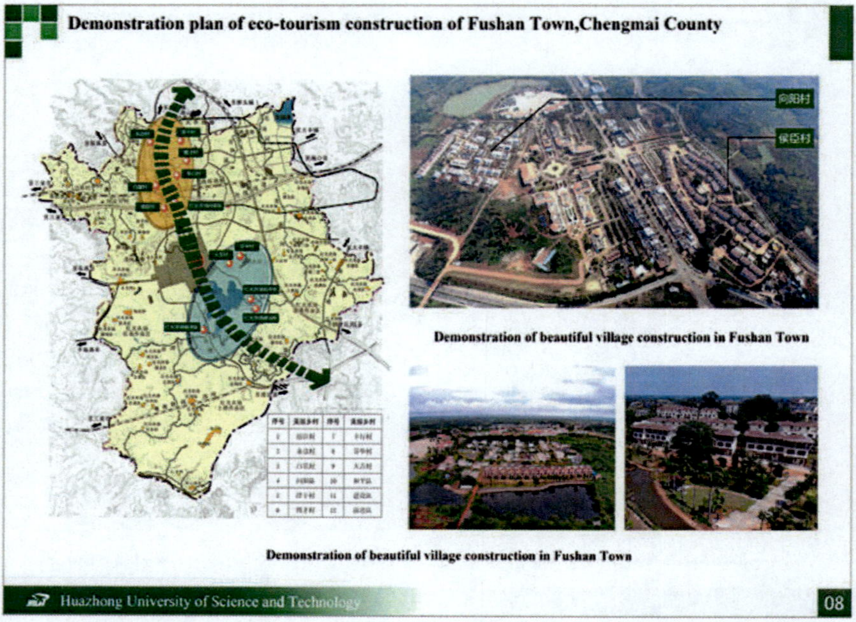

Fig. 5.19 Demonstration plan of beautiful village construction of Fushan Town, Chengmai County

(2) The implementation mechanism of the green development plan is the organizational guarantee of the implementation of the plan;
(3) The green development financing coordination mechanism is the project guarantee of the implementation of the plan;
(4) The demonstration and popularization mechanism of green development is the technical guarantee of the implementation of the plan; and
(5) The social participation mechanism of green development is the social security guarantee of the implementation of the plan.

References

1. Jin ZS, Lu WG (2010) Village plan, 3rd edn. Southeast University Press, Nanjing, p 19
2. Peng YG (1992) Landscape analysis of traditional village and town. China Building Industry Press, Beijing, pp 50–59
3. Wang XM, Peng WJ, Hua H et al (2016) Development and application of ecological niche theory of architecture. J Civ Eng Manag 33(03):1–8

Chapter 6
Green Village and Town Environmental Infrastructure Construction

Abstract Environmental infrastructure construction is the main component of green village and town construction. In this chapter, the key areas and construction technology methods of green village and town environmental infrastructure construction are analysed, including the construction of drinking-water source protection and facilities, domestic sewage disposal facilities, domestic garbage disposal facilities, clean energy facilities, greening and landscape facilities, and disaster prevention and reduction and early-warning facilities.

The environmental infrastructure of green villages and towns is a general term for material engineering facilities that provide environmental public services to ensure the green production of villages and towns and the green life of residents. The key objects of green village and town environmental infrastructure construction are drinking-water source protection and water-saving facilities, domestic sewage treatment facilities, domestic garbage treatment facilities, clean energy utilization facilities, greening and landscape facilities, and disaster prevention and mitigation and early-warning facilities.

6.1 Construction of Green Village and Town Drinking-Water Source Protection and Water-Saving Facilities

"Protecting water sources" and "saving water" are long-term development goals of green village and town construction, and the construction of drinking-water source protection facilities and water-saving facilities should be promoted simultaneously.

X. Wang and H. Hua, *Green Village and Town Construction in China*,
https://doi.org/10.1007/978-981-16-2098-0_6

6.1.1 Construction of Village and Town Drinking-Water Source Protection and Water-Saving Facilities

The construction of drinking-water source protection facilities is a project for supporting the green life of village and town residents. The construction focuses on distributed drinking-water sources and centralized drinking-water sources. At present, most rural areas in China have decentralized drinking-water sources, while administrative towns and most market towns have centralized drinking-water sources.

1. Decentralized drinking-water source protection and facilities

1) Decentralized drinking-water sources

This refers to existing, reserve and planned drinking-water sources that supply water at a certain small scale (generally supplying water to fewer than 1000 people). Distributed drinking-water sources may be protected by referring to the *Guidelines for Environmental Protection of Distributed Drinking Water Source Areas (Trial).*

2) Scope of drinking-water source protection

Protected surface water sources include the following: the protected area is not less than 1000 m upstream and not less than 100 m downstream of the water intake, and the depth of the two banks is not less than 50 m, not exceeding the rain collection range. A 200-m radius is maintained around lake reservoirs, not exceeding the rain collection range.

The water source protection range of cisterns includes the catchment area.

The source protection range of groundwater is 30–50 m around the water intake.

3) Water supply facility construction for decentralized drinking-water sources

(1) Water supply facility construction for surface water sources

① River and lake reservoir water sources. The water intake point should be as close as possible to the river central line and centre of lake reservoirs or situated far from the riverbank and lakeside. Common facilities for the construction and protection of river water and lake reservoir water resources are water intake ports (or gangplanks), diversion channels, reservoirs, submersible pumps, and sand filter wells (or sand filter tanks). The preliminary treatment of water sources includes coagulation sedimentation and basic disinfection.

② Cistern. A cistern is a special rainwater collection tank with a simple sedimentation and purification treatment facility near the collection pond. A drainage ditch should be installed around the collection tank to prevent surface runoff from contaminating the water source. In severely water-deficient areas, open areas should be chosen as much as possible as the catchment for the cistern, and areas with high background values of soil harmful factors should be hardened by the site.

③ Environmental protection requirements for surface water sources. The water quality of surface water sources should comply with the state regulations on the quality of drinking-water sources. When using surface water as the source of drinking water, the water quality should refer to *Environmental quality standards for surface water* GB 3838-2002.

(2) Water supply facility construction for groundwater sources

① Wells. Groundwater wells should have well bays, brandreths and manhole covers, and it is advisable to use relatively sealed wells; the bottom and walls of the well should ensure its health protection.

② Springs. In the vicinity of a spring water source, a diversion pool should be built. A fence and other isolation protection facilities should be built 100 m from the spring water and 500 m upstream. A simple diversion channel should be installed beside the spring to prevent rainwater or sewage from carrying large amounts of pollutants directly into the spring water.

③ Environmental protection requirements for groundwater sources. When using groundwater as the source of drinking water, the water quality should refer to the *Quality Standard for Ground Water* GB/T 14,848-2017.

4) Protection requirements for water sources without water purification treatment

The quality of water sources without water purification treatment should refer to *Standards for Drinking Water Quality* GB 5749-2006. When the water quality does not meet the national drinking-water source water quality regulations, it should not be used as a drinking-water source.

2. Protection and facility construction for centralized drinking-water sources

1) Centralized drinking-water sources

This refers to drinking-water sources that are supplied to users through a water pipe network on a certain scale (generally supplying water to more than 1,000 people). The protection of centralized drinking-water sources may be implemented by referring to the *Guidelines for Environmental Protection of Centralized Drinking Water Source Areas (Trial).*

2) Protection scope of the drinking-water source area

The demarcation of the centralized drinking-water source protection area should be based on the geographic location, topography, hydrogeological conditions, water supply, mining methods and pollution source distribution, combined with local landmarks or permanent buildings, in accordance with *Technical Guideline for Delineating Source Water Protection Areas* HJ/T 338-2018 or local regulations and standards.

The centralized drinking-water source protection area should be divided into the primary protection area and the secondary protection area, and quasi-protection areas should be designated if necessary. The division of centralized drinking-water source protection areas should meet the relevant standards and construction specifications, such as HJ/T 338-2018.

3) Water-quality requirements for centralized drinking-water sources

(1) Water-quality requirements for river drinking-water sources

For primary protection areas, the Class II water quality standards in *Environmental Quality Standards for Surface Water* GB 3838-2002 should be met.

For secondary protection areas, the Class III water quality standards in *Environmental Quality Standards for Surface Water* GB 3838-2002 should be met.

For prospective reserves, the water quality requirements for secondary protection areas should be referred to.

(2) Water-quality requirements for lake reservoirs

The primary protection area includes small reservoirs (total storage capacity V < 10 million m^3), all water areas below the normal water level of a small lake (water surface area S < 100 km^2) with a single water supply function, and large or medium reservoirs and lakes with part of the water area divided into the primary protection area. It's the water quality should meet the Class II water quality standards in *Environmental Quality Standards for Surface Water* GB 3838-2002.

The secondary protection area includes the water outside the boundary of the primary protection area of small lakes and small and medium-sized reservoirs. Large-scale reservoirs and large and medium-sized lakes with a radial distance of not less than 2000 m outside the primary protection zone are designated as secondary protection areas. The concentration of the main pollutants is reduced by meeting standard III to standard II in *Environmental Quality Standards for Surface Water* GB 3838-2002.

(3) Water-quality requirements for groundwater drinking-water sources

The water-quality indexes of groundwater drinking-water sources (including primary, secondary and quasi-protected areas) are not lower than Class III in the *Quality Standard for Ground Water* GB/T 14,848-2017.

4) Water supply facility construction of centralized water sources

(1) Contents of water supply facility construction of centralized water sources

The main contents of the water supply facility construction of centralized water sources in villages and towns are as follows:

① Selection and determination of water quantity and sources in the water supply system;
② Design and construction of water intake structures;
③ Design and construction of the water supply and distribution pipeline network;
④ Selection and construction of water treatment processes; and
⑤ Water supply system operational management.

(2) Construction requirements of water supply facilities for centralized water sources.

The design, construction and operational management of centralized water supply and water supply projects in villages and towns must be consistent with *Code for Design of Building Water Supply and Drainage* GB 50015-2019, *Structural Design Code for Special Structures of Water Supply and Waste Water Engineering* GB 50069-2002, *Standards for Drinking Water Quality* GB 5749-2006,*Quality Standard for Ground Water* GB/T 14848-2017, *Environmental Quality Standards for Surface Water* GB 3838-2002, and *Design Code for Rural Water Supply Engineering* SL 687-2014.

6.1.2 Construction of Village and Town Domestic Water-Saving Facilities

The construction of green village and town water-saving facilities refers to the construction of water-saving facilities related to the domestic water consumption of residents in built-up areas.

Rural residents' domestic water and water-saving facilities are often referred to as water-saving appliances. At present, domestic water appliances commonly used in Chinese villages and towns include water nozzles, toilets and toilet systems, showers, and washing machines. The selection of water-saving household appliances in green villages and towns should conform to *Domestic Water Saving Devices* CJ/T 164-2014.

When constructing water-saving facilities in villages and towns with a centralized water supply, attention should also be paid to the maintenance and protection of the water supply pipe network to avoid leakage. The leakage rate of a water supply networks in a green village or town should be not more than 1% of the repair value stipulated in the *Standard for Leakage Control and Assessment of Urban Water Supply Distribution System* CJJ 92-2016.

6.2 Construction of Green Village and Town Domestic Sewage Treatment and Facilities

6.2.1 Construction of Black-Water Treatment and Recycling Facilities

Domestic sewage disposal is the key content of green village and town infrastructure construction and environmental protection. Black-water treatment and toilet reconstruction are key targets for green village and town domestic sewage treatment.

1. Black-water treatment and facility construction

1) Black-water management strategy

In rural sewage, black water refers to faecal sewage, meaning domestic sewage containing faecal matter and toilet sewage containing water and faeces from flushing toilets. In addition, night soil refers to black water that contains no or a small amount of flushing water.

At present, the sewage treatment rate in China's vast rural areas is still quite low. By the end of 2016, the sewage treatment rate in China's rural areas was only 22%. Therefore, the Ministry of Housing and Urban-Rural Development clearly put forward the following governance objectives: in economically developed, densely populated, environmentally sensitive areas, rural sewage must be comprehensively treated. In other rural areas, however, black water must first be treated by 2020, and the proportion of black-water treatment should reach 70% by 2020 after the construction of toilets in rural areas is fully implemented.

2) Construction of black-water treatment facilities in villages and towns

The main village and town black-water treatment facilities are three-form septic tanks and biogas pools. Through rural toilet reform and strengthening late-use management so that toilet faeces and sewage are not discharged, black water from septic tanks can be used as a resource by being returned to the fields, and biogas tank-generated slurry and biogas residue are also high-quality farm fertilizers.

A three-form septic tank is inexpensive and easy to build. A small amount of money can solve the problem of disorderly discharge of black water. This is an important type of sewage treatment facility in rural areas, especially in villages and towns with slow economic development.

The construction of rural three-form septic tanks should be in line with the *Hygienic Standard for Rural Household Latrine* GB 19,379-2012. The Reference Standard for the Construction of Rural Household Biogas Digester is *Collection of Standard Design Drawing for Household Anaerobic Digesters* GB/T 4750-2002.

2. Toilet revolution and facility construction

Toilet renovation and up-to-standard treatment are the key components of the construction of black-water treatment facilities in green villages and towns. Black-water treatment should start from the source of toilet sewage. At present, toilet facilities in rural areas of China are still relatively underdeveloped. Even if water closets are adopted and sewage is washed away with clean water, most sewage is directly discharged without purification treatment due to imperfect sewage networks and the difficulty in treating excrement and urine. Direct discharge seriously pollutes local water bodies and exacerbates the problem of black and smelly water in rural areas.

The use environment of the toilet is directly related to the green life and environmental quality of a vast number of farmers. To change the long-term problems of rural toilet facility construction in a relatively short time, we must vigorously promote the "toilet revolution". The United Nations International Children's Emergency Fund

(UNICEF) proposes that the toilet revolution means developing innovative, sustainable designs; creating new markets; and mobilizing communities to establish new social norms and practices for toilets.

To promote the realization of China's rural "toilet revolution", eight departments, such as the Central Agricultural Office, Ministry of Agriculture and Rural Affairs, National Health Commission, Ministry of Housing and Urban Rural Development, in January 2019 jointly issued the *Guidance on Promoting the Special Action of "Toilet Revolution" in Rural Areas*. In accordance with the basic idea of "orderly promotion, overall promotion, equal emphasis on construction and management, and long-term operation", it is necessary to promote the standardization of management, marketization of operations and maintenance, and socialization of supervision of rural toilet construction. It is required that within approximately five years, toilet dung should be treated and recycled to improve the living environment in rural areas.

6.2.2 Construction of Village and Town Sewage Treatment Facilities

Green village and town sewage treatment facilities may adopt household treatment or centralized treatment according to different construction conditions, such as regional economic conditions, geographical climate and living customs.

1. Village and town sewage household processing facility construction

(1) Pre-precipitation device. A septic tank can be used as a pre-precipitation treatment unit.
(2) Sewage disposal facilities. Separate sewage disposal facilities should adopt an integrated device. The device standard is *Household Sewage Treatment Plant* CJ/T 441-2013.
(3) Comprehensive sewage treatment facilities. When the integrated device does not meet the requirements, natural biological treatment should be added, such as combination with an aerobic treatment device or a combination of anaerobic biofilm and artificial wetland.
(4) Household sewage treatment process. Village and town household sewage treatment technology can be selected by referring to *Household Domestic Sewage Treatment Equipment* CJ/T 441-2013.

The selection of the village and town household sewage treatment process is shown in Table 6.1.

A village and town combined household sewage treatment unit is shown in Fig. 6.1.

2. Village and town centralized sewage treatment facility construction

In town areas and economically developed village areas, centralized sewage treatment facilities should be built, or sewage should be brought to nearby municipal sewage treatment plants for comprehensive treatment.

Table 6.1 Selection of village and town household sewage treatment process

Number	Recommended biotechnology and combinations	Effluent water quality	
		Direct discharge	Irrigation water
1	Anaerobic biofilm + biological contact oxidation	√	
2	Hypoxia contact oxidation + aerobic biological contact oxidation	√	
3	Anaerobic digestion		√
4	Hypoxia/anaerobic biofilm method		√

Fig. 6.1 Village and town combined household sewage treatment plant

Commonly used sewage disposal technologies in villages and towns of China are as follows:

(1) Sequencing batch reactor (SBR);
(2) Anaerobic-oxic (A/O);
(3) Anaerobic-anoxic-oxic (AA/O);
(4) Oxidation ditch;
(5) Constructed wetland;
(6) Stabilization ponds;
(7) Land percolation; and
(8) Biological integration pool.

Facility construction types and methods can be flexibly selected according to the requirements and specific conditions of sewage treatment in different villages and towns.

6.2.3 *Construction Case of a Village and Town Sewage Treatment Facility*

This case takes the construction of village sewage treatment facilities in Yanhe Village, Gucheng County, Hubei Province, as an example.

1. Analysis of sewage characteristics of Yanhe Village

The sewage of Yanhe Village is domestic sewage with a low pollutant concentration and good biodegradability. The sewage contains certain amounts of ammonia nitrogen and phosphorus. The designed treatment process should therefore have the functions of phosphorus and nitrogen removal.

2. Principle of selecting sewage treatment process

In the design of a sewage treatment station, the characteristics of sewage in the village or town and the availability of original sewage treatment facilities and structures should be fully considered, and the following principles should be followed:

(1) Choose a plan with reliable technology, advanced equipment and reasonable operational costs;
(2) Use a suitable level of automated control system to ensure the treatment results, reduce labour intensity and reduce project investment and operational costs;
(3) Make full use of existing structures and equipment to reduce the occupied area and project investment; and
(4) Add necessary treatment facilities to ensure that the effluent quality meets the standard.

3. Selection of the sewage treatment construction scheme

Current domestic sewage treatment methods are mostly two-level bio-degradable treatment systems that use biological treatment as the core. Under biochemical treatment, there will be an ideal removal rate of organic substances with a lower daily operational expenditure and without using any chemicals (the removal rate of COD can reach 80–90%, and the removal rate of BOD is more than 90–95%).

According to the field survey results, there were four alternative construction plans for the sewage treatment project in Yanhe Village:

Scheme 1: A2/O process;
Scheme 2: Biology integration pool + ecological pool process;
Scheme 3: Sedimentation tank + constructed wetland process; and
Scheme 4: Anaerobic filter + constructed wetland process.

4. Optimization of the village sewage treatment scheme based on GA-BP

The BP network is also called a back-propagation neural network (BPNN), which is composed of an input layer, output layer and hidden layer.

The genetic algorithm (GA) has a strong global optimization ability and searches for the optimal weights of indexes in the process. First, it randomly generates a certain number of "chromosome" groups, namely, each evaluation index weight, and places the weights of each evaluation index in the sewage treatment plan optimization problem of the village or town. The fitness value of each index weight is calculated, from which the "chromosomes" with large fitness values are selected for replication, crossover and mutation operations to produce a new generation of "chromosomes" with high fitness values. With the development of a genetic algorithm, the superior individuals of the population are retained and combined to produce the optimal individual index weight.

When the GA is combined with the BP network, the weight of the BP network (GA-BP) can be obtained quickly.

The optimization method of the village sewage treatment scheme based on GA-BP is suitable for the sewage treatment scheme of Yanhe Village and for the sewage treatment method optimization of the original village. Additionally, it has a wider significance of popularization.

1) Evaluation index selection

The evaluation indexes of village and town domestic sewage treatment schemes are divided into three layers: target layer, criterion layer and index layer. The target layer is the comprehensive benefit, and the purpose is to evaluate the merits of the sewage treatment scheme; the criterion layer includes project economic evaluation, project technical evaluation and project operational management evaluation. The index layer includes 13 evaluation indexes, such as "capital construction investment", "COD and BOD removal rate" and "operational difficulty".

The optimal evaluation index and measurement method of village and town sewage treatment schemes are shown in Fig. 6.2

2) Evaluation index calculation

According to Fig. 6.2, the optimization evaluation index and measurement method of village and town sewage treatment schemes are proposed, and the evaluation indexes C1–C13 are calculated.

3) Evaluation reference value determination

Village and town sewage treatment schemes were divided into 5 grades from excellent to inferior, represented by I, II, III, IV and V, respectively.

The range of C1–C3 measurement reference values of the evaluation method is determined by selecting relevant data from 23 sewage treatment projects in different areas of the country, and the grade evaluation datum value is obtained by

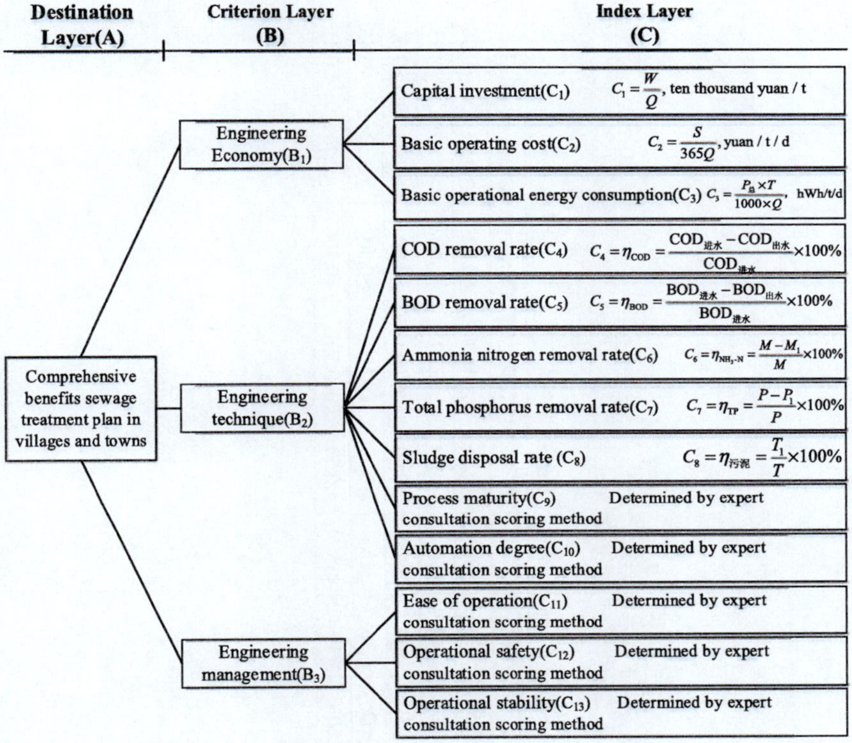

Fig. 6.2 Preferred evaluation indexes and measurement methods of village and town sewage treatment projects

the expert scoring method. The measurement reference value of C4–C7 was determined according to relevant national standards. The measurement reference value of C9–C13 was determined by the expert scoring method.

The classification index benchmark of the optimal selection evaluation of village and town sewage treatment schemes is shown in Table 6.2.

4) Optimal evaluation model of sewage treatment scheme based on GA-BP

(1) Classification of optimal evaluation grades

According to Table 6.2, the optimal score of a village and town sewage treatment scheme is level 5, and the score ranges from 0 to 5. It can be divided into the evaluation grades of optimal village and town sewage treatment schemes, as shown in Table 6.3.

(2) Modelling technique route based on GA-BP network

The optimal modelling technique path for the village and town sewage treatment scheme based on the GA-BP network is shown in Fig. 6.3.

Table 6.2 Grading index reference value of village and town sewage treatment plan optimization evaluation

Evaluation grade	C1	C2	C3	C4	C5	C6	C7	C8	C9	C10	C11	C12	C13
I	≤0.2	≤0.2	≤0.3	≥85	≥95	≥95	≥90	≥80	≥0.85	≥0.90	≥0.80	≥0.90	≥0.85
II	≤0.5	≤0.5	≤0.6	≥80	≥85	≥90	≥85	≥70	≥0.70	≥0.70	≥0.60	≥0.80	≥0.75
III	≤0.7	≤1.0	≤1.0	≥70	≥80	≥70	≥60	≥30	≥0.60	≥0.60	≥0.50	≥0.60	≥0.60
IV	≤1.0	≤1.2	≤1.5	≥60	≥60	≥60	≥30	≥15	≥0.50	≥0.30	≥0.30	≥0.50	≥0.45
V	>1.0	>1.2	>1.5	<60	<60	<60	<30	<15	<0.50	<0.30	<0.30	<0.50	< 0.45

Table 6.3 Optimal evaluation grades of village and town sewage treatment schemes

Grade	I	II	III	IV	V
Score	(0, 1]	(1, 2]	(2, 3]	(3, 4]	(4, 5]

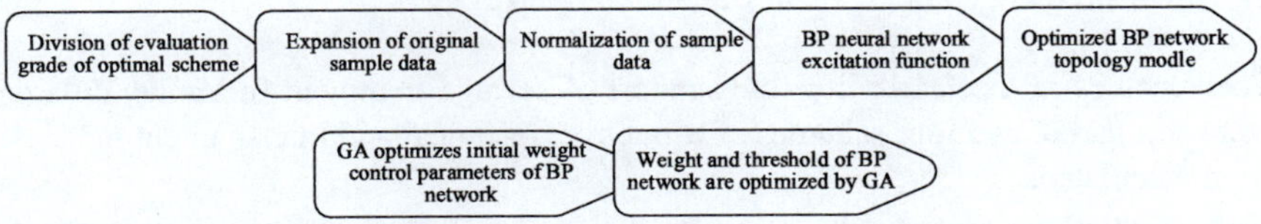

Fig. 6.3 Optimal modelling technique path of village and town sewage treatment scheme based on the GA-BP network

According to the modelling technique path shown in Fig. 6.3, the optimal modelling of the village and town sewage treatment scheme based on the GA-BP network was carried out.

5) GUI of the optimal model village and town sewage treatment scheme based on GA-BP

The graphical user interface (GUI) of the optimal model village and town sewage treatment scheme is established based on GA-BP. The user interface consists of a window, cursor, key, menu, dialog box, text description and other graphical elements. The user interface is established by the MATLAB R2010b GUIDE toolbox with a generic, flexible GA-BP network graphical user interface. See Fig. 6.4.

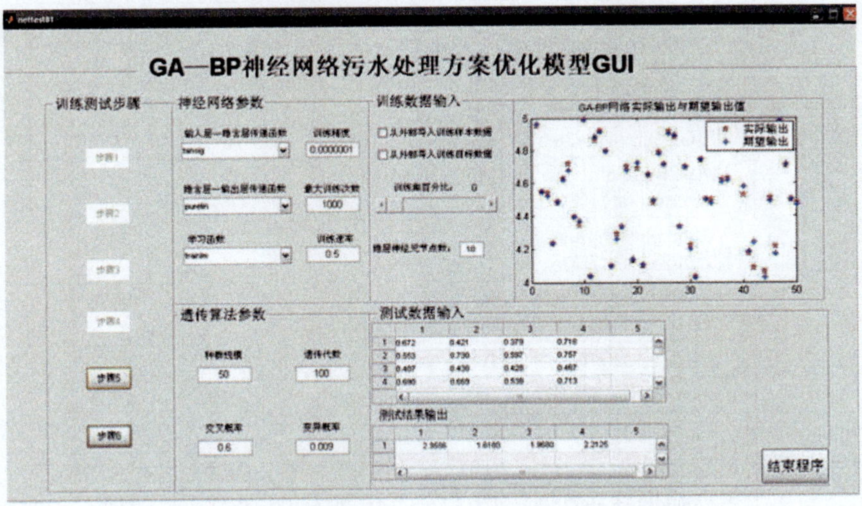

Fig. 6.4 Optimized results of GA-BP Yanhe Village sewage treatment scheme output

5. Construction of sewage treatment facilities of Yanhe Village

The above GA-BP method is applied to optimize and compare the sewage treatment schemes of Yanhe Village and determine whether the results are realistic.

1) Measurement comparison of evaluation indexes

The schemes in the feasibility study report of sewage treatment in Yanhe Village were compared, and four schemes of economic, technical and management indexes were calculated.

A comparison of the calculation results of the optimal evaluation indexes of various sewage treatment schemes in Yanhe Village is shown in Table 6.4.

2) Establishment of evaluation model and model-based CUI

Using the established evaluation model, the data in Table 6.4 are normalized by calculating Formula (6.1).

$$y' = 0.1 + 0.8(y - y_{\min})/(y_{\max} - y_{\min}) \tag{6.1}$$

Table 6.4 Comparison of calculation results of optimal evaluation indexes of sewage treatment schemes in Yanhe Village

Criterion layer	Index layer	Scheme 1	Scheme 2	Scheme 3	Scheme 4
Engineering economy	Capital investment (10 thousand yuan/t)	1.067	0.600	0.521	1.152
	Basic operating cost (yuan/t)	0.850	1.180	0.932	1.230
	Basic operational energy consumption (kWh/t)	0.761	0.820	0.813	0.912
Engineering technique	COD removal rate (%)	81	79	68	83
	BOD removal rate (%)	86	82	72	89
	Ammonia nitrogen removal rate (%)	70	80	65	75
	Total phosphorus removal rate (%)	75	85	70	80
	Sludge disposal rate (%)	95	90	85	85
	Process maturity	0.82	0.75	0.73	0.70
	Automation degree	0.90	0.80	0.70	0.78
Engineering management	Ease of operation	0.55	0.70	0.80	0.65
	Operational safety	0.83	0.92	0.95	0.75
	Operational stability	0.60	0.78	0.75	0.68

Table 6.5 Normalized results of the optimal evaluation index of the sewage treatment scheme in Yanhe Village

Criterion layer	Index layer	Scheme 1	Scheme 2	Scheme 3	Scheme 4
Engineering economy	Capital investment (10 thousand yuan/t)	0.672	0.421	0.379	0.718
	Basic operating cost (yuan/t)	0.553	0.730	0.597	0.757
	Basic operational energy consumption (kWh/t)	0.407	0.430	0.428	0.467
Engineering technique	COD removal rate (%)	0.690	0.669	0.538	0.713
	BOD removal rate (%)	0.739	0.693	0.578	0.773
	Ammonia nitrogen removal rate (%)	0.557	0.671	0.500	0.614
	Total phosphorus removal rate (%)	0.678	0.767	0.633	0.723
	Sludge disposal rate (%)	0.864	0.845	0.774	0.774
	Process maturity	0.742	0.680	0.662	0.636
	Automation degree	0.812	0.724	0.636	0.706
Engineering management	Ease of operation	0.502	0.636	0.726	0.592
	Operational safety	0.748	0.829	0.856	0.676
	Operational stability	0.549	0.708	0.681	0.620

where y is optimized raw data of sewage treatment plan and

y' is optimized data of the scheme after processing.

The normalized results of the optimal evaluation index of the sewage treatment scheme in Yanhe Village are shown in Table 6.5.

Take the index data in the A2/O process as GA-BP neural network test data a1:

a1 = [0.672 0.553 0.407 0.690 0.739 0.557 0.678 0.864 0.786 0.812 0.502 0.748 0.549]′;

Take the index data of the biological integration pool + ecological pool as test data a2:

a2 = [0.421 0.730 0.430 0.669 0.693 0.671 0.767 0.846 0.680 0.724 0.636 0.829 0.708]′;

Take the index data of the sedimentation tank + constructed wetland as test data a3:

a3 = [0.379 0.419 0.289 0.538 0.578 0.500 0.633 0.864 0.813 0.636 0.726 0.856 0.726]′;

Take the index data of the anaerobic filter + constructed wetland as test data a4:

$$a4 = [0.718\ 0.757\ 0.467\ 0.713\ 0.773\ 0.614\ 0.723\ 0.774\ 0.751\ 0.706\ 0.592\ 0.676\ 0.620]'.$$

Enter a1, a2, a3, and a4 into the GA-BP neural network graphical user interface panel table. Click on step 6 of the GUI interface to start the network test. The calculated results of the GA-BP model are sent to the test data output table of the GUI interface, and the calculated results are shown in Fig. 6.4.

Therefore, the optimal plan for Yanhe Village sewage treatment is plan 2, which is a biological integrated pond + ecological pool process.

After selecting and determining the optimal plan for sewage treatment, the author team worked closely with Wuhan Jincheng Environmental Engineering Company to design and construct the "integrated sewage processor-ecological pond" sewage treatment facility.

See Fig. 6.5 for the real scene of sewage treatment facility construction in Yanhe Village.

Fig. 6.5 Real scene of sewage treatment facility construction in Yanhe Village

6.3 Construction of Domestic Waste Disposal and Treatment Facilities of Green Villages and Towns

6.3.1 Facility Classification of Household Garbage Treatment of Villages and Towns

1. Treatment requirements for household garbage in villages and towns

The harmless and reductive disposal of household garbage is an important part of the environmental infrastructure construction of green villages and towns. "Household classification, village collection, town transfer and county treatment" should be implemented in collecting and transferring garbage in green villages and towns. The key points of the innocuous treatment of household garbage in villages and towns are classified collection, sealed transport and the separate collection, transport and treatment of hazardous garbage.

The *"13th five-year" national urban solid waste harmless treatment facilities plan* points out that by the end of 2020, the harmless disposal rate of domestic waste in China's towns will reach over 70%.

To protect agricultural land, sanitary landfills are generally not suitable for rural areas. If landfills are necessary, site selection, construction, operation, management and monitoring should be carried out in accordance with *Domestic Garbage Landfill Pollution Control Standards* GB 16,889-2008 and other relevant standards.

Town-level solid waste landfills should separate organic waste first. Waste with high organic and water content should be disposed of not by sanitary landfills but by composting. Sanitary landfill managers should ensure that the proportion of inorganic waste is more than 80% after classification.

2. Facility classification of household garbage in villages and towns

According to the *Classification and Evaluation Standard of Urban Household Garbage* CJJ/T 102-2004, garbage classification methods can be selected based on the characteristics and treatment methods of local garbage, and corresponding garbage treatment facilities can be built and provided. The main classification methods are as follows:

(1) The areas where sanitary landfills are used for garbage treatment should be classified according to recyclables, hazardous wastes, bulk wastes and other wastes;
(2) The areas used for composting should be classified according to recyclables, compostable garbage, harmful garbage, bulk garbage and other garbage; and
(3) According to the *Technical Specifications of Domestic Pollution Control for Town and Village* HJ 574-2010, organic waste such as agricultural fruits and vegetables, kitchen waste and faeces (which can enter household biogas digesters) and the remaining inorganic waste should be classified in villages and towns.

3. Setting of dustbins

Domestic waste in villages and towns is classified as a production source. Waste is classified as organic waste, recyclable waste, unrecyclable waste and harmful waste to realize the reduction classification of the sources of domestic garbage.

Trash cans are commonly used in village and town domestic waste classified collection: every house has 2 trash cans for organic and unrecyclable waste, and the capacity of each can should be 6–8 L. The service range per trash can in a town should be 50–80 m. Trash cans used for classification should be clearly marked and covered to avoid leakage.

4. Setting of waste collection sites

The service radius of waste collection sites in towns should be less than 100 m, and in villages, it should be less than 200 m.

5. Garbage collection vehicles

Domestic waste in villages and towns can be collected by box-type manual (or mechanical) collection vehicles; the capacity of a manual waste vehicle is 0.5–1 m^3, and the load is 0.2–0.5 t; the capacity of a mechanical waste vehicle is 2–5 m^3, and the load is 1–2 t.

6. Construction of waste transfer stations

The transportation of domestic waste in villages and towns is a process through which the waste collected at collection stations is pretreated, boxed and transported to urban waste treatment plants/factories or centralized waste treatment plants/factories. The construction of this project includes mainly waste transfer containers and vehicles. For waste transportation stations with a transportation distance greater than 5 km, a waste-compressing device suitable for reducing waste collection volume should be set up.

Waste transportation stations should be set up in towns and townships. The area of waste transportation stations should be more than 100 m^2, the service range should be less than 5 km, and all waste produced daily should be stored.

Small transportation stations used in cities and towns should be set up every 2–3 km^2. The designed transportation amount of small waste transportation stations should be $\leq$ 50 t/d or 50–150 t/d.

6.3.2 Main Disposal Methods of Household Garbage of Villages and Towns

1. Landfill disposal

(1) Sanitary landfills are generally not suitable for rural areas. However, if landfills are needed, site selection, construction, operation, management and monitoring should be carried out in accordance with the *Standard for Pollution Control on the Landfill Site of Municipal Solid Waste* GB 16,889-2008 and other relevant standards.
(2) Organic waste separation should be carried out first in landfill treatment at the town level. This waste contains a high percentage of organic waste or water and should not be disposed of in landfills; instead, it should be treated by composting. Sanitary landfill managers should ensure that the percentage of inorganic waste is more than 80% after classification.
(3) Villages that use local landfills should implement a stringent waste sorting system. The content of organic matter remaining in the inorganic waste after strict classification should be less than 10%. Wastes composed mainly of bricks, slag, sweeping ash and other inorganic matter can be used as landfills for rural abandoned pits and roads.
(4) Landfills should be treated with anti-seepage methods to prevent contamination of groundwater and surface water, and groundwater should be prevented from entering landfills. The anti-seepage system of the landfill should include leachate collection and treatment systems and dredging facilities.

2. Composting treatment

Simple high-temperature composting technology with a small scale, low mechanization and low investment and operating costs should be used in villages and towns. The garbage composting system should basically meet the following requirements:

(1) Organic matter content should be $\geq 40\%$;
(2) The temperature of the material in the reactor should be maintained at 55 °C for 5–7 days; and
(3) Residues of composting should be reused in farmland.

Domestic waste composting technology should be based on the current standard *Technical Code for the Composting of Municipal Solid Waste* CJJ 52-2014.

3. Biogas digester treatment

Organic farm waste can enter the household biogas digester for anaerobic treatment.

4. Resource recovery and harmless treatment

The utilization of domestic garbage resources is based on carrying out the "house classification" of garbage. Garbage that cannot be used is regularly transported and disposed of in a harmless manner, and other wastes are disposed of by means of manufacturing biogas, composting or resource recovery to realize the utilization of domestic garbage resources according to the principle of "reduction and harmless treatment".

Harmless treatment of domestic garbage refers to sanitary landfills, incineration and resource utilization (e.g., biogas production and composting). The harmless treatment rate of domestic garbage in green ecological villages should reach 100%. The harmless treatment rate of domestic waste in green ecological towns should be greater than or equal to 95%.

The calculation method for the harmless treatment rate of domestic garbage is as follows:

$$\text{Harmless treatment rate of domestic garbage} = \frac{\text{Innocent disposal of domestic garbage}(t)}{\text{Total amount of domestic garbage produced}(t)} \times 100\% \quad (6.2)$$

6.3.3 *Construction Case of Waste Disposal and Facilities of Villages and Towns*

We take Yanhe Village as an example. Since 2004, Yanhe Village has been conducting the classified collection and transfer of household garbage. Starting from the source of household garbage, each village household has two garbage cans that are collected daily by full-time cleaners using rickshaws. The organic waste is treated by the farmer him- or herself through the digester; inorganic garbage is sorted by the cleaners at garbage collection points. Recyclable garbage can be sold centrally to recycling stations, and the income is used as the labour income of the cleaners. The hazardous waste is collected separately and sent to the town for centralized disposal.

The garbage classification collection and evolution of Yanhe Village are shown in Fig. 6.6.

6.4 Construction of Clean Energy Applications and Facilities of Green Villages and Towns

6.4.1 *Facility Types of Clean Energy Utilization of Villages and Towns*

Clean energy utilization is an important indicator of the level of infrastructure construction and management in green villages and towns. At present, the clean energy utilization facilities commonly used in villages and towns in China are divided into six main types:

In 2007 In 2008 In 2009 In 2010

Fig. 6.6 Classification collection and evolution of household garbage in Yanhe Village

(1) Solar energy utilization facilities;
(2) Biomass energy utilization facilities;
(3) Wind energy utilization facilities;
(4) Water energy utilization facilities;
(5) Marine energy utilization facilities; and
(6) Geothermal energy utilization facilities.

The main clean energy facilities and utilization technologies of Chinese villages and towns are shown in Table 6.6.

6.4.2 Case of Innovative Mode of Biogas Utilization of Villages and Towns

1. General situation of biogas energy utilization in villages and towns of Gongcheng Yao Autonomous County

Gongcheng Yao Autonomous County is located in southeastern Guilin City, Guangxi Province. It is a National Sustainable Development Community and a National Ecological Demonstration Zone. The clean energy utilization of Gongcheng County includes wind energy and biogas utilization by large-scale farmers. Rural clean energy utilization in Gongcheng County has been recognized by the United Nations as "an apotheosis of rural ecological economic development in developing countries" and is known as the "Gongcheng pattern" for promotion in developing countries.

Table 6.6 Main clean energy facilities and utilization technologies of Chinese villages and towns

Clean energy utilization facility	Facility system classification	Clean energy utilization technology
Solar energy utilization	Solar water heating system	Vacuum tube solar water heater (95% domestic market share)
		Flat type solar water heater
	Solar cooker	Box-type solar cooker, flat-plate solar cooker, concentrating solar cooker, inner solar cooker
	Solar heating and cooling system	Solar energy greenhouse
		Solar house construction technology
		Solar heating technology
	Solar photovoltaic power generation system	Building attached photovoltaics (BAPV)
		Building integrated photovoltaics (BIPV)
Biomass energy utilization	Biogas system	Household biogas application technology
		Large and medium-sized biogas projects
		Domestic sewage purification biogas tank
	Straw energy utilization system	Straw biogas centralized gas supply technology
		Production of solidified straw fuel
		Centralized gas supply technology for straw pyrolysis and gasification
		Direct combustion power generation and straw distillation technology
		Soft straw carbonization production and processing technology
		Carbonization production and processing technology of hard straw

(continued)

Table 6.6 (continued)

Clean energy utilization facility	Facility system classification	Clean energy utilization technology
	Biomass power generation system	Agricultural and forestry waste direct combustion power generation, agricultural and forestry waste gasification power generation, waste incineration power generation, landfill gas power generation, and biogas power generation
Wind energy utilization	Small wind power	
	Wind energy extraction and drainage	
Water energy utilization	Micro-hydropower	
Marine energy utilization	Marine power generation (suitable for coastal and island villages)	
Geothermal energy utilization	Ground source heat pump system	Underground soil source heat pump, groundwater source heat pump, surface water source heat pump technology

Biogas utilization in Gongcheng County started in Huang Ling Village in 1979. By 2000, biogas digesters had spread to 9 towns and reached 73% of households. A three-in-one ecological agriculture industry chain has been formed with biogas utilization as the link, realizing three virtuous cycles of "ecological agriculture -ecological protection", "breeding-biogas-planting" and "economy-society". The household use rate of biogas in the county is over 89.5%, leading the country to utilize biogas in rural areas. Moreover, the county has taken the lead in exploring and establishing the "whole trusteeship" corporate operational mode of biogas in China and has achieved good results.

2. "Whole trusteeship" operational pattern of rural biogas in Gongcheng County

1) Establishment of the "whole trusteeship" operational pattern of biogas

We take the Langjiatun village group, Jianan Village, Limu Town, Gongcheng County, as an example. There are 53 households in the Lanjiatun group, and 53 biogas tanks have been built. Before the experimental period of the "whole trusteeship" operational pattern, 9 households stopped using biogas due to problems such as feed inlet cracking and damage to the fermentation system. At the same time, due to the lack of fermented raw materials, ageing of pipes, replacement of stoves, etc., none of the 53 households could use biogas normally. In 2015, Gongcheng County began piloting the "whole trusteeship" operational pattern for biogas use. The Lanjiatun group was identified as one of the pilot villages, and 50 households joined the biogas whole trusteeship pilot programme.

The biogas operational pattern in Gongcheng County includes three major "whole trusteeship" operations after the biogas digesters are completed. The model has been promoted in Gongcheng County.

(1) Whole trusteeship of input and output. The Lanjiatun group established a biogas service network to manage all incoming and outgoing materials. The service outlets reached an agreement with three large-scale farms and biogas users around Lanjiatun and are responsible for transporting biogas raw materials from the farms to 50 households in Lanjiatun and feeding them twice a month to ensure the provision of biogas materials. The outlets ensure the normal use of biogas. At the same time, the outlets are responsible for discharging for the farmers once a month and extracting slag for the farmers once a year to ensure efficient biogas operation.
(2) Whole trusteeship of maintenance. The network is responsible for the normal operation of the biogas facilities. It is responsible for overhauling and maintenance to solve the problems arising from biogas digesters, biogas stoves, biogas lamps, biogas transmission and distribution pipelines and components for the contracted households and providing long-term, thoughtful and honest services for farmers. Within 48 h of receiving notice of a problem, people will arrive at the scene to help troubleshoot.
(3) Implementation of consumption measurement. The person in charge of the outlet collects a deposit of 200 yuan from each household and installs a flow metre to implement consumption measurement. The farmers pay the outlet a fee of 2.0 yuan per cubic metre to ensure the daily energy consumption of "three meals a day".

2) Benefit evaluation of the whole trusteeship operation of biogas

(1) Social benefits. The pilot project of the whole trusteeship model fundamentally solves the problems of insufficient materials and maintenance technology in household biogas digesters and provides a successful example for the long-term operation and promotion of biogas in households. It also provides a beneficial exploration of the construction and management of clean energy utilization facilities in green villages.
(2) Environmental benefits. The outlets are the intermediary, and ensuring the normal operation of methane solves the problem for large farmers of environmental pollution caused by the difficulty of disposing of livestock droppings. In the process of normal biogas use, farmers can also effectively digest organic wastes such as livestock excrement in villages. Biogas slurry is used for agricultural production, which improves soil fertility, reduces the application of chemical fertilizers, reduces soil pollution, and plays a positive role in protecting rural living production environments and building a clean countryside.
(3) Economic benefits. In terms of the operational network, for a family of four, the average monthly gas consumption is approximately 30 cubic metres. Fifty users have signed up, and each household pays 60 yuan per month. The outlets can charge 3,000 yuan per month. For farmers, if they use approximately 1 canister of liquefied gas per month, the price is 125 yuan. The use of biogas

can save approximately 65 yuan per month, or 780 yuan per year. The total energy savings of the village are approximately 39,000 yuan per year. Thus, a win–win situation of economic benefits has been achieved.

3. "Trinity" Corporation Operational Patterns for Ecological Agricultural Development

1) "Replacing firewood by biogas" pilot village

Huangling Village is the first pilot village in Gongcheng County to carry out the pattern of "replacing firewood with biogas". Over more than 30 years of hard work, the utilization of biogas in Huangling Village has developed from the small "trinity" courtyard economic pattern of "pig-marsh-fruit" of pig manure and cow dung into the pool, biogas into the room for cooking and lighting, and biogas residue and biogas slurry into the orchard into a large "trinity" ecological agriculture development pattern and biogas utilization promotional pattern of "scale farming + corporate biogas whole trusteeship + intensive planting".

2) Corporation operational pattern

At present, Huangling Village has adopted a corporation operational pattern for the utilization of biogas. The main process is that the government introduced a biogas service company, and the company cooperates with large-scale farms to purchase livestock manure and carry out harmless treatment as raw material for the biogas tank. Farmers' biogas digesters implement "whole trusteeship" for feeding, discharging and maintenance. Farmers use gas only for swiping, and the biogas slurry and biogas residue produced by the biogas digesters are treated by the company as organic fertilizers for large-scale plantations and large planters. After the large "trinity" pattern for ecological agricultural development was implemented, a win–win situation was achieved for farmers' benefit and company development.

6.5 Construction of Greening and Landscaping Facilities of Green Villages and Towns

1. Construction content of greening and landscaping facilities of villages and towns

Greening and landscaping facilities are important components of infrastructure construction and environmental protection in green villages and towns.

(1) Greening facility construction of villages and towns includes green planting projects, public green space construction, green space construction for village and town protection, ancient and famous tree protection, and river (stream) flow landscape improvement.
(2) Landscaping facility construction of villages and towns includes the construction style and architectural style of villages and towns, protection of traditional

villages, construction of public spaces and landscapes, and construction of environmental sanitation facilities.

2. Construction requirements of greening and landscaping facilities of villages and towns

1) Included in town planning and village planning

Greening and landscaping facilities should be incorporated into green town planning and green village planning according to the characteristics of different regions.

2) Greening requirements of built-up areas in villages and towns

(1) Greening and planting in the built-up areas of villages and towns should be dominated by arbor trees combined with evergreen trees and deciduous trees with a focus on the combination of trees, irrigation and grass.
(2) Local plants should be chosen for rural green public areas to reflect the rural landscape. The necessary venues and fitness facilities will be constructed in the public green space, and public science and entertainment facilities can also be constructed in conjunction with public green space construction.
(3) The theme, form, materials and volume of sculptures and landscaping in the green areas of villages should be in harmony with the environment. Energy-saving lamps should be used for lighting in the green areas of villages and towns. Villagers are encouraged to build small landscape pieces and create rest spaces near their own courtyards or homes.
(4) Ancient trees and their growing environment in villages and towns should be properly protected. Fences or stone masonry can be built, and signs can be set up.

3) The construction requirements of public spaces and small landscapes in villages and towns

(1) The public spatial form of green villages and towns should be lively and free and convenient for villagers to carry out various public activities.
(2) The relevant facilities in the venue should be marked with a prominent logo. Various signs for public space settings should not impede pedestrian traffic and should coordinate with the landscape environment.
(3) Garbage bins should be reasonably set up in the public spaces of villages and towns to collect garbage.
(4) Eligible open space construction can be combined with emergency shelter construction to establish a safe passage.

4) Protection requirements of natural and historical cultural heritage in villages and towns

(1) Practical measures should be established to maintain the authenticity and integrity of the natural landscape of the village.
(2) The local folk customs and representative living and production landscape elements should be conserved and scientifically used.
(3) Ancient towns, ancient villages, and outstanding traditional houses should be renovated and protected.

5) Landscape construction requirements of rivers and streams in villages and towns

(1) The natural state of the river should be maintained, and the requirements of flood discharge and drainage should be met. Proper methods of slope protection should be selected according to different functional requirements. It is advisable to use local stones for stone masonry slope protection and planting bags (cages) or ecological bricks for ecological slope protection.
(2) Based on natural vegetation landscapes, artificial landscapes such as pavilions and suspension bridges should be used to integrate riverside landscapes into villages and towns to create an ecological corridor that is suitable for large numbers of people to take leisurely walks.
(3) For puddles and ponds, the focus should be on protecting the natural state and maintaining their complete ecosystem. They should not be over-utilized or landfilled.

6.6 Construction of Disaster Prevention, Reduction and Early Warning Facilities of Green Villages and Towns

6.6.1 Construction Requirements of Disaster Prevention and Reduction Facilities of Green Villages and Towns

Disaster prevention and reduction facility construction is necessary for environmental infrastructure construction in green villages and towns. These are important facilities to ensure the safety of the lives and production of residents and the sustainable development of the social economy in villages and towns.

1. Common disaster types of villages and towns

Disasters can be divided into man-made disasters and natural disasters according to their causes. Man-made disasters are all kinds of disasters caused mainly by human factors, such as natural resource depletion, environmental pollution, fire, traffic, epidemics, and war. Natural disasters are all kinds of disasters that harm human existence or damage the human living environment. According to the causes and mechanisms of natural disasters, they can be divided into geological disasters, meteorological disasters and environmental disasters.

2. Construction content of disaster prevention and mitigation facilities of green villages and towns

The construction of disaster prevention and mitigation facilities in green villages and towns is aimed at enhancing the ability of villages and towns to resist natural disasters. The key construction contents include fire control facilities, flood control facilities, earthquake resistance and control facilities, and wind prevention and disaster reduction facilities. Practical and effective disaster prevention and early-warning measures should also be established.

In constructing disaster prevention and reduction facilities in green villages and towns, special attention should be paid to the prevention and mitigation of disasters in villages. Green villages should have complete disaster prevention facilities and places to avoid natural disasters, a sound emergency plan management system for natural disaster relief, disaster relief volunteers and volunteer firefighters. Additionally, an emergency drill should be organized every 1–3 years.

6.6.2 Fire Control Facility Construction and Early-Warning Measures of Villages and Towns

1. Fire facility construction of villages and towns

The construction of fire safety facilities in green villages and towns should meet the requirements in the Standard for Planning of Town GB 50,188-2017, Code for Fire Protection and Prevention of Rural Area GB 50,039-2010 and Code for Fire Protection Design of Buildings GB 50,016-2014.

1) Fire lanes

Fire lanes in villages and towns should meet the requirements of the current national standard Code for Fire Protection Design of Buildings GB 50,016-2014:

(1) The net width and headroom of the lane should not be less than 4.0 m;
(2) The turning radius should meet the requirements for the turning of a fire engine; and
(3) There should be no obstacles, such as trees or overhead lines, that would hinder the operation of a fire truck between the fire lane and the building.

2) Rural building fire prevention

The fire resistance rating of rural buildings should not be lower than Grade I and Grade II and should meet the requirements of the current national standard Code for Fire Protection Design of Buildings GB 50,016–2014.

(1) Fire protection facilities for rural houses. Warehouses for storing materials such as firewood and agricultural implements and agricultural materials should be constructed independently; when combined with rooms for other uses, they should be separated by non-burning solid walls.
(2) Fire protection facilities for public buildings. The number of fire compartments and safety exits should meet the requirements of the current national standard Code for Fire Protection Design of Buildings GB 50,016–2014: not less than 2.

3) Fire point (station) setting

Villages and towns should set up fire stations and fire points according to scale, regional conditions, economic development and fire risk and other factors. The construction and equipment of the fire station should be implemented in accordance with the relevant fire station construction standards.

(1) Have fixed locations and buildings with clear marks;
(2) Be equipped with all or most firefighting equipment, such as fire trucks, hand-lifted motor pumps, water guns, hoses, fire extinguishers, and demolition tools; and
(3) Take insulation measures in cold areas.

4) Have a fire water source setting

Fire water sources should be set up in villages and towns. Villages and towns with water supply networks should be equipped with outdoor fire water supply systems, and those that do not have water supply networks should build fire pools or use natural water sources.

5) Outdoor fire hydrant setting

(1) When a village is within the protection scope of a fire station (point), the pressure of the outdoor fire hydrant bolt should not be lower than 0.1 MP; when a village is not within the protection scope of the fire station (point), the outdoor fire hydrant should meet the requirements of the unfavourable point of pressure and flow within the protection radius;
(2) The diameter of the fire water-supply pipe should not be less than 100 mm;
(3) The buried depth of fire water-supply pipelines should be determined according to climatic conditions, external loads, pipe properties and other factors;
(4) The distance between outdoor fire hydrants should not exceed 120 m; in rural areas where there are many buildings with fireproof Grade III and Grade IV materials, the distance between outdoor fire hydrants should not exceed 60 m;
(5) Outdoor fire hydrants in cold areas should adopt anti-freezing measures or use underground fire hydrants, fire water cranes or indoor fire hydrants; and
(6) Outdoor fire hydrants should be placed along the road and should be close to intersections, and the distance from the outer wall of a house should not be less than 2 m.

2. Fire warning measures of villages and towns

(1) Set up a fire alarm telephone and have reliable communication with the city fire command centre and water supply and power supply departments;
(2) Establish a full-time, obligatory or volunteer fire brigade;
(3) Carry out fire protection knowledge publicity and regular fire drills; and
(4) Formulate practical and feasible fire prevention and emergency response plans.

6.6.3 Flood Control Facility Construction and Early-Warning Measures of Villages and Towns

1. Flood control facility construction of villages and towns

The construction of flood control facilities and early-warning measures in villages and towns should meet the requirements of national standards such as the Standard for Planning of Town GB 50,188-2007 and Flood Control Standard GB 50,201-2014.

The main types of floods in villages and towns include river floods, sea tides, mountain torrents and mudslides.

Most villages and towns located in rural flood protection areas have no flood control capacity and need to take flood prevention measures to protect their safety. The rural protection zone should be divided into four protection grades according to the population or cultivated land area, and each grade corresponds to a type of flood control standard. The flood protection standard of Grade I is 100–50 years, that of Grade II is 50–30 years, that of Grade III is 30–20 years, and that of Grade IV is 20–10 years.

Flood control facilities in villages and towns include mainly cofferdams, safety platforms, water shelters, and flood shelters.

The safety value of the construction of flood control safety facilities such as cofferdams, safety platforms and water shelters should comply with the requirements of the Standard for Planning of Town GB 50,188-2007. Safety in the event of the greatest possible flood should be guaranteed.

Flood shelters in villages and towns should comply with the relevant provisions of the current national standard Technical Standard for Buildings in Flood Plains and Flood Detention Areas GB/T 50,181-2018.

Flood control and disaster reduction projects in villages and towns include river-bank remediation, river dredging, dam construction, and flood control projects. For villages and towns that are vulnerable to internal disasters, stagnation projects should be combined with drainage projects.

Flood prevention and rescue facilities in villages and towns include emergency evacuation points, medical rescue, material reserves and alarm devices.

2. Flood control early-warning measures of villages and towns

The flood warning measures of villages and towns should include the following:

(1) Establishing a flood warning mechanism;
(2) Conducting flood control and risk aversion drills; and
(3) Popularizing flood control information and education.

6.6.4 Earthquake Resistance Facility Construction and Early-Warning Measures of Villages and Towns

1. Earthquake resistance facility construction and earthquake prevention in villages and towns

The construction of earthquake and earthquake resistance facilities in villages and towns should comply with the requirements of Standard for Planning of Town GB 50,188-2007, Emergency Shelter for Earthquake Disasters-Site and Its Facilities GB 21,734-2008 and Code for Seismic Design of Buildings GB 50,011-2010.

1) Lifeline engineering and important facilities

(1) Facilities and food supplies for transportation, communications, water supply, power supply, energy, fire protection, medical care, etc. must be fully guaranteed;
(2) There should be multiple entrances and exits from different directions in densely populated areas and sections of villages and towns;
(3) A standby power supply should be set up for emergency command organizations for earthquake and disaster prevention in villages and towns; and
(4) Reasonable shelter and evacuation sites should be set up. Evacuation sites should be considered in conjunction with the construction of facilities such as squares and green spaces and comply with the provisions of the Standard for Planning of Town GB 50,188-2007.

2) Earthquake emergency shelters and support facilities

(1) The site selection is reasonable. Parks, green spaces, squares, stadiums, etc. can be selected as earthquake emergency shelters in villages and towns. Earthquake emergency shelters should have two or more evacuation roads connected to the outside world with different directions and should meet the safety requirements of the Emergency Shelter for Earthquake Disasters-Site and Its Facilities GB 21,734-2008.
(2) The facilities are fully equipped. Earthquake emergency shelters in villages and towns should have the necessary basic facilities, including emergency shelter facilities, medical assistance and sanitation facilities, emergency water supply and power supply facilities, emergency sewage systems and toilets, and emergency garbage storage and transportation facilities, and should meet the requirements of the Emergency Shelter for Earthquake Disasters-Site and Its Facilities GB 21,734-2008.

2. Early-warning measures of earthquake disasters in villages and towns

Early-warning measures for earthquake disasters in villages and towns should include the following:

(1) Establishment of earthquake prediction and early-warning mechanisms;
(2) Practice of earthquake prevention and risk avoidance; and
(3) Popularization of information and education about earthquake prevention.

6.6.5 Wind Disaster Prevention and Reduction Facility Construction and Early-Warning Measures of Villages and Towns

The construction of wind prevention and disaster reduction facilities in villages and towns should comply with the requirements of Standard for Planning of Town GB 50,188-2007 and Regulations on the Prevention of Meteorological Disasters, Order No. 570 of the State Council (2017).

1. Construction of wind-prone area facilities

Intensive shelter belts at the edge of the windward direction should be selected for villages and towns that are prone to becoming wind-struck areas and should meet the requirements of the Code for Design of Windbreak and Sand Fixation Forest GB 51,085-2015. Villages and towns with strong winds (sandstorms) and tornadoes should strengthen the construction of shelter belts and emergency shelters, and windbreaks and risk avoidance elements of buildings and structures should be regularly inspected.

2. Construction of typhoon-prone area facilities

Dams should be built in villages and towns in coastal areas and islands that are prone to be affected by typhoon disasters to enable them to withstand storm surges. The construction of seawalls, dikes, havens, windbreaks, sheltered anchors, and emergency shelters should be strengthened, and personnel transfer plans should be prepared.

3. Construction of storm drainage systems after wind disasters

Drainage systems in accordance with the maximum amount of precipitation should be built in villages and towns to ensure that rainstorm is eliminated timely after the wind disasters.

4. Emergency plans for wind disaster prediction

Emergency plans for wind disaster prediction in villages and towns should include the following:

(1) Establish a typhoon and gale forecast information network;
(2) Formulate contingency plans for windstorms; and
(3) Provide medical and rescue facilities.

6.6.6 *Construction Cases of Prevention and Mitigation of Disaster and Early-Warning Facilities of Villages and Towns*

The construction of disaster prevention and mitigation and early-warning facilities in Fushan Town, Chengmai County, Hainan Province is taken as a case study. For an overview of Fushan Town, see Sect. 5.3.2.

1. Plan of disaster prevention and mitigation of Fushan Town

The disaster prevention and mitigation plan of Fushan Town is divided into two parts: township and town area. The key contents of the plan are as follows:

(1) Establish an integrated prevention and control emergency command centre;
(2) Construct a prevention and control evacuation passageway and evacuation site;
(3) Set a floodwater level warning line;
(4) Ensure the availability of land for disaster prevention and treatment; and
(5) Protect key construction facilities, such as fire water pipelines, communication facilities, schools, hospitals, and sub-stations.

2. Construction of fire control facilities and early-warning measures in Fushan Town

(1) Define firefighting tasks and firefighting teams in accordance with the public safety industry standard Rural Fire Department Standard GA/T 998-2012. Implement various fire protection tasks.
(2) Strengthen the construction of fire control water sources and fire hydrants. Focus on the construction of fire water sources and hydrants in tourist reception areas and old neighbourhoods.
(3) Pay attention to the building of internal fire facilities. All kinds of newly built buildings in the town region must meet the requirements for fire-resistant grade and firefighting facilities stipulated by the state and Hainan Province.
(4) Ensure that the construction of fire lanes meets the relevant state requirements.
(5) Set up a fire alarm telephone and monitoring alarm system, formulate fire emergency plans, and set up obvious fire signs in tourist spots and communities.

3. Construction of flood control facilities and early-warning measures in Fushan Town

(1) Ensure that the flood control facilities in the town meet the fortification standards. Construction will be carried out according to the fortification standard once every 20 years and checked against the standard once every 20 years. The construction land in the town should be more than 2 m above the calibrated water level of the Fushan Reservoir (78.0 m).
(2) Pay attention to the maintenance and management of water conservancy and flood control facilities. The management of the main canals of the Fushan Reservoir and the Songtao Reservoir should be strengthened to ensure that the flood discharge channel and the river channel are unblocked so that floodwater can be discharged smoothly and in a timely manner in the case of flood disasters.

(3) Pay attention to drainage and waterlogging control. According to the national standard Code for Design of Outdoor Wastewater Engineering GB 50,014-2006, the drainage and waterlogging control in town should be based on the rainwater flow corresponding to a 10-min rainstorm every year.
(4) Set up construction and early-warning measures for other flood control facilities in Fushan Town in accordance with the requirements of Chengmai County, which meet the flood control requirements of the state and Hainan Province.

4. Construction of earthquake resistance facilities and early-warning measures in Fushan Town

(1) Pay attention to construction projects and village site selection. The basic intensity of a Fushan earthquake is in the range of 7 degrees, and the seismic acceleration is 0.15 g. When selecting the location of the new project and village, avoid the seismic fault zone and the seismic active area, and strictly follow the Code for Seismic Design of Buildings GB 50,011-2010.
(2) Pay attention to the seismic reinforcement of dangerous buildings in villages and towns. Guide the residents to construct houses that meet the requirements of earthquake resistance, and implement timely demolition and reinforcement measures for dangerous buildings.
(3) Pay attention to the construction of lifeline facilities for disaster prevention. Develop a comprehensive disaster prevention and emergency plan for infrastructures such as water supply, power supply, transportation, electricity, telecommunications, bridges and culverts as well as public service facilities and tourist reception facilities.
(4) Propagate scientific information about earthquake prevention and disaster reduction and improve the earthquake resistance of residential buildings in villages and towns and the self-rescue ability of residents.

5. Construction of windbreak and disaster reduction facilities and early-warning measures in Fushan Town

(1) Provide uniform arrangements for windbreak and disaster reduction facilities in Fushan Town.
(2) Prioritize the planning and construction of typhoon early-warning information communication systems. Through the analysis of typhoon trends, landing places and times, timely typhoon forecasts, typhoon warnings and emergency warnings can be released; through the establishment of an effective department linkage mechanism by the plan, typhoon disaster warning information can be transmitted to the safety personnel of all departments within 15 min after the release.
(3) Improve the emergency response capability to typhoon disasters. To ensure that typhoon disaster warning information can be disseminated to people in schools, construction sites, enterprises, villages and residential committees, and public places within 30 min after its release, safety publicity and inspection meetings should be held twice a year, and disaster prevention brochures should be distributed to each household to improve people's ability to actively defend against typhoon disasters.

Chapter 7
Construction and Renovation of Green Buildings in Villages and Towns

Abstract Green building construction and green reconstruction of existing rural buildings are the main components of green village and town construction. This chapter analyses the concepts and characteristics of green building construction of villages and towns; puts forward the basic requirements and energy-saving design methods for green building design of villages and towns; proposes seven key technologies of green building construction of villages and towns; analyses the properties of repair and renovation of existing rural buildings, the durable life of buildings and the types of repair and renovation construction; and describes the technical methods of green renovation of existing buildings.

The construction and renovation of green buildings are the main activities in green village and town construction. China's rural housing construction has shifted from the rapid increase in quantity to the improvement of building function and quality. In the construction and renovation of green building of villages and towns, the performance of energy savings, land savings, water savings, material savings and environmental protection over the whole life of a building should be improved, and the effects of monitoring and evaluating green construction methods should be reflected.

7.1 Design of Green Buildings in Villages and Towns

7.1.1 Concepts and Characteristics of Green Building Construction of Villages and Towns

1. Concepts of green building construction of villages and towns

Green buildings in villages and towns are rural buildings that save resources (energy, land, water and materials); protect the environment; reduce pollution; provide healthy, applicable and efficient space for farmers; and maximize the harmonious

X. Wang and H. Hua, *Green Village and Town Construction in China*,
https://doi.org/10.1007/978-981-16-2098-0_7

coexistence between humans and nature during the whole life of construction and use of buildings.

Green buildings of villages and towns are a specific type of rural sustainable building, and their design and construction should be guided by sustainable development concepts. Sustainable buildings are buildings built in accordance with the concepts and elements of sustainable development in the whole plan and design, including construction, management, demolition and recycling processes. The principles for assessing sustainable buildings include the principles of resource application efficiency, energy use efficiency, pollution prevention and environmental harmony.

2. Characteristics of green building construction of villages and towns

The construction methods of green buildings are different in villages and towns than in cities. Farmers in rural areas build mostly by means of rural collective planning, farmer or collective self-investment and self-organization. Rural green building design standards, construction methods and operational management are quite different from the uniform standard design, professional construction and centralized management of green buildings in urban areas. Therefore, the construction of green buildings in villages and towns should be adapted to different regional conditions, different levels of economic and social development and different construction techniques and methods in rural areas and be carried out according to local conditions.

The main objects of green building construction in villages and towns are as follows:

(1) Construction of green residential buildings for farmers, for example, farmer residences, farmhouses and farmhouse inns.
(2) Construction of green public buildings owned by collectives. These include village halls, farmer activity centres, and tourism and leisure buildings.
(3) Green renovation of various existing old buildings.

3. Elements of green building construction in villages and towns

(1) Land savings and outdoor environment include site selection, land use, outdoor elements, transportation and public services, and landscaping.
(2) Energy savings and energy utilization include the building and envelope, heating ventilation and air conditioning, lighting and electric aspects, clean energy and renewable energy utilization.
(3) Water savings and water resource utilization include the average daily water supply, leakage of pipe networks, use of water-saving appliances, and rational use of unconventional water sources.
(4) Material savings and material resource utilization include the building size coefficient, the amount of local building materials, and the amount of renewable and recyclable materials.
(5) Indoor environmental quality includes the sound environment, light environment, indoor temperature and humidity, and indoor air environment.

7.1.2 Basic Design Requirements of Green Buildings of Villages and Towns

1. Pay attention to the architectural design, layout and site selection

The building layout and site selection are the first steps in green building design. The following basic design requirements should be met according to the actual local conditions:

(1) Green building construction of villages and towns should be arranged and planned rationally and reflect local features and characteristics.
(2) Residential buildings should be constructed to harmonize with local conditions, save resources, protect the environment, and meet the requirements of energy savings, land savings, water savings and material savings.
(3) Residential land should be located downwind of the perennial minimum wind frequency of air pollution sources and upstream of water pollution sources, should be relatively concentrated, and should effectively use land and space resources.
(4) Residences should have good natural ventilation to adhere to the sunshine spacing coefficient stipulated by the local construction management department.
(5) In mountainous and hilly areas, the slope of the construction site should be under 25%. The southern, southeastern and southwestern slopes, exposed to the sun, should be preferred to avoid landslides, flood gullies and perennial wind gap. Flexible and diversified architectural design methods such as platform building, splintering, stacking and layered entrances should be adopted in the vertical treatment.

2. Meet construction requirements for green buildings

Construction is the key step in realizing green building functions. In the design, the characteristics of rural housing construction should be combined to promote the realization of green building functions:

(1) Control construction costs. Ensure that the positioning of living functions is reasonable, and the area is suitable. In a square building, full use should be made of the compact space, with reasonable layout of kitchen, bathroom, storage and traffic spaces and a reduced need for interior space.
(2) Reduce the housing cost. The building should receive sufficient sunshine in winter and have good orientation, natural ventilation and unblocked layout in summer. The kitchen and toilet should be naturally ventilated, exhaust gas should be removed, and clean sanitation methods should be achieved. Local building materials with good thermal insulation performance should be selected to improve thermal insulation performance and reduce energy consumption. Technologies such as solar water heating, small domestic sewage treatment and reuse, and methane can be adopted.

(3) Ensure that the construction is beneficial to farmers' production. According to the different industries engaged in by farmers in villages and towns, suitable production space and suitable storage space should be provided to promote the development of the family industry.
(4) Promote sustainable development. At the beginning of the design and construction of green buildings in villages and towns, there should be sufficient room for functional improvement and flexibility of renovation, which should not only reduce the cost of construction and use throughout the whole life of a building but also avoid the frequent demolition and construction of buildings in a short period of time, thus advocating the century-old rural housing construction concept.
(5) Realize energy-saving ecology. In terms of building materials, local building materials that are convenient, practical, environmentally friendly, recyclable and low in energy consumption can be selected; an appropriate layout and orientation can be arranged; and natural ventilation and lighting can be incorporated to reduce the energy consumption of the building.

3. Promote the organic integration of green functions

The core function of green buildings in villages and towns is "four savings and one environmental protection", namely, five factors: land savings and outdoor environment, energy savings and energy utilization, water savings and water resource utilization, material savings and material resource utilization, and indoor environmental quality. The design should provide overall consideration of these factors, and the optimal configuration of overall green functions should be realized through function combination and cost control.

4. Encourage farmers' participation in architectural design

It is important in the green building design of villages and towns to give full play to the role of construction communities and encourage local residents to actively participate in the design.

As shown in Fig. 4.2 and Table 4.1, the planning architect is an important member of the construction community of green villages and towns as the planner of a construction project, the designer of the construction scheme and a participant in the construction management process. Throughout the process of architectural design and construction management, the concept of green building should be promoted to farmers, green functional requirements of buildings should be solicited, and architectural design schemes should be jointly selected to guide the green construction of buildings and help village and town organizations to formulate green community management methods.

7.1.3 Energy-Saving Design Methods of Green Buildings of Villages and Towns

Energy-saving design is the core task of the green building design of villages and towns. Energy-saving design of green buildings should include energy savings in building objects, building equipment, material utilization, efficiency and other contents.

1. Energy-saving design of main structure

(1) Choose a reasonable building site. The location of the building site should fully consider the local climate, sunshine, wind direction and other factors to provide a good environment for residential energy savings.
(2) Determine the building shape coefficient. Under the condition that the design function of the subject is guaranteed to remain unchanged, the smaller the building shape coefficient is, the more energy saving the building is.
(3) Design an energy-saving building envelope that includes energy savings for the external walls, roof, floor, external doors and exterior windows. According to the actual situation in rural areas, different methods can be adopted for the energy-saving design of exterior walls, such as external wall external and internal insulation, or wall self-insulation. Roof design can use insulation materials and air insulation layers or plants on the roof to save energy. Floor energy savings can involve the addition of glass wool felt to ordinary floors, and outer doors and windows can be made of energy-saving materials according to local conditions. Outer windows should be double glazing windows. The ratio of windows to walls should not exceed the limit stipulated in the *Code for Thermal Design of Civil Building* GB 50176-2016.

2. Energy-saving design of building equipment

The principles of "suitable for local conditions" and "low cost, high energy efficiency" should be adopted according to the actual situation of the local countryside. Farmers should voluntarily choose energy-saving and water-saving construction equipment. The key points of the energy-saving design of building equipment are as follows:

(1) Energy savings for heating ventilation and air conditioning. For rural buildings in different climatic zones, high-efficiency energy-saving heating and ventilation methods are adopted in line with local characteristics and economic development level. Indoor temperature and humidity should meet the requirements of the *Code for The Thermal Design of Civil Building* GB 50176-2016. Energy-saving and environmentally friendly household air-conditioning equipment should be selected.
(2) Lighting and electrical energy savings Choose energy-saving lamps and household appliances.

(3) Clean energy and renewable energy utilization. Use clean biomass stoves, gas-fired stoves and biogas stoves; clean energy such as liquefied petroleum gas and natural gas; and renewable energy such as solar energy, geothermal energy and biomass energy.

3. Selection of building energy-saving material

(1) The principle of material selection involves good thermal insulation, non-toxicity, flame retardancy, low cost, and easy construction.
(2) Commonly used energy-saving materials include extruded Styrofoam sheets, vitrified hollow microbead fire insulation mortars, aerated concrete blocks, and felt; these materials have better thermal insulation performance and lower construction costs.
(3) Roof insulation materials include thermal insulation materials suitable for local climate conditions and durable planting roofs; these also have a good effect on heat preservation and energy savings and beautify the living environment.

4. Evaluation of building energy efficiency

(1) Meet the requirements of energy-saving design. The design of green buildings in villages and towns should comply with the relevant national and local building energy-saving design codes and standards. Compared with the same type of building without an energy-saving design, the energy-saving rate should not be lower than the requirements for local energy-saving buildings.
(2) Energy-saving calculation of building thermal engineering. The energy-saving calculation of thermal construction mainly verifies whether the specified index for each subitem meets the standard requirements, including the average heat transfer coefficient of the exterior walls, heat transfer coefficient of the roof, ground thermal resistance, window-wall ratio, visible light transmission ratio, airtightness of transparent curtain walls and exterior windows. If the specified index is to meet the requirements, the thermal performance of the building and the energy consumption of the building equipment should trade off in the calculation.

7.1.4 *Case of Energy-Saving Design of a Green Building in Yanhe Village*

This case is the design scheme of the green building energy-saving project of the Yanhe Village Reception Centre completed by the authors' team in 2009.

1. Overview of the energy-saving design of the village reception centre

1) Requirements of energy-saving design

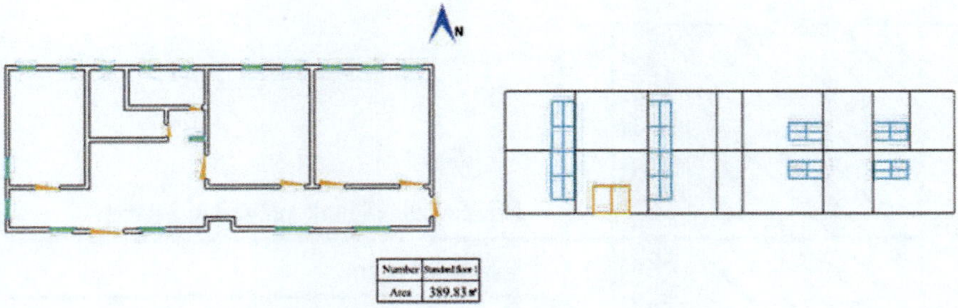

Fig. 7.1 Large sample of the reception centre

Yanhe Village, Wushan Town, Gucheng County, Xiangyang City, Hubei Province, is a typical Chinese hill village with hot summers and cold winters.

According to the regulations of building energy-saving design in Hubei Province in 2006, the energy-saving rate of new buildings in cities and towns of the province should reach 50%, and the energy-saving rate of new buildings in Yanhe Village was designed according to the above requirements.

2) Basic design data

The basic data of the energy-saving design of the Yanhe Village Reception Centre are as follows:

City: Gucheng (latitude = 32.3°, east longitude = 111.6°, altitude = 32.6 m); building orientation: south; building energy-saving calculation area: 779.66 m^2; building surface area: 1177.11 m^2; building volume: 3391.52 m^3; building shape coefficient: 0.347; building story number: 2; building height: 8.7 m.

A large sample of the reception centre is shown in Fig. 7.1.

Figure 7.2 shows the schematic diagram of the external wall of the reception centre.

The structural diagram of the energy-saving design of the reception centre roof is shown in Fig. 7.3.

The concept and design drawing of building energy savings and the actual scene after completion of the reception centre are shown in Fig. 7.4.

2. Energy-saving rate evaluation of reception centre

The energy-saving rate evaluation of the building is divided into 3 steps: to establish a building model; to determine the evaluation parameters, and to make the energy consumption simulation calculation.

1) Establish building model

For the evaluation case, 3 building models were established using the software Ecotect, and the simulation calculation and comparison of building energy consumption were carried out.

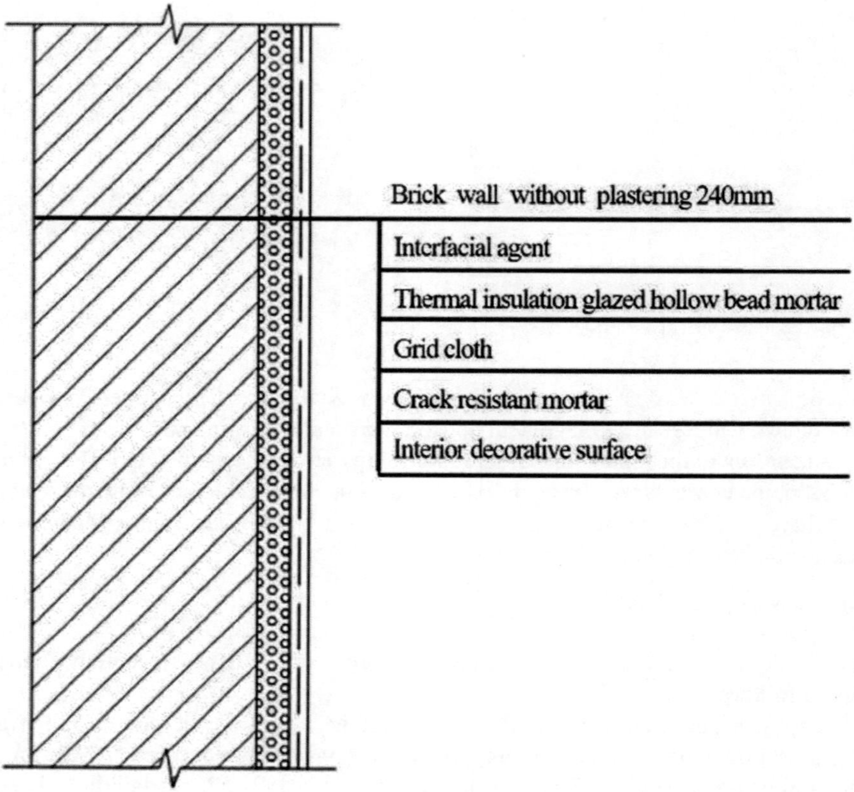

Fig. 7.2 Schematic diagram of the external wall insulation construction of the reception centre

Model I is the design building, which refers to Yanhe Village Reception Centre.

Model II is the reference building A, which is a hypothetical building for judging the thermal performance of the designed building envelope on the basis of the *Design Standard for Energy Efficiency of Public Buildings* GB 50189-2005;

Model III is the reference building B, which is a hypothetical building that does not use energy-saving design compared to the design building.

2) determine evaluation parameters

(1) Meteorological parameters

The evaluation data were obtained from the meteorological data (CSWD) of China's western Hubei region provided by the World Meteorological Organization (WMO) and the *Special Weather Data Collection for China's Building Thermal Environment Analysis*. The sunshine conditions of the design were determined according to the latitude and longitude of the village, namely, 32.3° north latitude and 111.4° east longitude.

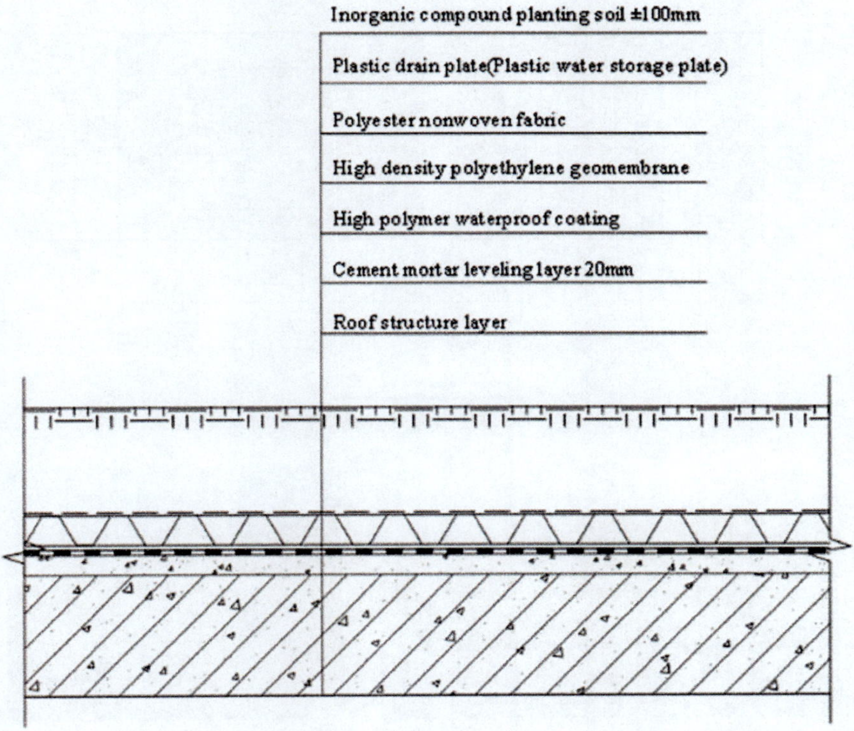

Fig. 7.3 Structure diagram of reception centre planting roof

(2) Thermal parameters

The main design thermal parameters of the 3 building models are shown in Table 7.1, Table 7.2 and Table 7.3 respectively.

3) Energy consumption simulation calculation

(1) Building evaluation model based on Ecotect

After the process of modelling, material assignment and parameter input, the energy consumption of the three building models was simulated.

The reception centre building model is shown in Fig. 7.5.

The building energy consumption simulation calculation process is shown in Fig. 7.6.

The data analysis of monthly energy consumption of the designed building, reference building A and reference building B calculated based on the building model is shown in Table 7.4, Table 7.5 and Table 7.6 respectively.

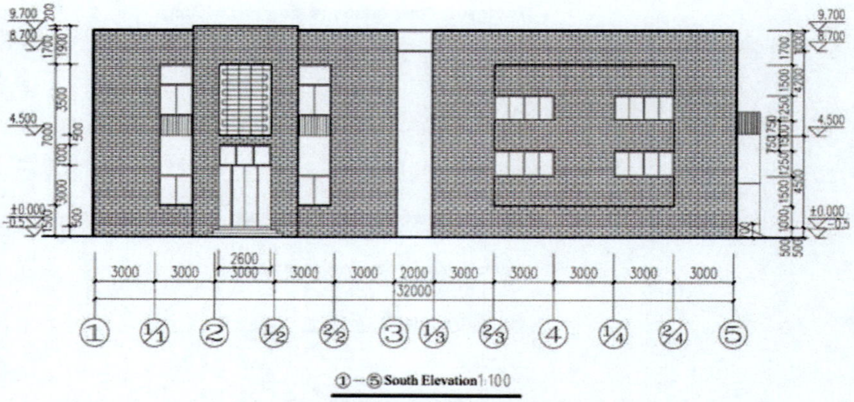

(a) Design drawing of reception centre

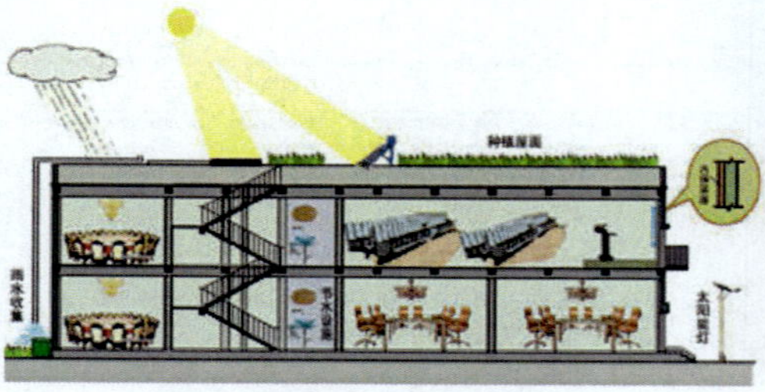

(b) Conceptual design for building energy savings of reception centre

(c) Actual scene after completion of reception centre

Fig. 7.4 Concept and design drawing of building energy savings and actual scenes after completion of reception centre

Table 7.1 Main design thermal parameters of roof and external walls

Envelope enclosure	Heat transfer coefficient K/(W/m^2 K)		
	Design building	Reference building A	Reference building B
Roof	0.58	0.7	4.17
External walls	1.35	1	2.37

(2) Evaluation of the thermal performance of the envelope of the designed building

① Calculation formula of the heat transfer coefficient of the envelope:

$$K = \frac{1}{R_0} \tag{7.1}$$

$$R_0 = R_i + R + R_e \tag{7.2}$$

Table 7.2 Main design thermal parameters of external windows

Envelope enclosure (external windows)		Main thermal parameters					
		Design building		Reference building A		Reference building B	
Direction	Window-wall ratio	Heat transfer coefficient $K/(\text{W/m}^2\ \text{K})$	Shading coefficient *SC*	Heat transfer coefficient $K/(\text{W/m}^2\ \text{K})$	Shading coefficient *SC*	Heat transfer coefficient $K/(\text{W/m}^2\ \text{K})$	Shading coefficient *SC*
East	–	–	–	–	–	–	–
South	0.20	0.27	0.94	4.7	1	4.7	0.95
West	0.11	0.27	0.94	4.7	1	4.7	0.95
North	0.22	0.27	0.94	3.5	1	4.7	0.95

Table 7.3 Main thermal parameters of bottom surface

Envelope enclosure	Thermal resistance R/[(m^2 K)/W]		
	Design building	Reference building A	Reference building B
Bottom surface	0.21	1.2	0.21

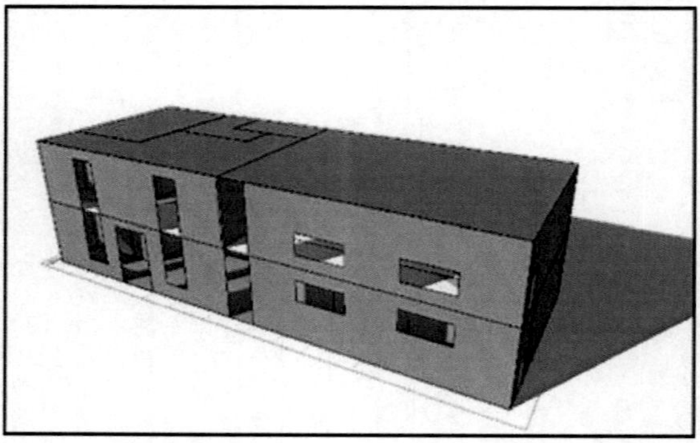

Fig. 7.5 Reception centre building model

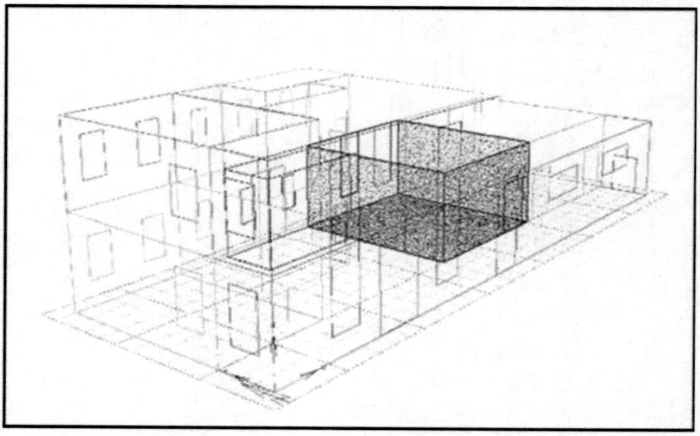

Fig. 7.6 Building energy consumption simulation calculation process

where K is the heat transfer coefficient of the envelope, W/m^2 K;
R_0 is the total thermal resistance of the envelope, m^2 K/W;
R_i is the inner surface heat transfer resistance, m^2 K/W;
R_e is the external surface heat transfer resistance, m^2 K/W; and
R is the heat transfer resistance of the envelope, m^2 K/W.

$$R_i = 0.11\,(\mathrm{m}^2 \cdot \mathrm{K/W}),\ R_e = 0.04\,(\mathrm{m}^2 \cdot \mathrm{K/W}).$$

Table 7.4 Monthly energy consumption of designed building

Month	Heating Wh	Refrigeration Wh	Total Wh
1	5629,567	0	5629,567
2	3542,061	0	3542,061
3	2602,230	0	260,2230
4	247,090	163,702	410,792
5	69,540	1108,599	1178,139
6	0	2682,796	268,2796
7	0	4066,986	406,6986
8	0	4755,627	4755,627
9	0	1246,168	1246,168
10	140,128	170,940	311,068
11	1766,546	476	1767,022
12	3712,764	0	3712,764
Total	17,709,926	14,195,294	31,905,220

Maximum heating power: 41856 W (January 12, 8:00)
Maximum cooling power: 34704 W (June 29, 14:00)

Table 7.5 Monthly energy consumption of reference building A

Month	Heating Wh	Refrigeration Wh	Total Wh
1	5754,240	0	5754,,240
2	360,3128	0	3603,128
3	263,3172	0	2633,172
4	239,173	210,466	449,,,639
5	68,176	1384,634	1452810
6	0	3152,402	3152,402
7	0	4731,523	4731,523
8	0	5496,716	5496,716
9	0	1489,077	1489,077
10	126,584	241,320	367,904
11	1784,701	14,838	1799,,539
12	3758,326	0	3758,326
Total	17,967,500	16,720,976	34,688,476

Maximum heating power: 41856 W (January 12,8:00)
Maximum cooling power: 34704 W (June 29, 14:00)

Calculation formula of the average heat transfer coefficient of the external wall:

$$K_m = \frac{K_P \cdot F_P + K_{B1} \cdot F_{B1} + K_{B2} \cdot F_{B2} + K_{B3} \cdot F_{B3}}{F_P + F_{B1} + F_{B2} + F_{B3}} \quad (7.4)$$

Table 7.6 Monthly energy consumption of reference building B

Month	Heating Wh	Refrigeration Wh	Total Wh
1	7019,424	0	7019,424
2	4568,682	0	4568,682
3	3409,452	0	3409,452
4	382,563	223,468	606,031
5	123,343	1496,847	1620,190
6	0	3444,020	3444,020
7	0	5118,496	5118,496
8	0	5912,018	5912,018
9	0	1609,483	1609,483
10	204,,355	200,,238	404,593
11	2373,664	15,188	2388,852
12	4706,068	0	4706,068
Total	22,787,551	18,019,758	40,807,,309
Maximum heating power: 56380 W (January 12,8:00) Maximum cooling power: 47625 W (June 29, 15:00)			

where K_m is the average heat transfer coefficient of the external walls, m^2 K/W;

K_p is the heat transfer coefficient of the main part of the outer walls (main walls), m^2 K/W, taking the calculated value;

F_p is the area of the main part of the outer walls (main walls), m^2;

K_{B1}, K_{B2}, and K_{B3} are the heat transfer coefficients of each thermal bridge part of the external wall (constructional column, ring beam, and floor protruding into the external wall, respectively), m^2 K/W, determined by calculation; and

F_{B1}, F_{B2}, F_{B3} are the areas of each thermal bridge part of the external walls.

② Envelope sub-index analysis of designed building

According to Formula (7.1), Formula (7.2) and Formula (7.3), the heat transfer coefficient or thermal resistance value of each part of the building model envelope can be calculated. By comparing the calculated results with the limits of the heat transfer coefficient and shading coefficient of the envelope in hot summer and cold winter areas in the *Design Standard for Energy Efficiency of Public Buildings* GB 50189-2005, the thermal performance of each part of the building model can be evaluated.

According to Table 7.4 and the calculation formulas above, each sub-index of the designed building envelope can be checked and analysed (see Table 7.7).

According to the analysis results shown in Table 7.7, the energy-saving index of the envelope of the designed building fails to meet the energy-saving standard in two aspects, namely, the average heat transfer coefficient of the external walls and the thermal resistance of the ground. The main reason is that the design of the reception centre needs to focus on the building facade modelling and function, which

Table 7.7 Checking the status of the sub-index of the envelope of the designed building

Building components	Up to standard or not
The average heat transfer coefficient of the exterior walls does not meet the standard requirements	×
The heat transfer coefficient of the roof meets the standard requirements	√
The thermal resistance of the ground does not meet the standard requirements	×
The window-wall area ratio, heat transfer coefficient and shading coefficient of the east window meet the standard requirements	√
The window-wall area ratio, heat transfer coefficient and shading coefficient of the south window meet the standard requirements	√
The window-wall area ratio, heat transfer coefficient and shading coefficient of the west window meet the standard requirements	√
The window-wall area ratio, heat transfer coefficient and shading coefficient of the north window meet the standard requirements	√
The visible light transmission ratio meets the standard requirements	√
The airtightness grade of the exterior windows and transparent curtain wall meets the standard requirements	√

leads to the thermal performance of the envelope failing to meet the requirements of energy-saving design standards. Therefore, according to the *Design Standard for Energy Efficiency of Public Buildings* GB 50189-2005, it is necessary to calculate the thermal performance trade-off of the overall designed building envelope.

③ The thermal performance trade-off calculation of the designed building

a. Select a reference building

The reference building in the trade-off calculation is a virtual building with the same function as that of the designed building and a similar shape that conforms to all the indexes in the energy-saving design standard. The shape, size, orientation, division of internal space and use functions of the reference building are completely consistent with those of the designed building, and parameters such as air conditioning system, lighting and electrical appliances and personnel density are also consistent with those of the designed building. Therefore, reference building A should be selected for the trade-off calculation in the evaluation.

b. Calculate the annual energy consumption of the designed building

According to the parameters of various building shapes and the parameters provided in Sect. 3.3.1 of the *Design Standard for Energy Efficiency of Public Buildings* GB 50189-2005, the annual energy consumption of the designed building is obtained, as shown in Table 7.8.

Table 7.8 Annual energy consumption of designed building

Energy type	Energy consumption (kWh)	Energy consumption per unit area (kW h/m^2)
Cooling power consumption	14,195	19.80
Heating power consumption	17,710	24.70
Total	31,905	44.50

Table 7.9 Annual energy consumption of reference building A

Energy type	Energy consumption (kW h)	Energy consumption per unit area (kW h/m^2)
Cooling power consumption	16,721	23.32
Heating power consumption	17,968	25.06
Total	34,689	48.38

c. Calculation of the energy consumption of reference building A

According to the parameters of all building shapes and the parameters provided in Sect. 3.3.1 of the *Design Standard for Energy Efficiency of Public Buildings* GB 50189-2005, the annual energy consumption of building A can be obtained, as shown in Table 7.9.

d. Evaluation results of building energy savings

According to the methods above, the simulated annual energy consumption results of the designed building and reference building A can be obtained.

See Table 7.10 for the comparison of the annual energy consumption calculation results between the designed building and reference building A.

See Fig. 7.7 for a comparative analysis of the annual energy consumption between the designed building and reference building A.

④ Trade-off calculation results of the energy-saving rate of the designed building

The annual energy-saving rate of the designed building is

$$\frac{34689 \times 2\text{-}31095}{34689 \times 2} = 55.18\%$$

Table 7.10 Comparison of annual energy consumption between designed building and reference building A

Calculation results	Designed building	Reference building A
Annual energy consumption (kW h)	31,905	34,689

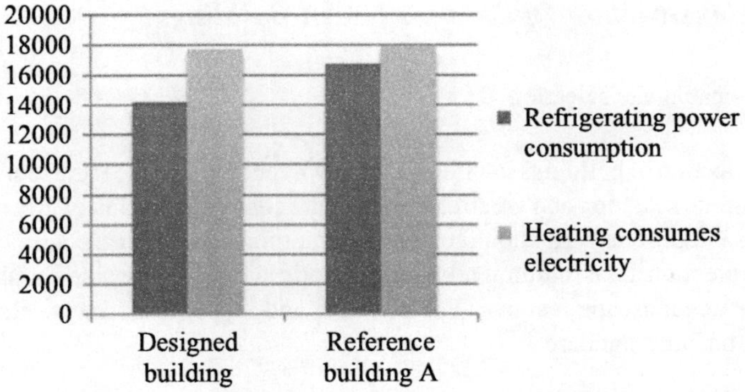

Fig. 7.7 Comparison of annual energy consumption between the designed building and reference building A

The analysis shows that the annual energy consumption of the designed building is less than that of reference building A, and the energy-saving rate is 55.18%. Therefore, according to Sect. 3.3.1 of the *Design Standard for Energy Efficiency of Public Buildings* GB 50189-2005, the energy-saving design of the building envelope meets the energy saving requirements.

7.2 Construction of Green Buildings in Villages and Towns

The main difference between the construction of green buildings in villages and towns and the construction of ordinary farm buildings is that attention should be paid not only to improving the energy savings, land savings, water savings, material savings, and environmental protection over the whole life of buildings but also to the monitoring and evaluation of the construction effect.

The construction of green buildings in villages and towns includes 7 key technologies: construction options for green buildings, building energy savings and clean energy utilization, water savings of buildings and water-saving facilities, building material savings and material resource utilization, improvement of the indoor environmental quality of buildings, green construction of buildings, and green operational management of buildings.

7.2.1 Construction Options for Green Buildings

1. Reasonable site selection

The site selection of buildings in villages and towns should first meet the requirements of building site selection and green elements of the construction plan of green villages and towns. Additionally, it should meet the requirements of construction control in various protection areas, cultural relics and historic site protection and should protect characteristic landscape features. The planning and layout of buildings should also meet the sunshine standard.

2. Site safety

The construction site should be free from the threat of natural disasters such as floods, landslides and debris flows and outside the range of industrial and other pollution sources. In addition, construction sites for disaster prevention and reduction facilities surrounding the construction should be ensured, such as using the central squares in villages and towns as public earthquake evacuation areas and establishing fire emergency channels.

3. Optimal land utilization

The land-use plan of farmhouse homesteads and building groups should reflect the economic and intensive use of land. Occupying basic farmland, water area and forest-land as construction land is strictly forbidden. Vacant land on slopes should be utilized as much as possible, and the scale of land use and approval should be considered. The land utilization should involve savings.

4. Ecological protection

It is necessary to maintain the original topography, to reduce high filling and deep excavation, and not to occupy local forestland or destroy vegetation to protect surface water. The building construction style should harmonize with the natural landscape.

5. Traffic services

Links with public transport and public services should be ensured. It is necessary to concentrate on the construction of residential areas and village groups, prioritize public transport, facilitate the sharing of public services between villagers and tourists, encourage green travel methods, and rationally plan and control parking lots.

6. Environmental beautification

Farmhouse courtyards should make full use of natural conditions and artificial elements to carry out garden greening and beautification and promote the vertical greening of farmhouse buildings.

7.2.2 Building Energy Savings and Clean Energy Utilization

1. Energy savings of building envelope

The shape, orientation, window-wall ratio and flat-face design of buildings should conform to current national or industrial standards. The thermal performance of the building envelope should meet or exceed the relevant current standards, and the use of thermal wall insulation materials and energy-saving doors and windows is encouraged.

2. Energy savings of heating, ventilation and air conditioning

High-efficiency and energy-saving heating and ventilation modes should be adopted for farmhouses in different climatic zones in line with local characteristics and the level of economic development. The indoor temperature and humidity should meet the requirements of the *Code for Thermal Design of Civil Building* GB 50176-2016. Construction should utilize local traditional technology and building materials and give full play to the natural lighting, cold-proofing, warmth-retaining, ventilation and windproof construction forms of traditional buildings, such as cave-type and straw-brick buildings, to achieve the self-saving effects of farmhouses. Green farmhouses should use energy-saving air-conditioning equipment.

3. Energy savings of lighting and electric

Farmhouse design should utilize natural lighting as much as possible and choose energy-saving lamps and other household appliances. For the lighting of common areas (such as halls, corridors, and stairwells), energy-saving control measures such as timing and induction should be adopted.

4. Clean energy and renewable energy utilization

Farmers should be required to use clean biomass stoves, gas cookers, biogas stoves, etc. and encouraged to use clean energy, such as liquefied petroleum gas and natural gas, and renewable energy, such as solar energy, geothermal energy and biomass energy.

7.2.3 Water Savings of Building and Water-Saving Facilities

1. Control of average daily water consumption in buildings

Farmers' water-saving behaviours should be cultivated, and water consumption should meet the water quota requirements of the *Standard for Water Saving Design in Civil Building* GB 50555-2010.

2. Avoidance of water-supply pipe leakage

Water-supply pipes and fittings should have good corrosion resistance and durability, and valves and other equipment should have good sealing performance. Deformation and leakage caused by rolling vehicles or heavy piles should be avoided for pipes buried underground. A graded water metre should be installed to detect leakage defects in the pipeline network so that they can be repaired in a timely manner.

3. Installation and use of water-saving appliances

The installation and use of water-saving appliances are required in green farmhouses, mainly including water-saving toilets, showers and washing machines; overflow alarms for tanks; and automatic linkage devices for inlet valves.

4. Rational use of non-traditional water sources

It is necessary to encourage the use of water systems in farmhouses design, scientifically treat and rationally use wastewater, build collective and domestic rainwater harvesting facilities in areas with low precipitation, control the setting of landscaped water bodies and reduce water use for car washing.

7.2.4 Building Material Savings and Material Resource Utilization

The selection of building materials is carried out in 4 stages, raw material use, material production, construction use and waste disposal, to minimize the environmental load and contribute to the health of villagers.

1. Reasonable selection of building size

The shape of buildings should conform to the current national standards. While retaining the traditional architectural style in the selection of regular building shapes, simplicity, generosity and material savings are advocated.

2. Preferred use of local building materials

Priority should be given to locally produced building materials. For example, construction should make full use of the rich resources of stones from hilly areas and adopt local environmental protection and energy-saving building materials such as waste brick, renewable wood or bamboo.

3. Use of reusable and recyclable materials

Farmhouse construction should use reusable and recyclable building materials and make the fullest possible use of old building materials and components. Construction materials and products prohibited and restricted by state and local authorities, such as clay bricks, should not be used to reduce environmental pollution, reduce waste landfills and save energy and resources.

4. Minimization of types of building materials

The design should avoid selecting different materials and adopt a structural design that is easy to disassemble and separate to facilitate the recycling, classification and reuse of building materials after maintenance or demolition.

7.2.5 Improvement of Indoor Environmental Quality of Buildings

1. Improvement requirements of the indoor environmental quality of buildings

(1) Improvement of acoustic environmental quality. The indoor noise level of the main functional room should meet the lowest standard of *Code for Design of Sound Insulation of Civil Buildings* GB 50118-2010. Low-noise ventilation and air-conditioning equipment, household appliances and drainage equipment should be used indoors, and attention should be paid to avoiding or reducing noise from traffic and outdoor social life.

(2) Improvement of light environmental quality. The main functional room should have a good outdoor view, and the lighting coefficient should meet the requirement of the *Standard for Daylighting Design of Buildings* GB 50033-2013. The quantity and quality of architectural lighting should meet the requirements of the *Standard for Lighting Design of Buildings* GB 50034-2013.

(3) Improvement of indoor air quality. The control of the concentration of indoor pollutants such as formaldehyde, benzene, ammonia, volatile organic compounds, and inhalable particulate matter should meet the requirements of *Indoor Air Quality Standard* GB/T 18883-2002. Green building decoration materials should be preferred.

2. Case of indoor environmental quality detection of buildings

The effects of improving the indoor environmental quality of green rural buildings in villages and towns should be verified and promoted through actual testing. To study and establish an indoor environmental quality testing method for green farmhouses in villages and towns, the authors' team worked with the National Sustainable Development Community of Baishan City, Jilin Province, and chose a group of grass-brick farmhouses as test objects in the northeast cold region in Shangdianzi Village, Baishan City. The group of grass-brick farmhouses are ecological farmhouses built as a demonstration of local straw grass-brick blocks as a building maintenance structure.

See Fig. 7.8 for an aerial view of Shangdianzi Village, Baishan City, Jilin Province.

For this section, rural buildings S_{31} and S_{32} in Shangdianzi Village are selected as examples.

1) Basic information of sample points

Sample points S_{31} and S_{32} are both grass-brick buildings built and put into use in 2002. The indoor heating methods are household small boiler and plumbing systems with heatable adobe sleeping platforms, and the energy sources used for heating and cooking are briquettes and firewood.

Sample point S_{31} has an area of 24 m^2, and sample point S_{32} has an area of 25 m^2. For the test, 1 test point is set in both S_{31} and S_{32}.

Fig. 7.8 Aerial view of Shangdianzi Village, Baishan City, Jilin Province

(a) Outdoor real scene of farmhouse S_{31} (b) Outdoor real scene of farmhouse S_{32}

Fig. 7.9 Real scene of indoor air environmental quality monitoring point of farmhouses

See Fig. 7.9 for the real scene of the indoor air environmental quality monitoring point of the farmhouse.

2) Monitoring data and evaluation

Table 7.11 shows the test data and results of the indoor air environmental quality testing of rural building S_{31} in Shangdianzi Village.

Table 7.12 shows the test data and results of the indoor air environmental quality testing of rural building S_{32} in Shangdianzi Village.

3) Analysis of test results

(1) Indoor air quality is poor

According to the test, the indoor air quality of the 2 farmhouses in Shangdianzi Village is poor, the main pollutant is inhalable particulate matter, and the pollution of benzene and volatile organic compounds is serious.

(2) Analysis of indoor air environment

① Although the indoor temperature of both buildings is not up to the standard, they are close to the lower limit of the standard value. Therefore, grass-brick buildings, as low-cost ecological buildings in rural areas, have good thermal insulation, energy savings and comprehensive straw utilization benefits. Under conditions of proper maintenance and repair, this type of construction is worth promoting for farmers' residences in rural areas of northeast China.

② The detection time was the coldest season in the local area. Buildings are usually insulated and windtight, which leads to poor ventilation and accumulation of pollutants.

③ Local farmers use briquettes and firewood as fuel for heating and cooking, and a large amount of particulate matter and benzene pollutants accumulate indoors.

④ Approximately 10 min before the test of sample point S_{32}, some people smoked in the test area. Under conditions of poor ventilation, inhalable particles from smoking accumulate in large quantities, causing serious indoor pollution.

Table 7.11 Test data and results of indoor air environmental quality testing of rural buildings in Shangdianzi Village

Sample number: S_{31}			Monitoring time: December 16, 2016				
Evaluation items	No.	Evaluation indexes	Scoring criteria	Measured values	Test results		
					Qualified (1)	Sub-standard (0)	Total score
Indoor air quality	1	Indoor temperature	22–28 °C (air conditioning in summer) 16–24 °C (heating in winter)	16	1		5
	2	Relative humidity	40–80% (air conditioning in summer) 30–60% (heating in winter)	61		0	
	3	Formaldehyde	Concentration $\leq$ 0.10 mg/m^3 (1-h average)	0.1	1		
	4	Benzene	Concentration $\leq$ 0.11 mg/m^3 (1-h average)	0.1	1		
	5	Ammonia gas	Concentration $\leq$ 0.20 mg/m^3 (1-h average)	0.3		0	
	6	Toluene	Concentration $\leq$ 0.20 mg/m^3 (1-h average)	0.1	1		
	7	Xylene	Concentration $\leq$ 0.20 mg/m^3 (1-h average)	0.3		0	
	8	TVOC	Concentration $\leq$ 0.60 mg/m^3 (8-h)	0.3	1		
	9	PM_{10}	Concentration $\leq$ 150 μg/m^3 (24-h average)	766		0	
	10	$PM_{2.5}$	Concentration $\leq$ 75 μg/m^3 (24-h average)	381		0	

Table 7.12 Indoor air environmental quality test times and test results in Shangdianzi Village

Sample number: S_{32}			Monitoring time: December 16, 2016				
Evaluation items	No.	Evaluation indexes	Scoring criteria	Measured values	Test results		
					Qualified (1)	Sub-standard (0)	Total score
Indoor air quality	1	Indoor temperature	22–28 °C (air conditioning in summer) 16–24 °C (heating in winter)	14		0	2
	2	Relative humidity	40–80% (air conditioning in summer) 30–60% (heating in winter)	60	1		
	3	Formaldehyde	Concentration ≤ 0.10 mg/m^3 (1-h average)	0.08	1		
	4	Benzene	Concentration ≤ 0.11 mg/m^3 (1-h average)	0.8		0	
	5	Ammonia gas	Concentration ≤ 0.20 mg/m^3 (1-h average)	0.7		0	
	6	Toluene	Concentration ≤ 0.20 mg/m^3 (1-h average)	0.8		0	
	7	Xylene	Concentration ≤ 0.20 mg/m^3 (1-h average)	2.0		0	
	8	TVOC	Concentration ≤ 0.60 mg/m^3 (8-h average)	1.9		0	
	9	PM_{10}	Concentration ≤ 150 $\mu g/m^3$(24-h average)	>1000		0	
	10	$PM_{2.5}$	Concentration ≤ 75 $\mu g/m^3$(24-h average)	>500		0	

7.2.6 Green Construction of Buildings

1. Site environmental protection

Construction measures include:

(1) The protection of cultural relics and historical features. It is necessary to take protective measures for on-site cultural relics and ancient famous trees;
(2) The protection of the water environment. It is necessary to protect the original groundwater around the site and reduce the extraction of groundwater; and
(3) The control of construction dust. Vehicles carrying soil and muck should be covered or sealed, and shielding, dust suppression and other measures should be taken in construction work.

2. Material savings of construction

Construction measures include:

(1) Proper disposal of construction waste. Construction waste should be collected and centralized, and gravel and earthwork can be used as foundation and roadbed landfill materials;
(2) Earthwork backfilling. Earth excavated in construction can be reasonably backfilled and utilized;
(3) Material savings. Templates and scaffolding should use renewable and reusable materials, such as bamboo; and
(4) Recycling of resources. Temporary construction facilities can utilize existing buildings and surrounding roads to recycle building packaging materials.

3. Water savings of construction

Construction measures include:

(1) Adoption of water-saving measures. It is necessary to reasonably arrange the water supply and drainage systems, use water-saving appliances and install water metres for on-site domestic water use, and adopt water-saving construction technology to prevent water leakage; and
(2) Water resource utilization measures. Tap water should not be used for pavement spray. Non-official water resources can be used for the construction site.

4. Energy savings of construction

Construction measures include:

(1) Energy savings for temporary construction facilities. Energy-saving equipment and automatic control devices should be adopted for temporary electricity use. Energy-saving lighting fixtures should be used on the construction site;

(2) Energy savings for construction machinery. Effective and energy-saving construction machinery and equipment should be adopted, and construction equipment and products that are explicitly prohibited by the state and local governments should not be used;
(3) Energy savings for material transportation. Nearby building materials can be chosen to reduce the transportation distance and minimize night work and winter construction time; and
(4) Renewable energy utilization. Solar energy, geothermal energy, wind energy, biomass energy and other renewable resources should be rationally utilized.

5. Land savings of construction

Construction measures include:

(1) Reduction of construction land occupation. The construction site layout should be as compact as possible;
(2) Protection of construction land. During construction, it is necessary to protect the surrounding cultural landscape and prevent soil erosion. Mountain wasteland can be used as sites for soil abandonment. After construction, the vegetation should be restored; and
(3) Protection of underground resources. Underground cultural relics discovered during construction should be effectively protected.

The abovementioned green construction technology should be combined with the characteristics of rural construction projects and should be applied according to local conditions.

7.2.7 Green Operational Management of Buildings

Green operational management of buildings in villages and towns refers to green management in the use of buildings after completion, which is an indispensable and important aspect of green building construction. The green operation and management of buildings in villages and towns includes 5 aspects:

(1) Formulate and implement management systems for energy savings, water savings, material savings and greening;
(2) Implement strict garbage management systems, rational planning of garbage logistics, domestic garbage collection, and garbage container settings;
(3) Meet discharge standards for pollutants, such as sewage generated during the use of the building;
(4) Achieve energy savings in the building and normal operation of water-saving facilities and provide complete records of heating, air conditioning, lighting, etc.; and

(5) Organize farmers to actively participate in the planning, design and construction of buildings and motivate them to consciously participate in green operation and management in the process of building use.

7.3 Green Renovation of Rural Existing Buildings

7.3.1 Overview of Green Renovation of Rural Existing Buildings

1. Significance of green renovation of existing rural buildings

An existing building is a building that has been in use for more than 2 years. There is a huge number of existing rural buildings in Chinese villages and towns; they not only constitute enormous wealth and valuable resources of rural economic and social development but also inherit the long history and cultural characteristics of Chinese villages and are important objects for the construction and protection of green rural buildings. According to the *Statistical Yearbook of China's Urban-Rural Construction 2017*, in 2017, a total of 29.282 billion m^2 of rural housing was built, including 25.585 billion m^2 of residential buildings, 1.562 billion m^2 of public buildings and 2.135 billion m^2 of production buildings.

The proportion of existing buildings in the total building area of rural China in 2017 is shown in Fig. 7.10.

2. Renovation problems of rural existing buildings

For historical reasons, most existing buildings in villages and towns have problems such as large energy consumption, poor construction quality, and imperfect functions. A large number of rural buildings need to be rebuilt every year. To create a

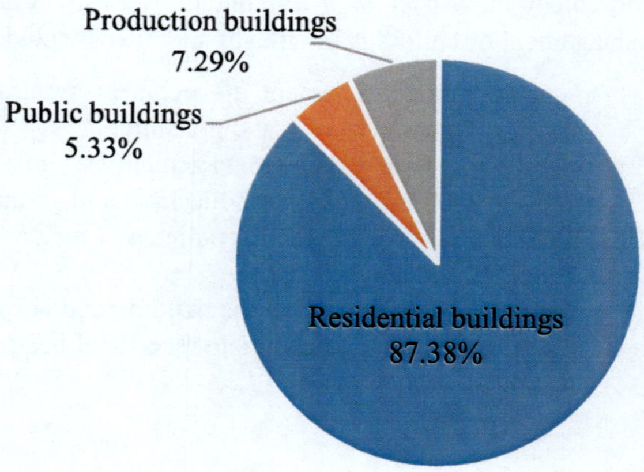

Fig. 7.10 Proportion of existing buildings in the total building area of rural China in 2017

beautiful countryside and inherit rural history, culture and characteristics, a large number of rural buildings with historical and cultural protection value must be renovated and updated. Large-scale demolition and construction are not allowed in these cases because of the resulting damage to the environment, civilization and social resources. At the same time, it is necessary to improve the use functions, energy-saving efficiency and appearance of existing buildings.

3. Content of green renovation of rural existing buildings

Green renovation refers to the maintenance, renewal and reinforcement of existing buildings with the goal of saving energy, improving the living environment and improving the use functions.

Green renovation of existing rural buildings refers to the maintenance, renewal and reinforcement of those buildings with the goal of saving rural energy resources, improving the rural living environment and improving the use of rural buildings and activities associated with them.

Renovation of existing rural buildings should follow the principles of adapting to local conditions and emphasizing energy savings and environmental protection functions during the whole life of buildings. The plan and design scheme, use of structures and materials, equipment and implementation transformation, construction and operational management of building renovation projects should be implemented in combination with the types and functions of the buildings while simultaneously taking into account the geographical and climatic, resource and environmental, economic and social, historical and cultural characteristics of the area. The *Assessment Standard for Green Retrofitting of Existing Building* GB/T 51141-2015 can be taken as a reference.

7.3.2 Property and Types of Building Maintenance and Renovation

To scientifically formulate a green renovation plan for existing rural buildings in villages and towns and rationally select the applicable technology for green renovation of existing buildings, an understanding of the basic properties and types of maintenance and renovation of existing buildings is required.

Repair and renovation activities for existing buildings include building quality inspection and appraisal, organization and management of building repair activities, assessment of housing defects and causes, selection of repair and renovation types and repair construction [1]. The process involves decision parameters and decision evaluation methods such as the maintainability, maintenance time, maintenance standard, maintenance costs and maintenance accessibility of existing buildings [2]. It also includes the evaluation of the energy efficiency of existing building renovation [3].

1. Properties of building maintenance and renovation

The construction activities of buildings can be divided into 2 major stages: the new construction stage and maintenance-renovation stage. New construction creates the physical form and usage value of the building through planning, design, construction and other activities. Through maintenance and repair, transformation and renovation, demolition and reconstruction activities, etc., the value of buildings can be improved. The new construction is the basis of maintenance and reconstruction, and maintenance and reconstruction supplement and continue the new construction. The two stages together constitute the whole process of the material change of the building.

Building maintenance and renovation are meant to reduce and prevent the performance loss of the building by restoring, repairing and renovating the existing building within the specified service life to restore and improve its original functions, ensure its normal use, and properly extend its service life. With a small maintenance and renovation investment, the maximum comprehensive benefits can be obtained.

2. Types of building maintenance

Generalized building maintenance includes 3 types: maintenance, retrofitting and renovation. Therefore, it is also called building maintenance and renovation in engineering practice.

1) Building maintenance

Building maintenance refers to the maintenance or restoration of the components, decoration and equipment in the building to meet the user's requirements and extend the service life of the building.

The relationship among the maintenance, functional deterioration and life span of a building is shown in Fig. 7.11.

In the figure, point *A* represents the perfect initial functional value of the building after completion (with the functional value set at 100). With increased service time, without maintenance, the building will be damaged naturally, and its function will gradually decrease to point *C*, as shown in the figure (with the functional value set at 0). At this time, the actual service life of the building is the *CB* value, and point *B* is the allowable lower limit of its durability. When the building is damaged to a certain extent, such as the *K* value for repair, its function can recover to *D*. In this way, the building can be damaged and repaired several times. Each repair extends the durability of the building, but the original function cannot be completely restored until the building's service function drops to the allowable lower limit *C* and the building loses its maintenance value and is scrapped. The life durability limit *CH* of the building after maintenance will be greater than the natural damage durability limit *CB* under the condition of no maintenance.

The maintenance of existing rural buildings is generally carried out by farmers using buildings in accordance with loss and defects, irregular maintenance and repair.

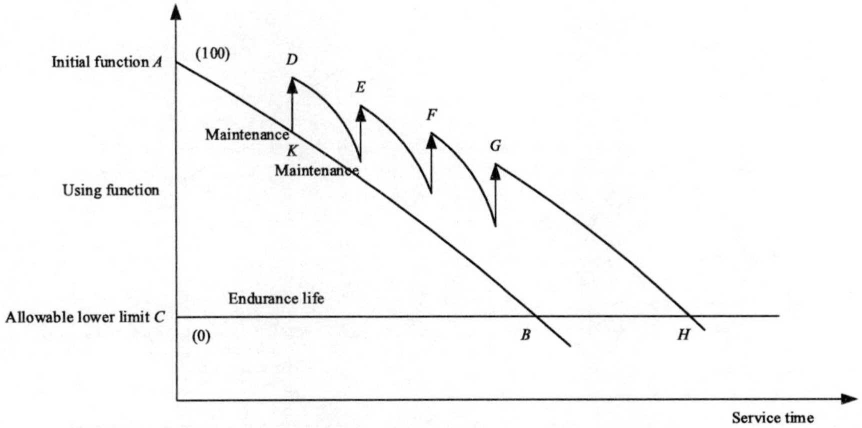

Fig. 7.11 Relationship among the maintenance, functional deterioration and life span of a building

2) Building renovation

Building renovation is undertaken to improve the original function, construction standards and environmental quality of the building.

In a building renovation project, based on the premise of basically retaining the main structure of the building, its usage function and component facilities are improved, the living standard and environmental quality of the old building are brought close to the standard for a new building, and the service life of the building is prolonged.

The main measures for building renovation include changing the layout; expanding the area and adding layers; adjusting the effective use area; improving the building decoration standards; increasing or improving the equipment in the kitchen and bathroom; improving the water supply, gas supply and HVAC equipment; and adding modern equipment and facilities according to the users' demand. In a broad sense, building renovation is a type of building maintenance activity.

The renovation of existing rural buildings in villages and towns is generally undertaken by hired craftsmen or construction teams after 15–20 years of building use.

3) Building replacement project

Building replacement refers to the demolition of old buildings and the construction of new buildings on the original base. Rebuilding the original building in the same place is also called building redevelopment or replacement. When the basic structures of the old building reach a durable age and are on the verge of being scrapped and unable to continue in use or need to be eliminated for other reasons, such as social, economic, planning and construction reasons, demolition should be considered. After the old building is demolished, the original foundation can be utilized. The new foundation

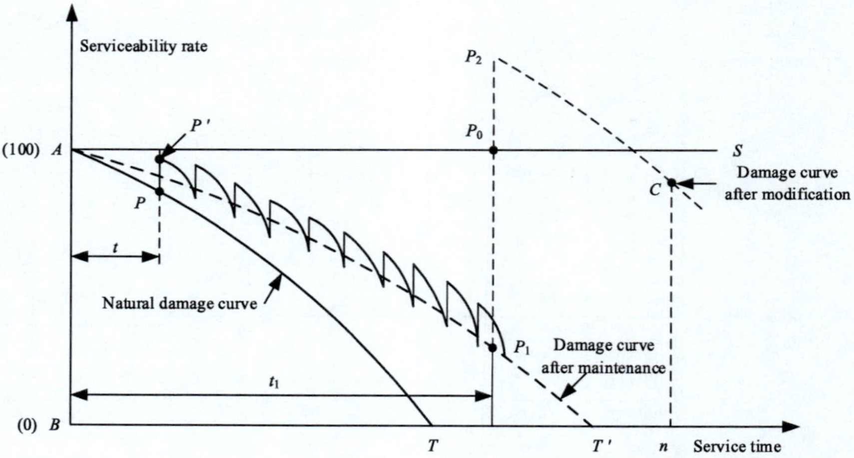

Fig. 7.12 Relationship among repair, renovation and building damage

and partial structural components and materials can be constructed on the original site.

The renovation of existing rural buildings in villages and towns is usually done by combining maintenance, reconstruction and replacement.

The relationship among repair, renovation and building damage is shown in Fig. 7.12.

In Fig. 7.12, line *AS* represents the initial intact functional state of the building. If no maintenance is carried out, the natural damage process and life span of the building will be shown in curve *AT*, and all damage will be complete by year *T*. If repairs are carried out at year t (e.g., point *P* on damage curve *AT*), the function of the building returns to point P_1. As the building continues to be maintained, its use function will decline slowly, and the life of the building will be extended to year *T'*.

When the building is over-repaired, retrofitted or renovated in year t_1, not only will its integrity rate return to the original function P_0, but the actual function of the building will improve to the level of P_0 ~ P_2. P_2C represents a new round of the damage curve of the building after renovation, indicating that its service life under normal functional conditions can be extended by *n* years after renovation.

The renovation or reconstruction of existing buildings in villages and towns is similar to the construction of new buildings. The renovation of farmhouses is generally undertaken by artisans or construction teams hired by the farmers. In the process of the development of town areas, village rectification and beautiful village construction, village renovation or reconstruction is often planned by the town government or village committee.

7.3.3 Green Renovation Method of Rural Existing Buildings

The fundamental purpose of the green transformation of existing rural buildings is to continuously improve and enhance the functions and qualities of rural buildings and to prolong the durability of rural buildings under the premises of safety, economy, application and aesthetics by following the principle of "energy savings, land savings, water savings, material savings, and environmental protection".

To carry out the green transformation of existing buildings, we must first master the durability analysis method of existing buildings and then specifically select the green transformation technology method.

1. Durability analysis of existing buildings

1) Concept of building durability

Building durability refers to the performance of all kinds of components, decoration and equipment of a building that can maintain its normal working state over a certain period of time. The durability of a building is determined by many factors related to its physical properties, chemical properties, and economic performance. Therefore, the durability of various components and materials should be determined by the causes of damage and several adverse factors, and the durability of the building can then be calculated.

The durability of residential buildings in villages and towns [4] refers to the ability of those buildings to maintain their safety, normal use and acceptable appearance under normal use and maintenance conditions during a certain service life.

2) Analysis of building service life

Building service life refers to the average service life of the building as specified in the design and is the statutory durability period for the operation and use of the building. Within the normal durability period of the building, building maintenance and renovation must be carried out to fully or partially restore the lost function of the building and reasonably extend its service life.

The building service life can be calculated by the "ageing curve". According to the durability and economic and technological factors of each component of the building, the minimum allowable lower value limit of a certain function is stipulated. When the functional loss of the main components (such as load-bearing structure) of the building reaches or falls below this lower limit, the durability of the building is considered to have ended.

Theory and practice show that existing rural areas can reasonably extend the normal service life of buildings and effectively improve their function and energy efficiency through comprehensive maintenance, renovation and function improvement and other green construction methods.

Table 7.13 Classification of service life of civil buildings

Category	Design life (years)	Example
1	5	Temporary building
2	25	Easy-to-replace building structure
3	50	Ordinary building and structure
4	100	Monumental building and particularly important building

3) Grade standard of building durability

The grade standard of building durability is an important basis for green reconstruction designs and construction schemes for existing rural buildings. At present, there is no specific class standard for the durability of buildings in villages and towns in China, but the current *Code for Design of Civil Buildings* GB 50352-2019 can be followed on the basis of the classification of service life of civil building design and in combination with the actual conditions in different rural areas.

See Table 7.13 for the classification of designed service life in terms of the importance grade of civil buildings.

With the development of the rural social economy and the continuous improvement of building function quality, the inevitable development demand of green village and town construction in China is to promote the "hundred-year farmhouse" system and formulate the green renovation standard of existing rural buildings.

2. Green renovation technology and method for rural existing buildings

According to the rural characteristics of different areas and the actual needs of existing rural buildings, the corresponding green renovation technology for existing rural buildings can be chosen.

1) Historical feature protection of rural building renovation

When planning and constructing green renovations of rural historic buildings, historic cultural blocks and historic villages and towns, the relevant provisions of the state and local governments concerning the protection of historical culture and heritage should be followed. Specific reference can be made to *Code of Conservation Planning for Historic Cities* GB 50357-2018 and *Regulations on the Protection of Famous Historical and Cultural Cities, Towns and Villages* promulgated by the State Council in 2018.

Rural historic buildings are rural buildings or structures with historical, scientific and artistic value. They can reflect the historical features of the countryside, architectural styles of different periods and local characteristics, such as an old street or an old well. They are valuable and non-renewable architecture resources.

When rural buildings are renovated, the ecological environment and natural landscape around the existing building should be protected. Protection efforts should focus on green areas, water systems, roads, and ancient trees with protection values around rural buildings. It is necessary to adhere to the original protection and utilization, rationally retain the use of structures and construction facilities, and prevent large-scale and unnecessary demolition and destruction.

2) Performance improvement in rural building renovation

(1) Improve the style and coordination of rural building groups. In the process of transformation, do not add a large number of decorative components or building types that do not coordinate with the original style.
(2) Improve the energy-saving performance of rural buildings. The key point is to transform the roofs, exterior walls, exterior doors and windows, etc.; adopt passive energy-saving measures as much as possible; reduce the energy consumption of heating, refrigeration and air conditioning; and achieve or approach the current national energy-saving design standard for buildings. The renovation should also meet the requirements of the current *Standard for Daylighting Design of Buildings* GB 50033-2013 and cannot reduce the sunshine standards of surrounding buildings.
(3) Improve the clean energy utilization performance of farm buildings. Combined with the renovation project, we strongly advocate the use of clean energy such as solar energy and biogas.
(4) Improve the environmental health of buildings. The focus is on the renovation of toilets, kitchens, domestic sewage facilities and water-saving facilities.
(5) Improve the seismic disaster prevention of rural buildings. In areas where seismic fortification is required, seismic reinforcement should be carried out on rural buildings that have no seismic structure. *Technical Specification for Seismic Strengthening of Building* JGJ 116-2009 can be taken as a reference. Renovated rural buildings should meet the national seismic requirements to ensure safety.
(6) Improve the repairable and updatable performance of rural buildings. Component materials that are easy to maintain, replace and recycle should be reasonably selected, flexible and changeable space should be reserved and adjustable plane layout should be implemented to realize the sustainable transformation of existing rural buildings.

3) Site safety of rural building renovation

(1) The site where the existing rural building is located should be safe. There should be no risk of natural disasters, such as floods, landslides, and mudslides, and no sources of pollution exceeding the standards.
(2) There should be no noise pollution around rural buildings. Noise can be divided into indoor noise and outdoor noise. Indoor noise comes from ventilation and air-conditioning equipment and household appliances. Outdoor noise comes from surrounding traffic, daily life, industrial production, etc. The noise-control standard can be implemented by referring to *Code for Design of Sound Insulation of Civil Buildings* GB 50118-2010.

4) Optimization of structural materials for rural building renovation

(1) Make full use of local materials. In the green renovation of existing rural buildings, materials and products prohibited by the state and local authorities cannot be used. In the renovation, the utilization rate of the original components should not be less than 70%. Recycled materials and recyclable materials such as local stone and bamboo should be used as much as possible for new buildings.
(2) Save component materials as much as possible. During the renovation, materials should be saved. Under the premises of safety, reliability, economy and environmental protection, the original structural components, such as beams, boards, columns, and walls, should be fully utilized to avoid unnecessary removal or replacement and to minimize damage to the original structure and components.

5) System transformation of equipment and facilities

(1) Energy-saving renovation of equipment and facilities. Energy-saving diagnosis of the building can be carried out in villages and towns where conditions permit. According to the local climate and natural resource conditions, renovations can make full use of renewable energy such as solar and geothermal energy to provide domestic water heating and energy for air conditioning. Central heating is advocated in villages and towns in cold northern regions. In the renovation of rural electrical systems, energy-saving lamps and controls should be used. The energy-efficiency level of lighting, ballast and distribution transformers cannot be lower than the requirements of the *Standard for Daylighting Design of Buildings* GB 50034-2013.
(2) Water-saving renovation of water system. In the green renovation of water systems in rural buildings, the status quo of water resource utilization should be evaluated before the renovation plan is formulated. The key points are the effect of water and energy savings, the rationality and economy of technology, the impact on the surrounding environment and buildings, and whether the water demands of farmers and future visitors can be met. Water system green renovation of buildings should use a water-saving sanitation apparatus and equipment.
(3) Use of rational non-traditional water sources. This is the most direct and effective way to save water, and farmers should be encouraged to use rainwater, reclaimed water and seawater for greening irrigation, landscape watering, and road and vehicle washing. Water-saving behaviours should be advocated. A rainwater utilization infrastructure should be set up by using farmhouses and site spaces, including roof greening, rainwater gardens, sunken green spaces, shallow gullies with vegetation, rainwater ponds, and rainwater throttling and infiltration facilities. Paved areas and parking spaces should be paved with grass brick and pervious tiles.
(4) Renovation of sewage treatment facilities. In the green renovation of rural buildings, sewage collection and treatment facilities should be built, and the sewage treatment and emissions rates should be up to 100%.

6) Green construction and management of rural building renovation

(1) Construction management of green renovation. Measures include dust-control measures such as setting up dust nets, sprinkling water and covering. Low-noise, low-vibration construction equipment should be used, and sound insulation and noise reduction construction technology should be adopted. Measures should be taken to reduce waste from construction, formulate and implement waste recycling plans, select nearby sources of building materials for renovation and repair, and select factory-processed components and materials as far as possible to reduce waste from construction sites. Water-saving and water-use plans should be formulated and implemented for construction.
(2) Operational management of green renovation. After the green renovation of rural buildings, it is necessary to implement energy-saving, water-saving, material-saving and greening management systems for buildings in villages and towns. It is necessary to implement systems for the storage, collection and harmless reduction of domestic waste for buildings in villages and towns. It is also necessary to establish and implement a normal operational records supervision system for relevant facilities and encourage all residents to actively participate in green operational management.

References

1. Wang XM (1994) Building maintenance management. China Building Materials Industry Press, Beijing
2. Wang XM (2004) Research on maintainability and maintainability decision of residential buildings. J Huazhong Univ Sci Technol (Urban Science Edition) 21(1):5–9
3. Hua H, Li W, Wang XM, et al (2017) Comparative Study on energy consumption measurement and energy efficiency of adobe buildings. J Civil Eng Manag 6:17–22
4. Yang J, Wang XM, Wang Y (2011) Economic analysis of the durability of rural residential buildings. J Civil Eng Manag 28(2):82–86

Part III
Evaluation of Green Village and Town Plans and Construction

Chapter 8
Evaluation of Green Village and Town Plans

Abstract Evaluating the implementation effects of green village and town plans is an important way to guarantee the planning and construction goal realization of green villages and towns and optimize the benefits of planning and construction projects. This chapter analyses the basic requirements for evaluating green village and town plans and establishes 6 specific evaluation methods. Yanhe Village in Hubei Province is taken as a case study for the application of the evaluation method of green village and town plans.

The evaluation of green village and town plans refers to the evaluation of the completion of the plan index and the effects of planning and construction in the process of plan implementation. It is not only an analysis and summary of the implementation in the process of planning but also an important basis of plan revision and measure implementation optimization.

Evaluation of green village and town plans involves a comprehensive judgement based on the analysis and comparison of the implementation effects of green village or town plans or special plans using certain technical methods according to the current corresponding standards and specifications. It is an important guarantee in the stage of plan implementation. It has 6 aspects: evaluation of site selection and functional division optimization, evaluation of niche suitability plans, evaluation of water resource utilization and sewage disposal plans, evaluation of waste disposal plans, evaluation of clean energy utilization plans, and evaluation of plan implementation and ecological environmental restoration.

The authors have carried out long-term research and construction practice in the green ecological Yanhe Village, Hubei Province. This chapter and Chaps. 9–11 discuss Yanhe Village as an evaluation case for green village and town plans and construction.

X. Wang and H. Hua, *Green Village and Town Construction in China*,
https://doi.org/10.1007/978-981-16-2098-0_8

8.1 Basic Evaluation Requirements of Green Village and Town Plans

1. Purpose of evaluation

The purpose of evaluating green village and town plans is to evaluate the green construction effects of village or town plan implementation and the impact of construction activities on the ecological environment. The key points are to evaluate the ecological effects and sustainability of a village and town plan to promote the comprehensive optimization and implementation of the plan, verify the guidance and implementation of the plan, and evaluate the greenness and effectiveness of planning and construction projects in villages and towns.

2. Basis of evaluation

The main resources for the evaluation of green village and town plans are as follows:

(1) *Standard for Planning of Town* GB 50188-2007;
(2) *Guidelines for the Construction of Beautiful Village* GB/T 32000-2015;
(3) *Technique Code for Village Rehabilitation* GB 50445-2008;
(4) *Current Land Use Classification* GB/T 21010-2017;
(5) *Code for Water Resources Planning* GB/T 51051-2014;
(6) *Code of Practice for Water Resources Bulletin* GB/T 23598-2009;
(7) *Technical Guideline for Planning Environmental Impact Assessment: General Principles* HJ 130-2019;
(8) *Guideline for the Town-level General Land Use Planning* TD/T 1025-2010;
(9) *Technical Code for Ecosystem Status Assessment* HJ 192-2015;
(10) *Standard for the Assessment of Regional Biodiversity* HJ 623-2011; and
(11) *Technical Specification for Village Landscape Environmental Engineering* CECS 285-2011, etc.

3. Content of evaluation

The main content of the evaluation of green village and town plans is the special evaluation of green construction aspects of the plan. It includes evaluation of site selection and functional division optimization, evaluation of niche suitability, evaluation of water resource utilization and sewage treatment, evaluation of waste treatment, evaluation of clean energy utilization, and evaluation of eco-environmental remediation in the implementation of green village and town plans.

The evaluation of green village and town plans can include all the above special projects or only one or several of them according to the actual needs.

4. Process of evaluation

The evaluation programme for green village and town plans is generally divided into 3 stages: preliminary preparation, analysis and evaluation, and report compilation.

1) Preliminary preparation stage

(1) Establish a clear evaluation scope and purpose. Clarify whether this evaluation is a comprehensive evaluation of the green village or town plan implementation or an evaluation of the implementation effect of one or several special projects.
(2) Collect the information needed for evaluation. Collect relevant laws, regulations, policies and planning documents and materials related to green village planning; conduct on-site investigation, reconnaissance and testing of the evaluation area to obtain relevant basic data; evaluate the economic, social and ecological environment of the region to conduct a preliminary analysis; and collect remote sensing data and other geographic information.
(3) Formulate the evaluation technical scheme. Analyse and summarize the collected data, and formulate a suitable technical evaluation scheme for the project.

2) Analysis and evaluation stage

(1) Basic content of the evaluation. Evaluation of green village and town plans should include an overview of the plans, evaluation indexes and methods, evaluation conclusions and recommendations and other main contents.
(2) Technical requirements of the evaluation. A comprehensive evaluation of the implementation of green village and town plans should be based on the establishment of the evaluation index system to determine the appropriate index weight. For a special evaluation of the implementation of a programme, corresponding evaluation indexes should be formulated according to different evaluation contents. The appropriate evaluation level should be based on current evaluation criteria.
(3) Conclusions and suggestions of evaluation. According to the results of the evaluation, a clear evaluation conclusion should be drawn and specific improvement suggestions put forward.

3) Report compilation stage

The report compilation of the evaluation of green village and town plans requires concise writing, a complete structure, sufficient arguments, prominent points, comprehensive charts, informative data, clear conclusions and feasible suggestions.

5. Compilation of evaluation report

The compilation format of the evaluation report of green village and town plans includes 5 parts as follows:

1) General

Summarize the origin of the task, and clarify the evaluation basis, purpose and principle of the evaluation, evaluation scope (with pictures), and evaluation focus.

2) Analysis and evaluation

Outline the development objectives, positioning, scale, layout, structure, timing, and main content of the plan. Summarize the investigation of the status quo of the village or town; briefly describe the natural geographical situation, social and economic situation, resource occurrence and utilization, environmental quality and ecological status of the village or town; analyse and evaluate the problems existing in the construction and development of the village or town and environmental and ecological protection needs; and clearly propose factors restricting the implementation of the plan.

3) Establishment of evaluation index system

To determine the evaluation objectives of different planning periods, establish an evaluation index system should be established and provide a specific evaluation index number. Use appropriate evaluation methods and established evaluation index systems to evaluate the implementation and effect of village and town plans.

4) Evaluation conclusions

Summarize the work results of the evaluation, provide clear conclusions, and put forward improvement suggestions for the plan or opinions regarding rectification and optimization of the plan.

5) Evaluation attachments

Attach the necessary maps and tables to the evaluation, such as the current status survey scope, monitoring point distribution maps and other figures or tables.

8.2 Evaluation of Site Selection and Functional Division Optimization of Green Villages and Towns

Saving construction land resources and realizing the functional division optimization of land use are the primary tasks and implementation emphases of green village and town plans as well as the technical difficulties of evaluation. This section proposes a systematic evaluation method for the site selection and functional division optimization of green villages and towns that can be selected and used according to the different characteristics and technical conditions of the evaluated villages and towns.

8.2.1 Evaluation Object and Basic Method

1. Scope of evaluation

The evaluation scope of the site selection of green villages and towns is the township area of the town or the village area of the administrative village.

The evaluation scope of the functional division optimization of green villages and towns is the township area of the town or the built-up area of the administrative village.

2. Content of evaluation

1) Evaluation of site selection of green villages and towns

(1) Evaluation basis. Evaluation of site selection of villages and towns is based on *Standard for Planning of Town* GB 50188-2007. The basis for selecting the evaluation index is *Current Land Use Classification* GB/T 21010-2017.
(2) Evaluation content. The essence of site selection evaluation is to evaluate the compliance of land use in the process of plan implementation. The main evaluation contents include the normative audit of "classification of land use" and "calculation of land use" in the current town or village plan and the implementation degree of the key index in the process of plan implementation.
(3) Evaluation focus. The key point of site selection evaluation is to evaluate whether the land use of the town or village is consistent with the current construction plan and whether it reflects the purpose of saving land.

2) Evaluation of functional division optimization of green villages and towns

(1) Evaluation basis. The selection of construction land for villages and towns should comply with the requirements of "classification of construction land" and "standard of planning and construction land" in the *Standard for Planning of Town* GB 50188-2007.
(2) Evaluation content. The evaluation of the functional division optimization of villages and towns includes the status evaluation of the implementation of the functional division plan of green villages and towns and the optimization evaluation of the functional division during the implementation of green village and town plans.
(3) Evaluation focus. There should be an evaluation of the consistency between the current functional division status and the current functional division plan of towns or villages and an optimization evaluation of functional division during the implementation of village and town plans. According to the evaluation results, suggestions on plan revision or construction activity rectification should be put forward for villages or towns that deviate from the plan requirements.

3. Method of evaluation

1) Review of current plan and related information

The review focuses are the "classification of land use", "calculation of land use" and functional division of land use in the current town or village plan.

2) Survey of the current situation of land-use plan and implementation

According to the survey requirements and restrictions of village and town plans, the space-ground integration measurement (S-GIM) method can be adopted to survey the land-use status of villages and towns. By using GPS and UAV surveys, remote sensing data analysis methods, basic data extraction of land-use changes and other field survey methods, the plan implementation status of the evaluated area can be measured and evaluated.

3) Consistency evaluation of plan implementation space

According to field survey results, the space consistency analysis of the land-use plan implementation is carried out from the perspective of land type map locations, and the implementation of the site selection plan is evaluated.

4) Optimization evaluation of land-use functional division

By using remote sensing data, GPS and UAV measurement methods, the status quo of land-use functional division of built-up areas of towns or villages is analysed, and the optimization evaluation of land-use functional division of built-up areas of villages and towns is carried out through multi-objective decision-making and other evaluation methods.

4. S-GIM method and application

1) S-GIM integrated measurement method

The S-GIM method is a data acquisition method that combines 3 measurement methods: remote sensing data analysis, unmanned aerial vehicle measurement and global positioning system measurement.

In general, the data and images of the status quo of the evaluated area can be obtained separately by using any of the above methods. However, because the scope of this evaluation is a village or town, there are often limiting factors that might affect the accuracy of the evaluation data.

When the S-GIM method is applied, the evaluation data can be obtained by combining 3 or 2 measurement methods according to local conditions.

2) Advantages and limitations of using a single measurement method

(1) Single use of remote sensing data

Satellite remote sensing data analysis is effective for evaluating site selection and functional division optimization in villages and towns. It has the characteristics of accurate data, convenient calculation, and continuous data for the evaluated area that can be obtained when data sources are sufficient. However, the evaluation scope is a village or town. Compared with the big data of satellite remote sensing, the evaluation area is very small. At the same time, limited by data sources, the data that can be accurately used for quantitative analysis in the evaluation are still insufficient. Therefore, the single use of remote sensing data analysis is likely to produce accuracy differences between big data and small areas.

(2) Single use of UAV data

UAVs can be used for low-altitude photography to immediately obtain accurate imaging data for the evaluated area. However, the acquisition of the image of the assessed area by the UAV method is generally suitable only for qualitative analysis. Therefore, it is difficult to meet the quantitative analysis requirements of this evaluation by using UAVs to collect data.

(3) Single use of global positioning system data

GPS is a commonly used method of field measurement; it includes a base station and a mobile receiver that can be used to measure the topographic data of the evaluation range through the association of the base station and receiver. However, this method will be affected to a certain extent by the topography of the evaluated area; for example, if the base station and the receiver are blocked by ridges in the measurement range, the accuracy of the measurement data will be affected.

3) Advantages of the S-GIM integrated method

In the S-GIM method, the regional geographic information data of the time specified for the quantitative analysis of the evaluated area can be obtained by remote sensing data analysis, the images needed for qualitative analysis can be obtained by UAVs, accurate terrain data can be obtained by GPS, and more accurate evaluation data can be obtained by data analysis.

The S-GIM method combines 3 or 2 measurement methods by using the advantages of each measurement method while avoiding the defects of evaluation data caused by the single use of a certain method to obtain more accurate data for the current measurement of site selection and functional division of villages and towns and achieve more accurate evaluation conclusions.

5. Method of evaluation data acquisition

1) GPS and UAV measurement method

(1) Through GPS measurement, the evaluation data of residential land, public service facility land, production facility and storage facility land, green space and public square land and other construction land in built-up areas are obtained.
(2) Through UAV measurement, graphical data such as the current land-use situation in built-up areas, villages and towns were obtained.

2) GPS image processing

(1) Determination of the reference points of measurement

Through graphic analysis and field investigation, the representative reference points of the surveyed area are determined.

(2) Graph superposition and comparison analysis

① The satellite remote sensing image data before and after the plan implementation are superimposed and compared with the established reference points to extract the changes in map sites that reflect the construction activities of villages and towns.
② Geographic information system (GIS) technology is used to obtain quantitative statistics for the area and scope of changes in map sites.

③ Qualitative analysis and storage management are carried out for the land-use attribute information of the changed map site and compared with the village and town planning data to determine the accurate location of changed map sites on the village and town planning map.

④ Whether changes in the land area and property of the changed map site are in accordance with the village plan is determined.

6. Remote sensing image information processing

The processing of remote sensing image information includes three main steps: image pre-processing, classification processing and classification result processing. Correlation analysis was performed using erdas, eCognition and ArcGIS software.

1) Remote sensing image pre-processing

This step mainly includes band fusion, projection and coordinate conversion, image cutting, geometric correction and image enhancement. The above operations are completed using erdas software.

2) Classification and processing of remote sensing images

This step is divided into remote sensing image segmentation, classification system definition, classifier insertion, sample object definition, feature space optimization, image classification, classification result evaluation and other contents. The above operations are completed using eCognition software. The specific operations are as follows:

(1) Remote sensing image segmentation

To establish an evaluation project in eCognition software, it is necessary to rank the importance of remote sensing images and add specific segmentation rules (such as segmentation algorithms and segmentation parameter-setting information) to perform image segmentation. In this evaluation, multi-scale segmentation is adopted for remote sensing image segmentation.

(2) Classification of land use

Combined with the land requirements of the planning and construction of green villages and towns, this evaluation classifies village land according to six categories: cultivated land, forestland (including garden land and grassland), residential land, traffic land, water area and unused land.

(3) Classifier insertion

Defining the feature space increases the degree of identification of different features and the application of classification in the feature space to the definition of the classification system; the inserted eCognition software provides oversight of the classifier. The nearest classifier is used in this evaluation.

(4) Sample object definition

eCognition uses the adjacent classification method, similar to supervised classification in traditional image analysis software, to define the training area on behalf of the type of land use. The defined types of land-use samples are selected, and the samples are initialized according to the standard nearest neighbour classification.

(5) Feature space optimization

The contribution of different features is calculated by the spatial feature optimization tool to identify the optimal feature space combination for classification, and the optimized feature space combination is applied to the classification system.

(6) Image classification

After the sample object is selected, relevant tools are used to conduct preliminary classification operations for the image. On the basis of the preliminary classification results, misclassified ground object samples are adjusted, correct categories are assigned, classification operations are conducted, and so on until the results are satisfactory.

(7) Evaluation of classification results

eCognition software provides four evaluation classification accuracies in the statistical method, and a classification evaluation, if the classification accuracy is not ideal, can modify the sample object, optimize the feature space, perform classification, and repeat the process until satisfactory results are obtained.

3) Classification results statistics and processing

The classification results are converted into raster data in ArcGIS software, and basic information on land use is collected to generate the land-use classification map.

8.2.2 Evaluation of Site Selection and Plan Space Consistency of Green Villages and Towns

The core of the consistency evaluation of site selection and plan space in green villages and towns is to compare the planning objectives, the plan implementation process and the results from the "base year" to the "completion year" of the plan.

1. Object of evaluation

Based on the *Standard for Planning of Town* GB 50188-2007 and *Current Land Use Classification* GB/T 21010-2017, this evaluation method integrates green village and

Table 8.1 Classification and elements of land use in green villages and towns

Number	Land category	Meaning of land use
1	Farmland	Refers to land on which crops are grown, including orchards and tea gardens
2	Forestland	Refers to land where trees, bamboo, and shrubs grow
3	Residential land	Refers to land with residential buildings, public buildings and ancillary facilities, including residential land, commercial land, and public management and public service land
4	Traffic land	Refers to land used for transportation of ground lines, stations, etc.
5	Waters	Refers to river water, lake water, pit pond water, ditches and other water area
6	Unutilized land	Refers to land that has not been utilized

town land into six categories: cultivated land, forestland, residential land, transportation land, water area and unused land (pastures in livestock areas can be attributed to forestland). In general, industrial and mining land are not included in this evaluation.

The classification and meaning of land use in green villages and towns are shown in Table 8.1.

2. Time limit of evaluation

The time limit for this evaluation includes the whole process of plan implementation, namely, the plan from the "base year" to the "completion year".

3. Method of evaluation

(1) Determine the number of "assessment years" during the plan implementation period, from "base year" to "completion year".
(2) Carry out on-the-spot investigation and obtain relevant evaluation data for the evaluated area in the determined evaluation year, and carry out spatial compatibility calculation and current status evaluation during plan implementation for green villages and towns.
(3) Calculation formula

The calculation formula for the spatial compatibility of site selection and plan implementation in green villages and towns is

$$P_i = \frac{X_i - C_{bi} - C_{wi}}{C_i} \tag{8.1}$$

where p_i is the calculated value of the spatial compatibility of land i from the evaluation year to the completion year;

i is land i;

X_i is the current actual area of land i in the evaluation year, hm^2;

C_{bi} is the area of the plot that does not meet the plan during changes in land i in the evaluation year, hm^2;

Table 8.2 Consistency evaluation standard of site selection and planning in green villages and towns

Evaluation items	Level I	Level II	Level III	Level IV	Level V
Level of conformity with land map site	$P_i = 1$	$0.95 \leq P_i < 1$	$0.90 \leq P_i < 0.95$	$0.85 \leq P_i < 0.90$	$P_i < 0.85$

C_{wi} is the area of the plot *where* land plan i has not been implemented in the evaluation year, hm^2; and

C_i is the total area of the *i* land plan, hm^2.

(4) Standard for evaluation level

This evaluation adopts 5 evaluation criteria levels to evaluate the spatial compatibility of site selection and plan implementation in green villages and towns.

The consistency evaluation standard of site selection and planning in green villages and towns is shown in Table 8.2.

When $P_i = 1$, the level is I, indicating that the plan implementation results are in full compliance with the planning objectives. In general, it is possible to achieve the level I standard only at the end of the plan implementation period.

When $P_i < 1$, the level is II–V, which means that the plan implementation results are not consistent with the plan objectives, and the current situation of the site selection of the village (town) from levels II to V differs from the plan.

8.2.3 Evaluation of Functional Division Optimization of Green Villages and Towns

1. Purpose of evaluation

The functional division optimization of land use in green villages and towns refers to the combination of natural land environmental conditions to improve the existing land-use function; reasonably allocate existing land resources to villages and towns to achieve rational and effective spatio-temporal implementation of land resources; promote rural ecological environmental improvement, economic development and social progress of people's livelihoods; and improve rural planning and land use in different functional areas to obtain environmental, economic and social benefits.

The core of evaluation of the division optimization of land use in green villages and towns is judging whether the comprehensive benefits of land use in the evaluated villages and towns reach the maximum values.

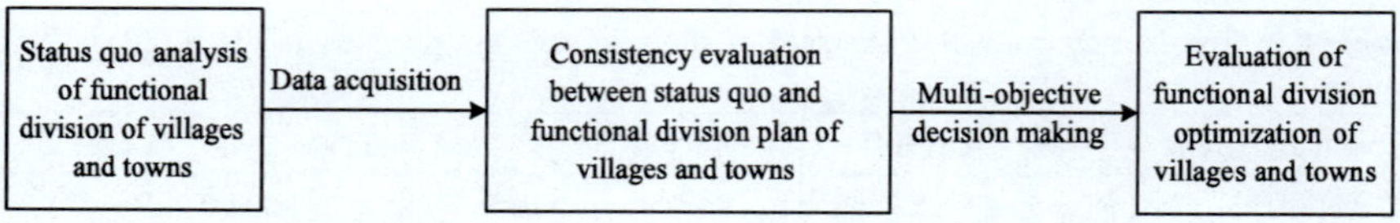

Fig. 8.1 Evaluation process of functional division optimization in green villages and towns

2. Content of evaluation

The evaluation of functional division optimization in green villages and towns includes the following main contents:

(1) Analysis and evaluation of the status quo of the functional division of villages and towns;
(2) Consistency evaluation between the status quo and the functional division plan of villages and towns; and
(3) Use of the multi-objective decision-making method to evaluate the functional division optimization of villages and towns.

3. Process of evaluation

The evaluation process of functional division optimization in green villages and towns is shown in Fig. 8.1.

4. Current situation evaluation of the functional division of villages and towns

1) Data acquisition

The S-GIM method is used to obtain the current situation data for the functional division of a village or town. For example, UAV aerial photography is used to obtain a qualitative analysis picture of the current situation of the evaluated region. GPS measurement technology is used to obtain the terrain data of the current situation of the functional division of the village or town and create a map of the current situation of the functional division.

2) Status analysis

Correlation analysis software is used to analyse the current situation data of the functional division of the village or town obtained by measurement and to obtain the status map of the functional partition of the evaluated region. By comparing the current situation data and maps of the functional division of the village or town with the current plan, the difference between the current situation and the functional division plan of the village or town can be analysed and evaluated.

5. Evaluation of consistency between the current situation and the plan of a village or town

1) Evaluation focus

The functional division evaluation of land use in villages and towns includes two main types of consistency:

(1) Consistency with land-use division in the upper-level plan. According to the *Guideline for the Town-level General Land Use Planning* TD/T 1025-2010, the land in a town is divided into basic farmland protection area, general agricultural area, construction area, independent industrial and mining area, scenic tourism land area, ecological environmental safety control area, natural and cultural heritage protection area, forested area, animal husbandry area and other areas.

(2) Consistency between industrial development plans and land-use functional division. According to the development of industry in villages and towns, functional division of villages and towns will be evaluated by regional leading industrial use to determine the type of division and the implementation of land-use control.

2) Evaluation content

The functional division evaluation of land use in villages and towns has 5 aspects:

(1) Dividing ecological land-use area. Land with ecological security functions, such as forestland and water sources, is divided into ecological land-use area.
(2) Dividing agricultural development area. Cultivated land, garden land, economic forestland, and pond water surfaces are divided into agricultural development area.
(3) Dividing village construction area. Residential areas, industrial and mining land, public facility land, transportation land and other lands are divided into village construction area.
(4) Dividing tourism area. Local tourism land plans are used to identify the tourism area.
(5) Drawing a functional division map and analysing it. According to the division results, a functional division map of the village or town is drawn and analysed with ArcGIS software.

3) Standard of evaluation level

This evaluation adopts 5 evaluation criteria levels to evaluate the spatial compatibility of functional division and plan implementation in green villages and towns.

The evaluation standard is shown in Table 8.3.

Table 8.3 Evaluation standard of consistency of status quo of functional division with plans of green villages and towns

Evaluation items	Level I	Level II	Level III	Level IV	Level V
Conformity of functional division	$P_i = 1$	$0.97 \leq P_i < 1$	$0.95 \leq P_i < 0.97$	$0.90 \leq P_i < 0.95$	$P_i < 0.90$

When $P_i = 1$, the level is I, indicating that the plan implementation results are in full compliance with the planning objectives. In general, it is possible to achieve level I only at the end of the plan implementation period.

When $P_i < 1$, the evaluation level is level II–V, indicating that the plan implementation results are not consistent with the planning objectives. In addition, from Level II to Level V, the consistency of functional division in the village (town) is significantly different from the plan.

6. Evaluation of functional division optimization of villages and towns

This evaluation uses a multi-objective decision-making analysis method to analyse and evaluate the functional division of villages and towns. The main steps are as follows:

1) Selecting decision variables

The decision variables in this evaluation are land-use types, including cultivated land, garden land, forestland, residential land (commercial land, residential land, public management and public service land), transportation land, water area and unused land.

2) Determining constraints

Three aspects of constraints are determined: land economic benefit, social demand and ecological environmental requirements. The calculation is as follows:

$$\begin{cases} \sum_{j=1}^{n} a_{ij}x_j = (\geq, \leq)b_i & (i = 1, 2, \ldots, m) \\ x_j \geq 0 & (j = 1, 2, \ldots, n) \end{cases} \tag{8.2}$$

where x_j is area of various types of land;

a_{ij} is the constraint factor; and

b_i is the constraint constant.

3) Setting the objective function

Land use in green villages and towns is a multi-objective socially constructed activity that requires harmony among economic, social and ecological benefits. This evaluation sets 3 aspects of the objective function: economic benefit, ecological benefit and social benefit.

(1) Economic benefit objective function

The primary basis of land-use activity evaluation is economic benefit. The economic benefit objective function is evaluated by the output benefit per unit area of various lands. The calculation is as follows:

$$\max f(x) = \sum_{i=1}^{n} c_i x_i \tag{8.3}$$

where c_i is output benefit per unit area of various lands;
x_i is area of various lands; and
n is the number of various lands.

(2) Ecological benefit objective function

For this evaluation, the green equivalent is measured. The forest coverage rate to a certain extent can reflect other ecological status factors, such as per capita green space, soil erosion and atmospheric environmental quality [1]. This approach uses the minimum forest coverage rate as a recommended evaluation index to establish an ecological objective function model based on green equivalents:

$$\max h(x) = \frac{\sum_{i=1}^{n} x_i g_i}{S_{\text{forest}}} \tag{8.4}$$

where g_i is the average green equivalent of land of class i; and
S_{forest} is the forested area in the region at the best (lowest) forest coverage rate.

According to the principle of green equivalents, traditional land is divided into land with green equivalents, land with implicit green equivalents, and land without green equivalents.

Land with green equivalents includes cultivated land, garden land, and forestland.

Land with implied green equivalents is used mainly for water regulation, and its green equivalent is generally set to half that of forestland.

Land without green equivalents includes mainly commercial land, industrial and mining storage land, residential land, public management and public service land, and transportation land, and its green equivalent is 0.

(3) Social benefit objective function

Since it is difficult to quantify the social benefit to form the maximum or minimum objective function, the technology converts the social benefit objective functions into constraints for processing and computing.

4) Results analysis

According to the constraint conditions and the objective function, the optimal solution is obtained by a computer program, and the economic, ecological and social benefits are analysed under the optimal solution.

8.2.4 Evaluation Case of Site Selection and Functional Division Optimization of Green Villages and Towns

1. Implementation evaluation of site selection plan

1) Review of basic data for the plan

According to *The Construction Plan of Yanhe Village, Wushan Town, Gucheng County (2008–2020)*, the land plan of Yanhe Village is divided into 8 types of use: residential land, land for public facilities, farmland, tea planting base, fungus planting base, road land, greening land and water area. The plan of the functional division of Yanhe Village contains 5 functional zones: residential living area, agricultural sightseeing area, tea sightseeing area, tourist sightseeing area and natural ecological protection area.

The functional division plan of Yanhe Village is shown in Fig. 5.9.

The land-use plan of Yanhe Village is shown in Fig. 5.10.

2) Survey and analysis of the implementation status quo of the land-use plan

(1) Status quo of land use in 2008. Qstar8, iRTK2, Hi-Net Server and other measurement instruments were used to conduct on-the-spot measurements and analysis of the land use of Yanhe Village, and through the analysis, the total area of the village and the occupation of the classified land can be known. The total area of Yanhe Village is approximately 1164 hm^2, including 139.63 hm^2 of cultivated land, 971.23 hm^2 of forestland, 10.05 hm^2 of residential area, 18.55 hm^2 of traffic area, 15.63 hm^2 of water area and water facilities, and 8.91 hm^2 of unused land.

The status analysis of the land-use structure in Yanhe Village is shown in Fig. 8.2.

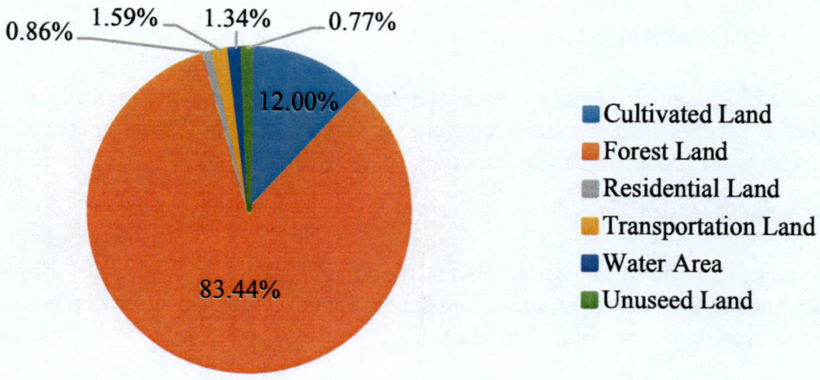

Fig. 8.2 Status analysis of land-use structure in Yanhe Village

(2) Status quo of road system. The road system in Yanhe Village is divided into 3 levels: the main road, secondary roads and branch roads. The total road system in Yanhe Village covers 18.55 hm^2, accounting for 1.59% of the village area.
(3) Status quo of ecological land. The ecological land in Yanhe Village includes mainly forestland and water area. The forestland includes mainly shrubland and other forestland, with a total area of 971.23 hm^2, accounting for 83.44% of the village area. Water area and water conservancy facilities cover 15.63 hm^2, accounting for 1.34% of the village area.
(4) Status quo of agricultural land. The agricultural industrial development area of Yanhe Village covers 39.7 hm^2, and the total cultivated land area is 139.63 hm^2, accounting for 12.00% of the village area.
(5) Status quo of residential land. The residential construction land in Yanhe Village includes mainly homesteads, land for public management and public service facilities, land for commercial use and other construction land. The area is approximately 10.05 hm^2, accounting for 0.86% of the village area.

3) Analysis of remote sensing data of land-use classification changes

(1) Remote sensing image pre-processing. In erdas software, 4 groups of remote sensing image data from 2008, 2012, 2014 and 2016 (the "evaluation year" selected during the plan implementation process) were processed with image cutting, geometric correction and image enhancement.

(2) Remote sensing image classification

① The 4 groups of remote sensing image data from Yanhe Village from 2008, 2012, 2014 and 2016 were imported into eCognition software for image segmentation. The segmentation scale and shape parameters were set at 120 and 0.5, respectively, and the segmentation images were obtained.
② Definition of classification system. According to the survey results, the cultivated land and garden land of Yanhe Village are collectively classified as cultivated land. There is no grassland in Yanhe Village. In the "Class Hierarchy" panel, the land system of Yanhe Village is classified as cultivated land; forestland; residential, industrial and mining land; transportation land; water area; and unused land.
③ Classifier insertion. The average values of 3 layers of remote sensing images were used as feature space in the defined classification system, and the nearest classifier was selected for insertion into the classification description.
④ Definition of sample. Samples of cultivated land; forestland; residential, industrial and mining land; transportation land; water area and unused land were selected for initialization.
⑤ Feature space optimization. Feature space optimization was carried out, and the system automatically calculated the optimal classification feature combination based on the samples.
⑥ Image classification. A classification program in the "Process Tree" panel was created, and the land classification was activated. The misplaced feature samples were adjusted to the correct category, and the classification calculation was then carried out. Iterations were performed until the results were satisfactory.

The processing process of remote sensing images of Yanhe Village by eCognition is shown in Fig. 8.3.

⑦ Evaluation of classification results. "Accuracy Assessment" was selected from the "Tools" menu, and "Error Matrix Based on Samples" was selected as the statistical method for evaluating the classification accuracy. The statistical results are shown in Fig. 8.4.

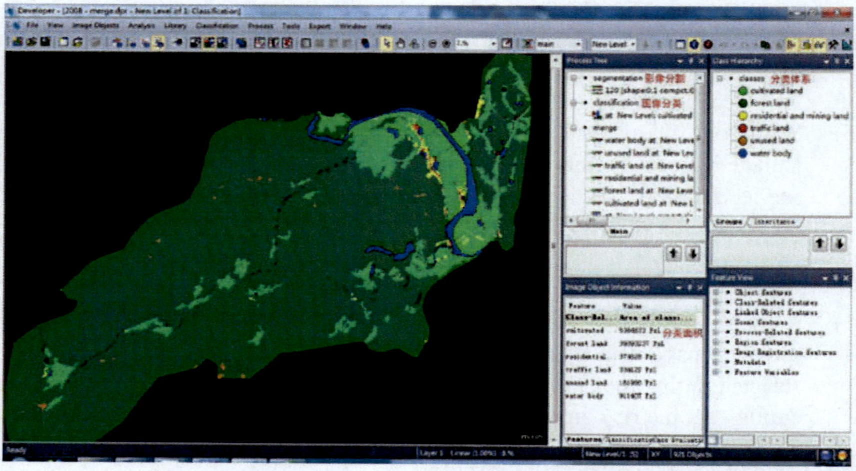

Fig. 8.3 Remote sensing image processing process

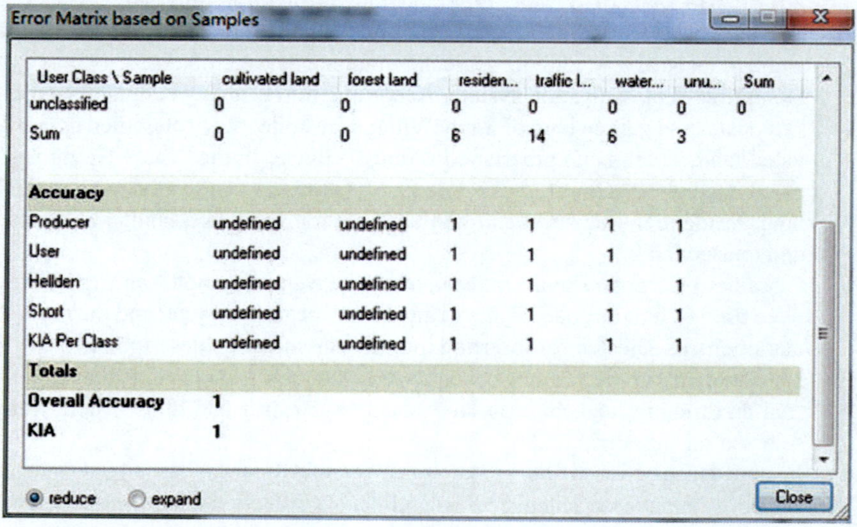

Error Matrix based on Samples

User Class \ Sample	cultivated land	forest land	residen...	traffic l...	water...	unu...	Sum
unclassified	0	0	0	0	0	0	0
Sum	0	0	6	14	6	3	
Accuracy							
Producer	undefined	undefined	1	1	1	1	
User	undefined	undefined	1	1	1	1	
Hellden	undefined	undefined	1	1	1	1	
Short	undefined	undefined	1	1	1	1	
KIA Per Class	undefined	undefined	1	1	1	1	
Totals							
Overall Accuracy	1						
KIA	1						

reduce expand Close

Fig. 8.4 Classification accuracy evaluation statistical results

The overall accuracy and KIA values in the figure are both 1, indicating that all selected training samples were classified correctly with satisfactory classification accuracy.

4) Statistics and processing of classification results

First, the classification results of 4 groups of remote sensing image data for Yanhe Village from 2008, 2012, 2014 and 2016 were converted into raster data in ArcGIS software, and the basic information of various land types was incorporated to obtain the changes in all kinds of land-use areas in Yanhe Village. Then, the classification map of the land in Yanhe Village was generated.

The land areas used for the classification of Yanhe Village in 2008–2016 are shown in Table 8.4.

The land-use classification comparison of Yanhe Village in 2008–2016 is shown in Fig. 8.5.

5) Implementation analysis evaluation of site selection plan

(1) Basic data processing

According to the construction plan of Yanhe Village, the base year for the implementation evaluation of site selection is 2008, and the completion year of the plan is 2020. For this evaluation, 2008, 2012, 2014 and 2016 were chosen as evaluation years for the implementation.

Using ArcMap to intersect the land-use classification map obtained from the eCognition analysis data, the implementation status of land-use plan in different evaluation years of Yanhe Village can be obtained through field measurement, and the results are compared with the set value of land-use plan in 2020 respectively, the coincidence degree and deviation reasons are analyzed, and the improvement countermeasures are put forward.

The implementation analysis of the land-use plan in Yanhe Village is shown in Table 8.5.

(2) Consistency evaluation of site selection and plan implementation space

① Spatial coincidence degree calculation and grade classification of site selection for land use

Table 8.4 Areas of land used for the classification of Yanhe Village in 2008–2016

Index	Land category	Land area (hm^2)			
		2008	2012	2014	2016
1	Arable land	132.14	158.40	139.28	139.63
2	Woodland	986.57	945.48	974.02	970.47
3	Residential land	9.37	9.58	10.44	10.05
4	Traffic land	8.35	13.12	15.89	18.55
5	Water area	22.76	16.94	15.48	15.63
6	Unutilized land	4.05	19.70	8.12	8.91

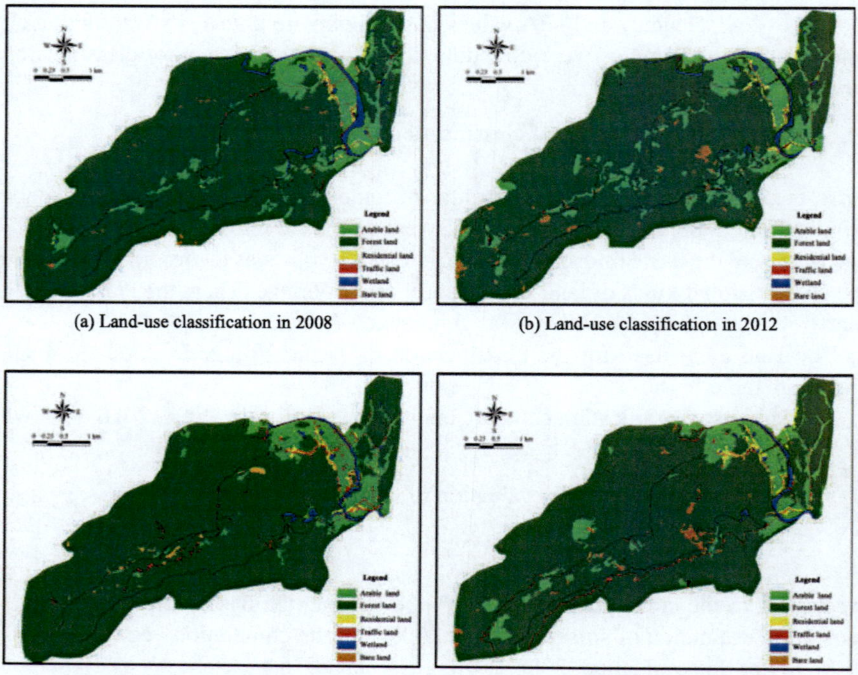

(a) Land-use classification in 2008

(b) Land-use classification in 2012

(c) Land-use classification in 2014

(d) Land-use classification in 2016

Fig. 8.5 Land-use classification comparison of Yanhe Village in 2008–2016

Using the "intersect" in ArcMap to superimpose the land classification maps of the base year (2008) and the evaluation year (2012, 2014, 2016), the area of the changing land that does not conform to the planned area in 2008–2012, 2008–2014, 2008–2016 can be obtained.

Formula (8.1) is used to calculate the spatial coincidence degree of the land maps in the implementation of land-use plan in Yanhe Village from 2008 to 2012, 2008 to 2014 and 2008 to 2016.

The analysis of the spatial coincidence between the land plots in each evaluation year and the land areas in the plan year is shown in Table 8.6.

② Spatial coincidence degree evaluation of land-use plan and implementation in Yanhe Village

According to Table 8.6, the evaluation level of the spatial coincidence degree of the maps for Yanhe Village did not reach level I, indicating that from the planning base year to the assessment year, all types of land in Yanhe Village did not fully meet the expected targets. The spatial coincidence degrees of cultivated land and forestland are higher than 0.9, which means that cultivated land and forestland are planned according to the programme, and other land types are adjusted.

Table 8.5 Implementation analysis of land-use plan in Yanhe Village (Unit: hm^2)

Evaluation time	Measurement and comparison	Cultivated land	Forestland	Residential land	Transportation land	Unutilized land	Water area
Planning base year	Actual statistics in 2008	132.14	986.57	9.37	8.35	4.05	22.76
Planning completion year	Plan setting points in 2020	108.50	1012.50	10.40	8.70	5.70	18.20
2012	Measured Value in 2012	158.40	945.48	9.58	13.12	19.7	16.94
	Difference by points in 2020	−49.90	67.02	0.82	−4.42	−14	1.26
2014	Measured Value in 2014	139.28	974.02	10.44	15.89	8.12	15.48
	Difference by points in 2020	−30.78	38.48	−0.04	−7.19	−2.42	2.72
2016	Measured Value in 2016	139.63	970.47	10.05	18.55	8.91	15.63
	Difference by points in 2020	−31.13	42.03	0.35	−9.85	−3.21	2.57

The variation trend analysis of the spatial coincidence of land types during the implementation of the plan for Yanhe Village is shown in Fig. 8.6.

(3) Evaluation conclusion of site selection and plan implementation in Yanhe Village

Figure 8.6 shows that 2014 was a turning point in the plan implementation.

Before 2014, except for residential land and transportation land, the coincidence degree of land types showed an increasing trend, indicating that cultivated land, forestland, unutilized land and water area were planned according to the programme during 2008–2014.

After 2014, Yanhe Village began to vigorously construct the Bairi Mountain Scenic Spot and expand the uphill road so that the 2 evaluation indexes of "residential land" and "traffic land" fell in line with the plan.

From 2014 to 2016, in addition to that of forestland, the coincidence degree of other land types declined. The main reason was that the conversion of some forestland into cultivated land did not meet the objectives, resulting in a decrease in the consistency of cultivated land. The expansion of the construction land in the Bairi Mountain Scenic Spot also reduced the coincidence degree of land use for the residential area, and the construction of the mountain road caused a decline in the coincidence degree of traffic land. The water area remained unchanged.

Table 8.6 Analysis of spatial coincidence between land plots in each evaluation year and land areas in plan year in Yanhe Village

Land types	Plan setting points in 2020/hm^2	2012			2014			2016		
		Measured value/hm^2	Spatial coincidence	Rating	Measured value/hm^2	Spatial coincidence	Rating	Measured value/hm^2	Spatial coincidence	Rating
Cultivated land	108.50	158.40	0.90	III	139.28	0.94	III	139.63	0.91	III
Forestland	1012.50	945.48	0.86	IV	974.02	0.91	III	970.47	0.92	III
Residential land	10.40	9.58	0.84	V	10.44	0.76	V	10.05	0.73	V
Transportation land	8.70	13.12	0.92	III	15.89	0.91	III	18.55	0.83	V
Unutilized land	5.70	19.70	0.82	V	8.12	0.87	IV	8.91	0.80	V
Water area	18.20	16.94	0.69	V	15.48	0.69	V	15.63	0.70	V

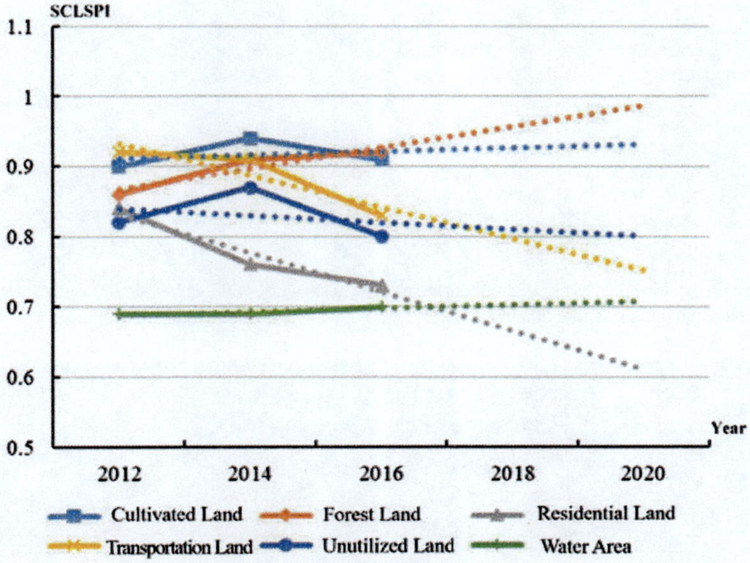

Fig. 8.6 Variation trend analysis of the spatial coincidence of land types during plan implementation (SCLSPI) for Yanhe Village

2. Implementation evaluation of functional division plan

1) Status quo analysis of functional division

(1) Data acquisition

Unmanned aerial vehicle photography is used to qualitatively analyse the current status of the functional division of Yanhe Village. Global satellite positioning system (GPS) on-site measurement technology is used to quantitatively analyse the current status of the functional division of Yanhe Village.

The UAV aerial photography analysis of Yanhe Village functional division is shown in Fig. 8.7.

The data analysis method of the status of the functional division of Yanhe Village is shown in Fig. 8.8.

(2) Evaluation of the spatial coincidence degree of the functional division plan

① Calculation of spatial coincidence degree and grade classification of functional division planning

According to Formula (8.1), the spatial coincidence degrees of functional division in the evaluation planning process of Yanhe Village from 2008 to 2012, 2008 to 2014 and 2008 to 2016 are calculated.

The analysis of the spatial coincidence between the functional areas of Yanhe Village in each evaluation year and the planned year is shown in Table 8.7.

(a) New village residential functional area

(b) Bairi Mountain Scenic Spot

(c) Water area

(d) Tea garden production area

Fig. 8.7 UAV aerial photography analysis of functional division of Yanhe Village

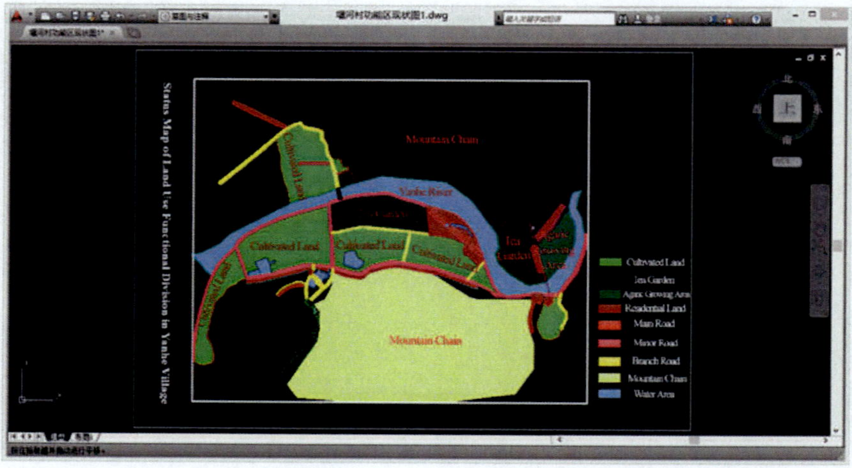

Fig. 8.8 Data analysis method of the status of the functional division of Yanhe Village

Table 8.7 Analysis of the spatial coincidence between the actual functional areas of Yanhe Village in each evaluation year and the planned year

Land functional division	Plan setting points in 2020/hm^2	2012			2014			2016		
		Measured value/hm^2	Spatial coincidence	Rating	Measured value/hm^2	Spatial coincidence	Rating	Measured value/hm^2	Spatial coincidence	Rating
Natural ecological reserve	1036.4	982.12	0.85	V	997.62	0.91	IV	995.00	0.91	IV
Basic farmland area	84.9	134.80	0.87	V	115.68	0.92	IV	116.03	0.88	V
Tea production area	20.8	20.8	1.00	I	20.8	1.00	I	20.8	1.00	I
Fungus production area	2.8	2.8	1.00	I	2.8	1.00	I	2.8	1.00	I
Residential area	19.10	22.70	0.88	V	26.33	0.83	V	28.60	0.78	V

② Analysis of the spatial coincidence degree of the functional division plan

According to the calculation results shown in Table 8.7, the planning and implementation of the functional zones such as the natural ecological reserve, the basic farmland area, and the residential area of Yanhe Village do not meet level I. This finding indicates that the 3 functional zones have not fully met the expected targets between the planning base year and the assessment years. The spatial coincidence degrees of the tea production area and fungus production area are both 1, indicating that these 2 functional areas in Yanhe Village are planned in accordance with the programme.

According to Table 8.7, the trend of the spatial coincidence degree of functional divisions in the implementation process in Yanhe Village during 2012–2016 can be analysed.

The variation trend of the spatial coincidence of functional division during the implementation of the plan for Yanhe Village is shown in Fig. 8.9.

(3) Evaluation conclusion of functional division plan optimization for Yanhe Village

Figure 8.9 shows that before 2014, the spatial coincidence degrees of the basic farmland area and the natural ecological protection area in Yanhe Village showed an increasing trend, indicating that the two functional areas were planned according to the programme during 2008–2014.

After 2014, Yanhe Village began to construct Bairi Mountain Scenic Spot, expand the mountain road and occupy part of the forestland and farmland, resulting in a decrease in the spatial coincidence degree of the basic farmland area and the natural

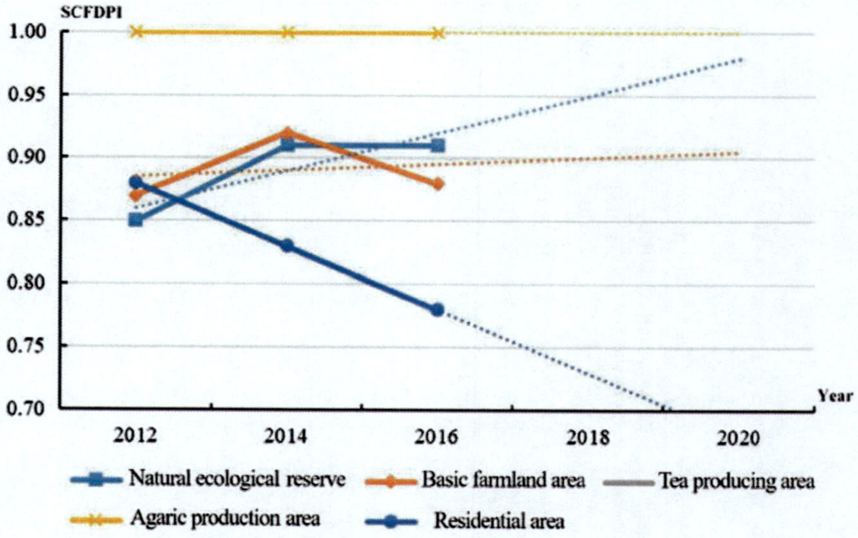

Fig. 8.9 Variation Trend of the Spatial Coincidence of Functional Division during Plan Implementation (SCFDPI) for Yanhe Village

ecological reserve. At the same time, the construction of scenic spots and the expansion of roads increased the residential area, resulting in a decreasing trend in the spatial coincidence degree of the residential area.

3. Evaluation of functional division plan optimization

The evaluation methods of functional division plan optimization in green villages and towns are as follows: selecting optimization decision variables, determining development constraints, setting optimization objective functions, calculating optimization solutions for multiple objective functions, and determining future optimal development plans of functional division of villages and towns.

1) Selection of decision variables

Based on the characteristics of the land resources and social and economic development requirements of Yanhe Village, this evaluation considers the determinacy of the objective function benefit coefficient and sets 5 variables.

The variable settings in the multi-purpose plan model are shown in Table 8.8.

2) Determination of constraints

According to the characteristics of rural land resource utilization, the requirements of the overall land-use plan and the comprehensive consideration of the operability of relevant data, the following 6 constraints are set: total land constraint, total population constraint, ecological protection constraint, cultivated land reserve constraint, total scale of construction land constraint, and decision variable non-negative constraint.

(1) Total land constraint

The sum of all kinds of land types in Yanhe Village is equal to the total land area of Yanhe Village. Therefore, the total land constraint value for Yanhe Village is

Table 8.8 Variable settings in the multi-purpose plan model of Yanhe Village

Variable	Functional area	Current area (hm^2)	Proportion (%)
Z_1	Natural ecological reserve	995.00	85.53
Z_2	Basic farmland area	116.03	10
Z_3	Tea production area	20.80	1.78
Z_4	Fungus production area	2.80	0.24
Z_5	Residential area	28.60	2.45

$$S_{\text{total}} = \sum_{1}^{n} Z_i = 1163.23\,\text{hm}^2$$

(2) Total population constraint

The natural population growth method directly calculates the future population based on population growth in the base year. The calculation formula for the population growth estimation is

$$p = p_0(1 + k)^n \tag{8.5}$$

In the formula, *P* is the estimated total population of the evaluated area in the planned completion year, people;

P_0—The actual total population of the evaluated area in assessment year, people;

K—Natural population growth rate, ‰;

n—Assessment years.

According to the population natural growth method, this evaluation predicts the population of Yanhe Village in 2020. According to the change in the population of Yanhe Village over the years, the population growth rate of Yanhe Village is set as 2.4‰, so it can be predicted that the total population of Yanhe Village in 2020 will be 1,183. Therefore, the total population constraint value of Yanhe Village until 2020 is 1183.

(3) Ecological protection constraint

The natural ecological reserves in the semi-mountainous area of western Hubei, where Yanhe Village is located, are forestland and water area. Therefore, this evaluation selected forestland and water area as the natural ecological reserve of Yanhe Village and referred to the relevant national standards or the current planning of the assessed area to determine the value of ecological protection constraints.

① Forestland. *Guidelines for the Construction of Beautiful Village* GB/T 32000-2015 stipulate that forest and grass coverage rates should be up to or pass 80% of the total area. Therefore, the protected area of the forestland in Yanhe Village is

$$1163.23 \times 80\% = 930.584\,\text{hm}^2$$

② Water area. The water area planned in Yanhe Village should not be less than 18.2 hm^2. Therefore, the ecological protection constraint value Z_1 is

$$Z_1 \geq 930.584 + 18.2 = 948.784\,\text{hm}^2$$

(4) Cultivated land reserve constraint

According to the statistics in Table 2.5, the per capita arable land area in China in 2016 was 0.086 hm^2. In the same year, the per capita arable land in Gucheng County, where Yanhe Village is located, was less than 0.07 hm^2, which was even lower than the national per capita cultivated land level.

This evaluation classifies the basic farmland area Z_2, tea production area Z_3, and fungus production area Z_4 as cultivated land.

① The cultivated land area in Yanhe Village should be no less than the national per capita cultivated land area, so

$$Z_2 + Z_3 + Z_4 \geq 0.086 \times 1183 = 101.74\,\text{hm}^2$$

② The cultivated land area in Yanhe Village should not be lower than the target in the 2020 plan, so

$$Z_2 + Z_3 + Z_4 \geq 108.50\,\text{hm}^2$$

Therefore, the constrained value of the cultivated land reserve in Yanhe Village is

$$Z_2 + Z_3 + Z_4 \geq 108.50\,\text{hm}^2$$

(5) Total scale of construction land constraint

① Residential land-use constraint. Residential land use includes residential land, commercial land, public management and service land, for which the values should not be lower than the plan values.

Therefore, the land-use constraint value of the Yanhe Village residential area is 10.05 hm^2.

② Traffic land constraint. Traffic land affects the economic development of Yanhe Village, and the value should not be lower than the current value, which is 18.55 hm^2.

Therefore, the constraint value of the total scale of construction land in Yanhe Village is

$$Z_5 \geq 10.05 + 18.55 = 28.6\,\text{hm}^2$$

(6) Decision variable non-negative constraint

The optimization evaluation model of the functional division of Yanhe Village requires that all variables be non-negative. This means that $Z_i > 0$, where Z_i is the area of type i land use in the plan completion year.

3) Setting of the target function

(1) Objective function of ecological benefit of Yanhe Village

① Determine the minimum woodland area of Yanhe Village. *Guidelines for the Construction of Beautiful Village* GB/T 32000-2015 stipulate that the coverage rate of forest and grass in mountainous areas should be up to and beyond 80%. Therefore, the minimum woodland area of Yanhe Village is

$$S_{forest} = 1163.23 \times 80\% = 930.58\,\text{hm}^2.$$

② Calculate the green equivalent of cultivated land and forestland in Yanhe Village. The average green equivalent of cultivated land and forestland in Yanhe Village is set as $g_1 = 0.5$ and $g_1 = 1$ [1].

③ Determine the ecological benefit objective function of Yanhe Village. According to Formula (8.4), the objective land-use function optimization for Yanhe Village is

$$\max h(Z) = \frac{Z_1 + 0.5(Z_2 + Z_3 + Z_4)}{930.58} \tag{8.6}$$

where $\max h'(Z) = 930.58 \max h(Z)$, and when $h'(Z)$ is at the maximum value, $h(Z)$ is also maximized. At this time, the independent variables of the two functions have the same value. To make the mathematical analysis more convenient, the ecological benefits in this evaluation are calculated and analysed according to $\max h'(Z)$.

Therefore, the objective ecological benefit function of Yanhe Village is

$$\max h^{'}(Z) = 930.58 \max h(Z)$$

(2) Objective function of economic benefit in Yanhe Village

According to the survey, the per capita income of Yanhe Village was 12000 yuan in 2016, and the total revenue was approximately $1.2 \times 1050 = 12.6$ million yuan. The income comes mainly from tourism, tea, flowers and seedlings. The tourism income is approximately 6 million yuan, the tea cultivation and processing income is approximately 3 million yuan, the flowers and seedlings income is approximately 500000 yuan, and the black fungus cultivation income is approximately 200000 yuan. The incomes of the basic farmland area, tea production, edible fungus production, and residential area have a ratio of 5:30:2:60.

According to Formula (8.3), the objective function of land-use structural optimization economic benefit for Yanhe Village is

$$\max f(Z) = f(Z) = 0.56Z_2 + 18.74Z_3 + 9.28Z_4 + 27.25Z_5 \tag{8.7}$$

4) Multi-objective function solving

The linear weighted sum method was used to convert the multi-objective function into a single objective function, and the weight of both the economic benefit objective function and the ecological benefit objective function was set as 0.5. The objective function can be converted into

$$\max g(Z) = 0.5Z_1 + 0.53Z_2 + 9.62Z_3 + 4.89Z_4 + 13.625Z_5 \quad (8.8)$$

The model pair (8.4) was established by using Lingo software to obtain the optimal solution.

The optimal solution of multi-objective decision making for functional division optimization of Yanhe Village is shown in Table 8.9.

5) Functional division optimization scheme for Yanhe Village

By optimizing the functional division of Yanhe Village based on the optimization model established in this evaluation, the functional division optimization scheme of Yanhe Village under the constraints of ecological protection, cultivated land reserve and construction land can be obtained.

The optimization scheme for the functional division utilization of Yanhe Village is shown in Table 8.10.

The optimized land-use functional area of Yanhe Village is substituted into Formula (8.2) to calculate the maximum ecological benefit of 1.07.

The optimized land-use functional area of Yanhe Village is substituted into Formula (8.3) to calculate the maximum economic benefit of 3495.

Table 8.9 Optimal solution of multi-objective decision making for functional division optimization of Yanhe Village

Variable	Optimal solution
Z_1	948.78
Z_2	84.90
Z_3	9.74
Z_4	0.00
Z_5	119.81

Table 8.10 Optimization scheme for functional division utilization of Yanhe Village (Unit: hm^2)

Variable	Functional area	Status quo area	Optimized area	Area change
z_1	Nature reserve	995.00	948.78	−46.22
z_2	Basic farmland area	116.03	84.90	−31.13
z_3	Tea producing area	20.80	9.74	−11.06
z_4	Fungus producing area	2.80	0.00	−2.8
z_5	Residential area	28.60	119.81	91.21

6) Evaluation conclusion of the functional division plan for Yanhe Village

(1) Under the constraints of ecological protection, cultivated land reserves and construction land, the natural ecological protection area, basic farmland area, tea production area and fungal production area in Yanhe Village were reduced by 46.22 hm^2, 31.13 hm^2, 11.06 hm^2 and 2.8 hm^2, respectively.
(2) If the residential area increases by 91.21 hm^2, the economic and ecological benefits of the land-use functional area of Yanhe Village can be maximized simultaneously.
(3) The optimized economic benefit will reach 34.95 million yuan, an increase of 22.35 million yuan over the current value of 12.6 million yuan.
(4) The optimized ecological benefit decreased, and the green equivalent also decreased from 1.14 to 1.07.

4. Comprehensive evaluation of site selection and functional division plan consistency

According to the above calculation of the spatial coincidence degree of various land-use types and functional zones in Yanhe Village, the comprehensive value of the spatial coincidence degree for site selection and functional division plan implementation in Yanhe Village was calculated by the weighted average method.

Through calculation, the comprehensive evaluation conclusion of the spatial coincidence of site selection and functional division plan implementation in Yanhe Village can be obtained as follows:

(1) The comprehensive value of the spatial coincidence of site selection plan implementation in Yanhe Village is 0.815, and the evaluation grade is V. Based on the evaluation criteria shown in Table 8.2, the anastomosis degree is quite different.
(2) The comprehensive value of the spatial coincidence of the functional division plan implementation in Yanhe Village is 0.914, and the evaluation grade is IV. Based on the evaluation criteria shown in Table 8.3, the anastomosis degree is quite different.

8.3 Niche Suitability Evaluation of Green Village and Town Plan

8.3.1 Evaluation Object and Basic Method

1. Concept of niche

(1) The niche, also known as the ecological niche, refers to the position of an organism in the environment.
(2) The niche dimension refers to the environmental impact factor of the niche, and one impact factor constitutes one dimension.
(3) A hypovolume niche refers to a niche space with no less than three niche dimensions.

(4) In ecology, niche breadth is an indicator of the diversity of bioavailable resources. Resources are divided into several levels in a certain way. Through investigation, the number of different resource levels of each species are recorded, and the niche breadth of the species is obtained through appropriate calculation. The niche breadth varies from 0 to 1; 0 indicates no use, and 1 indicates equal use at all levels.

2. Niche suitability green village and town plan

Niche suitability refers to the matching relationship or closeness degree between the actual niche of a species in a certain dimension and the optimal niche and to the degree of influence of an activity on ecological factors in a certain region.

Niche suitability as used in this evaluation refers to the match between the green village and town social and economic development plan, the main factors affecting local environmental quality, and the main factors affecting the ecological environment.

An evaluation of green village and town plan niche suitability can effectively indicate whether the current conditions of the village or town environment meet the development needs of the plan and propose directions for improvement and optimization.

The value of green village and town plan niche suitability ranges from 0 to 1.

3. Eco-environmental impact factors of green village and town plans

1) Eco-environmental impact factors and classification

The environmental impact factors of green village and town plans are divided into two categories: environmental quality and ecological impact.

(1) Basis for determining the environmental impact factor

This basis includes *Standard for Planning of Town* GB 50188-2007, *Technique Code for Village Rehabilitation* GB 50445-2008, *Guidelines for the Construction of Beautiful Village* GB/T 32000-2015, *National Ecovillage Construction Standards (Trial)*, *National Ecological Town Construction Indicators (Trial)*, *National Ecological Civilization Construction Demonstration Village and Town Indicators (Trial)*, and *Technical Guideline for Planning Environmental Impact Assessment: General Principles* HJ 130-2019.

(2) Basis for determining ecological impact factors

This basis includes*Technical Code for Ecosystem Status Assessment* HJ 192-2015, *Standard for the Assessment of Regional Biodiversity* HJ 623-2011, *Standard for Planning of Town* GB 50188-2007, *Guidelines for the Construction of Beautiful Village* GB/T 32000-2015, and *Technical Specification for Village Landscape Environmental Engineering* CECS 285-2011.

The analysis of eco-environmental impact factors and categories of green village and town plans is shown in Table 8.11. The "–" in the table indicates that there are no data.

Table 8.11 Factors impacting the eco-environment and classification of green village and town plans

Category	Environmental impact factors					
	Water	Air	Solid waste disposal	Sound	Ecological impact	Rural landscape
Environmental quality	√	√	√	√	–	–
Ecological conditions	–	–	–	–	√	√

2) Analysis of eco-environmental impact factors

(1) Water environment

According to the relevant current standards, the content of green-ecological village and town water environmental plan includes mainly surface water environmental quality, groundwater environmental quality, drinking-water quality, water pollutant discharge, and seawater quality.

The evaluation standards and methods for a green village and town water environmental quality plan are shown in Table 8.12.

(2) Air environment

The main evaluation scope of the green village and town air environmental plans is the second-class environmental air function division in the seat of government of a town or village.

The evaluation standards and methods of a green village and town air environmental quality plan are shown in Table 8.13.

(3) Solid waste disposal

The evaluation of a green village and town solid waste disposal plan includes mainly solid waste disposal in village and town enterprises and village and town domestic garbage disposal.

The evaluation standards and methods of a green village and town solid waste disposal plan are shown in Table 8.14.

(4) Acoustic environment

With the continuous development of the rural social economy and transportation, the openness of villages and towns in China has been increasing, and the control of the acoustic environment has become an important factor in village and town plan.

The evaluation standards and methods of a green village and town acoustic environmental quality plan are shown in Table 8.15.

Table 8.12 Evaluation standards and methods of green village and town water environmental quality plan

Impact factors	Evaluation content	Evaluation indexes	Evaluation standards
Surface water environmental quality	Rivers, lakes	pH, dissolved oxygen, COD_{cr}, BOD_5, NH_3-N, TP, TN	*Environmental Quality Standards for Surface Water* GB 3838-2002
Groundwater environmental quality	General groundwater	pH, total hardness, Solubility total solid, NH3-N, Permanganate index	*Quality Standards for Groundwater* GB/T 14848-2017
Drinking-water quality	Water quality and quantity of drinking-water source	Total number of colonies, arsenic, fluoride, nitrate (in N), colour, turbidity, pH, total dissolved solids, total hardness, COD_{Mn}, Fe and manganese, chloride, sulfate	*Standards for Drinking Water Quality* GB 5749-2006
Water pollutant discharge	Rural sewage, sewage disposal facility drainage	pH, COD, BOD5, Animal and vegetable oil, NH3-N, TN, TP	*Integrated Wastewater Discharge Standards* GB 8978-2002 *Discharge Standard of Pollutants for Municipal Wastewater Treatment Plant* GB 18918-2002
Seawater quality (suitable for coastal areas)	Offshore water quality	Colour, odour, taste, suspended matter, coliform, faecal coliform, pathogens, water temperature, pH, dissolved oxygen, COD, BOD5, inorganic nitrogen (in N), non-ionic nitrogen (in N), active phosphate (in P)	*Seawater Quality Standards* GB 3097-1997

Table 8.13 Evaluation standards and methods of green village and town air environmental quality plan

Impact factors	Evaluation scope	Evaluation indexes	Evaluation standards
Ambient air quality	Village and town second-class functional area	SO_2, NO_2, PM_{10}, $PM_{2.5}$, TSP	*Ambient Air Quality Standards* GB 3095-2012

Table 8.14 Evaluation standards and methods of green village and town solid waste disposal plan

Impact factors	Evaluation content	Evaluation indexes	Evaluation standards
Domestic garbage disposal in villages and towns	Domestic garbage	Garbage production, garbage classified collection and transportation rate, garbage innocuous disposal rate	*Method to Calculate and Forecast Municipal Domestic Refuse Output* CJ/T 106-2016 *Technical Specifications for Collection and Transportation of Municipal Solid Waste* CJJ 205-2013
Solid waste disposal in village and town enterprises	Industrial solid waste	Solid waste production, main pollutant disposal rate, hazardous waste disposal rate, solid waste generation, classified collection rate, classified storage rate, perishable garbage self- disposal rate	*Standards for Pollution Control of Storage and Disposal Sites for General Industrial Solid Waste* GB 18599-2001 Other relevant standards
	Catering industry solid waste Tourism solid waste		*Specification for Environmental Protection of Catering Trade* HJ 554-2010 *Technical Code for Food Waste Disposal* CJJ 184-2012

Table 8.15 Evaluation standards and methods of green village and town acoustic environmental quality plan

Impact factors	Evaluation content	Evaluation indexes	Evaluation standards
Fixed sound source	Village and town industry transportation	Daytime equivalent sound level (L_d) Nighttime equivalent sound level (L_n) Maximum A sound level ($L_{\max}$)	*Technical Guidelines for Noise Impact Assessment* HJ 2.4-2009
Mobile sound source	Noise generated by vehicles passing over village roads, railways, and waterways	Daytime equivalent sound level (L_d) Nighttime equivalent sound level (L_n) Maximum A sound level ($L_{\max}$)	

Table 8.16 Evaluation standards and methods of green village and town ecological status plan

Impact factor	Evaluation content	Evaluation indexes	Evaluation standards
Habitat quality	Woodland and grassland	Woodland grassland area	*Technical Code for Ecosystem Status Assessment* HJ 192-2015
	Wetland	Wetland area	
	Cultivated land	Cultivated area	
	Construction land	Construction land area	
	Un-utilized land	Un-utilized area	
Biological abundance	Biological abundance	Wildlife abundance index, wild vascular bundle abundance index	*Standard for the Assessment of Regional Biodiversity* HJ 623-2011

(5) Ecological status

The ecological status impact assessment includes natural resource impact factors and ecosystem impact factors in the region, including two major impact factors: habitat quality and biological abundance.

The evaluation standards and methods of a green village and town ecological status plan are shown in Table 8.16.

(6) Village and town landscape

Landscape environmental impact factors include the appearance of the seat of government of a town, the appearance of a village and so on. To simplify the evaluation method, this evaluation incorporates village and town green plan and historical and cultural protection plan evaluations into the landscape evaluation indexes.

The evaluation standards and methods of a green village and town landscape plan are shown in Table 8.17.

4. Evaluation process of green village and town niche suitability

The evaluation process of green village and town plan niche suitability is as follows:

(1) Field investigation includes a water environmental status test, air environmental status test, solid waste discharge and treatment status test, acoustic environmental status test, ecological protection status survey, and landscape status survey.
(2) Data processing includes evaluation standard inquiry, measurement and calculation of the water, air and acoustic environments, solid waste disposal, ecological protection status, and village landscape status.
(3) Niche suitability evaluation includes the determination of the evaluation threshold, determination of the evaluation index of each impact factor, and

Table 8.17 Evaluation standards and methods of green village and town landscape plan

Impact factor	Evaluation content	Evaluation indexes	Evaluation standards
Township appearance	Landscape spatial layout	Reasonable spatial layout, prominent local characteristics, images, style proportion and colour of buildings, structures appearance coordination	*Standard for Planning of Town* GB 50188-2007
	Landscape ornamental layout	Simple form, colour harmony, easy to identify	
	Greening of main roads	The proportion reaches 100%	
	Ratio of public green space	The proportion is at least 12%	
	Cultural heritage	An overall pattern of outstanding historical and cultural relics and the protection of the spatial scale, the maintenance and improvement of buildings, structures and environmental elements	
Village appearance	Architectural form, courtyard space	The architectural form and courtyard space conform to the local traditional style	*Guidelines for the Construction of Beautiful Village* GB/T 32000-2015 *Technical Specification for Village Landscape Environmental Engineering* CECS 285-2011
	Public space and Ornamental landscaping	The public space is lively and free, which is convenient for villagers' public activities. The ornamental landscape reflects the local characteristics and coordinates with the surrounding environment. The location of signs is reasonable	
	Ratio of public green space	The proportion is at least 4%	

(continued)

Table 8.17 (continued)

Impact factor	Evaluation content	Evaluation indexes	Evaluation standards
	Environmental hygiene	The fronts and backs of houses are clean, with no sewage overflow, no scattered garbage, no open garbage or straw burning, clean water, no odour, etc.	

comprehensive evaluation of the ecological green village and town plan niche suitability.

8.3.2 *Evaluation Model of Green Village and Town Plan Niche Suitability*

1. Green village and town niche dimension

The eco-environmental impact factors in a green village and town plan mainly include the water environment, air environment, solid waste disposal, acoustic environment, ecological impact and landscape.

The green village and town niche is a multi-dimensional super-volume niche. In the evaluation of a village and town niche plan, the area has been determined, and the climatic conditions can be regarded as basically stable within a certain time limit (except for extreme weather), with a usual planning time limit of 5–20 years. Therefore, the eco-environmental impact factors will change over time within the planning time and constitute a response relationship.

2. Green village and town niche suitability

Green village and town construction is an organism, and its survival and development are closely related to the environment. By using the niche suitability method, it is possible to effectively evaluate whether the planning of the organism matches the current conditions of the village or town environment and propose an optimization programme.

Green village and town plan niche suitability ranges from 1 to 0. When the actual carrying capacity of the eco-environmental impact factor fully meets the planning development needs, the niche suitability is 1. When the actual carrying capacity of the eco-environmental impact factor does not meet the planning development needs, the niche suitability is 0.

Table 8.18 Classification of green village and town eco-environment impact factors

Number	Classification	Environmental quality impact factor	Ecological impact factor
1	First category S_{1i}	Water environmental quality S_{11}	–
2		Air environment S_{12}	–
3		Solid waste S_{13}	–
4		Acoustic environment S_{14}	–
5	Second category S_{2i}	–	Ecological conditions S_{21}
6		–	Landscape S_{22}

3. Evaluation model of green village and town niche suitability

1) Estimation of ecological suitability for 2 types of eco-environmental impact factors

The 6 types of impact factors of the green village and town ecological environment are divided into 2 categories. The first category is environmental quality factors, denoted X_{1i}, and the second is ecological impact factors, denoted X_{2i}. Generally, an environmental quality factor is a parameter that can be measured and quantitatively analysed by relevant national standards. However, there is a suitable interval for the measurement of ecological impact factors, and a semi-quantitative method is often adopted to analyse the data.

The classification of green village and town eco-environmental impact factors is shown in Table 8.18. The "–" in the table indicates that there are no data.

The following 2 methods can be used to estimate the suitability of the above 2 different types of eco-environmental impact factors.

(1) Assessment of environmental quality impact factors

The suitable assessment method for environmental quality impact factors is shown in Formula (8.9).

$$X_{1i} = \begin{cases} 0 & S_{1i} < D_{1i,p,\min} \text{ or } S_{1i} > D_{1i,p,\max} \\ R_{1i} & D_{1i,p,\min} < S_{1i} \le D_{1i,p,\max} \\ 1 & S_{1i} = D_{1i,p} \text{ or } R_{1i} = 100\% \end{cases} \tag{8.9}$$

where X_{1i} is niche suitability of the i-type environmental quality impact factor;

S_{1i} is current status measure of the i-type environmental quality impact factor;

$D_{1i,p}$ is the planning requirement value of the i-type environmental quality impact factor, usually the corresponding national standard value;

$D_{1i,p,\min}$ is the minimum of the i-type environmental quality impact factor standard;

$D_{1i,p,\max}$ is the maximum of the i-type environmental quality impact factor standard; and

R_{1i} is the guarantee rate of the *i*-type environmental quality impact factor, usually measured by the qualified rate.

(2) Assessment of ecological status impact factors

The suitable assessment method for ecological status impact factors is shown in Formula (8.10).

$$X_{2i} = \begin{cases} 0 & S_{2i} \le D_{2i,\min}, S_{2i} \ge D_{2i,\max} \\ \frac{\sum_{i=1}^{n} W_{2i} \cdot S_{2i}}{\sum S_j} \cdot R_{2i} & D_{2i,\min} < S_{2i} \le D_{2i,\max} \\ \frac{S_{2i}}{S_{2i,\max}} \cdot R_{2i} & D_{2i,\min} < S_{2i} < D_{2i,\max} \\ S_{2i} \cdot R_{2i} & 0 < S_{2i} < D_{2i,\max} \end{cases} \tag{8.10}$$

where X_{2i} is the niche suitability of the *i*-type environmental quality impact factor;

$D_{2i,\min}$ is the minimum of the *i*-type ecological impact factor;

$D_{2i,\max}$ is the maximum of *i*-type ecological impact factors;

W_{2i} is the weight of *i*-type ecological impact factor;

$S_{2i,\max}$ is the maximum measured value of the *i*-type ecological impact factor; and

R_{2i} is the guarantee rate of the *i*-type ecological impact factor, which can usually be measured by the planning index of the *i*-type impact factor in the current plan:

This indicator exists in the plan and is in line with national standards, $R_{2i} = 100\%$;

This indicator exists in the plan but has no specific value, $R_{2i} = 50\%$; or

There is no such indicator in the plan, $R_{2i} = 0$.

The remaining symbols have the same meaning as in Formula (8.9).

2) Processing method of data reduction

The evaluation of green village and town plan niche suitability involves many impact factors, and the data acquisition and processing methods for each impact factor differ, which makes it difficult to determine the field measure value (S_i) of each impact factor. To facilitate the evaluation, the following simple methods are proposed.

(1) Determination of the field measure value of the *i*-type impact factor S_i

When the *i*-type impact factor has different measurement points within the evaluation range, the mean value of the impact factor is calculated according to Formula (8.11).

$$S_i = \frac{\sum_{i=1}^{n} S_i'}{n} \tag{8.11}$$

where S_i' is the detection value or normalized value of the *i*-type impact factor and

N is the number of points in indicator *j* of the *i*-type impact factor in the evaluation range.

(2) Standardized processing of monitoring data

The following standardized value processing method for different detection data of various impact factors can be adopted (8.12),

$$S_i' = \frac{S_{ij} - S_{j,\min}}{S_{j,\max} - S_{j,\min}} \tag{8.12}$$

where S_i' is the normalized value of the detection value of the i-type impact factor;

S_{ij} is the mean value of the j-th indicator of the i-type impact factor at different evaluation points within the evaluation range;

$S_{j,\min}$ is the minimum measured value of indicator j of the i-type impact factor of different measurement points within the evaluation range; and

$S_{j,\max}$ is the maximum measured value of indicator j of the i-type impact factor of different measurement points within the evaluation range.

(3) Estimation of niche suitability indicator of multi-dimensional eco-environmental impact factors

Green villages and towns are multi-dimensional super-volume niches. For an ecological space with multi-dimensional impact factors, as long as one of the impact factors cannot meet the minimum demand, the niche suitability of the entire ecological space is 0. Therefore, the niche suitability assessment for green eco-village planning can be carried out using Formula (8.13).

$$X = X_1 + X_2 \tag{8.13}$$

where X is the green village and town plan niche suitability indicator;

X_1 is the environmental quality impact indicator,

$$X_1 = W_{1i} \cdot X_{1i} = [W_{11}, W_{12}, \ldots, W_{14}][X_{11}, X_{12}, \ldots, X_{14}]^T$$

X_{1i} is the niche suitability indicator of the i-type environmental quality impact factor;

W_{1i} is the evaluation weight of the i-type environmental quality impact factor;

X_2 is the ecological status impact indicator,

$$X_2 = W_{2i} \cdot X_{2i} = [W_{21} \cdot W_{22}][X_{21} \cdot X_{22}]^T$$

X_{2i} is the niche suitability indicator of the i-type ecological status impact factor; and

W_{2i} is the evaluation weight of the i-type ecological status impact factor.

Table 8.19 Evaluation grade of green village and town plan niche suitability

Grade	Very suitable	Comparatively suitable	Basically suitable	Not very suitable	Unsuitable
Score *X*	1.0–0.80	0.79–0.60	0.59–0.40	0.39–0.20	0.19–0

4. Evaluation grade of green village and town plan niche suitability

According to the analysis of the evaluation model, the green village and town plan niche suitability *X* value is between 0 and 1. According to the obtained scores, it is divided into 5 evaluation grades: very suitable, comparatively suitable, basically suitable, not very suitable, and totally unsuitable, and converted into corresponding scores.

The evaluation grade of green village and town plan niche suitability is shown in Table 8.19.

8.3.3 Implementation Evaluation Case of Green Village and Town Plan Niche Suitability

1. Data investigation of niche suitability

In this evaluation, 2012 was used as the year for the evaluation of the niche suitability of the implementation of the Yanhe Village plan.

1) Conduct field investigation

The field investigation of Yanhe Village included acquisition of the current plan preparation and implementation and monitoring of the status quo of the eco-environmental impact factors of Yanhe Village.

2) Determination of monitoring indexes

The monitoring indexes for eco-environmental quality factors in the niche suitability evaluation of the Yanhe Village plan include the following:

(1) Water environmental quality—surface water quality, groundwater, domestic wastewater, wastewater treatment facilities, discharge water and other indexes.
(2) Air environmental quality—SO_2, NO_x, PM_{10}, $PM_{2.5}$ and other indexes.
(3) Solid waste disposal status—household garbage classification collection and fixed-point clearance rate, household garbage harmless disposal rate and other indexes.
(4) Acoustic environmental quality—acoustic environment of Yanhe Village built-up area and other indexes.

(5) Eco-environment status—vegetation coverage, water area ratio and other indexes.
(6) Landscape status—architectural form & courtyard and other indexes.

2. Niche suitability indicator calculation

1) Calculation of niche suitability indicator X_{1i} of environmental quality impact factors

The environmental quality factors for niche suitability evaluation are the "category I environmental quality factors" in Table 8.18, including water environmental quality, air environmental quality, solid waste disposal and acoustic environmental quality.

The basis of niche suitability assessment is the current national standards for the quality control of environmental quality factors. The basic evaluation data are obtained by laboratory analysis through field testing or field sampling.

According to the green ecological village and town plan niche suitability model established by this evaluation, the environmental quality impact factors of niche suitability indicator X_{1i} are determined according to Formula (8.9), and the guarantee rate of environmental quality impact factor R_{1i} is the standard rate of monitoring data of different monitoring points for the same evaluation project. The average value of the quality impact factor S_{1i} is calculated according to Formula (8.11).

(1) Water environmental quality niche suitability indicator X_{11}

The key point of the of the water environmental quality niche suitability indicator calculation for Yanhe Village is to test water environmental quality parameters such as pH, ammonia, nitrogen, COD, total nitrogen, total phosphorus and BOD_5 on site and to obtain basic data.

The calculation results of the water environmental quality niche suitability indicator for Yanhe Village are shown in Table 8.20.

(2) Air environmental quality niche suitability indicator X_{12}

The basic data of the air environmental quality niche suitability indicator calculation for Yanhe Village are $PM_{2.5}$, PM_{10}, SO_2 and NO_x detected on site at different monitoring points at Yanhe Village.

The calculation results of the air environmental quality niche suitability indicator for Yanhe Village are shown in Table 8.21.

(3) Solid waste disposal status niche suitability indicator X_{13}

There are no enterprises producing industrial pollution in range of Yanhe Village, so the main parameters of the solid waste disposal environmental quality niche suitability indicator are the fixed-point removal rate and the harmless disposal rate of household garbage. The calculation results of the basic data were obtained through field investigation and testing.

Table 8.20 Calculation of water environmental quality niche suitability index for Yanhe Village

Monitoring content	Monitoring value															
	Water quality in rivers and streams S_{111}				YanTang water quality S_{112}				Well-water quality S_{113}				Wastewater treatment facility drainage water quality S_{114}			
	S_{111i} (mg/L)	R_{111i} (%)	X_{111i}	X_{111}	S_{112i} (mg/L)	R_{112i} (%)	X_{112i}	X_{112}	S_{113i} (mg/L)	R_{113i} (%)	X_{113i}	X_{113}	S_{114i} (mg/L)	R_{114i} (%)	X_{114i}	X_{114}
pH/dimensionless	7.39	100	1.00	0.71	7.56	100	1.00	0.60	7.25	100	1.00	0.83	6.55	100	1.00	0.57
NH_3-N	0.27	100	1.00		2.00	100	1.00		0.03	0.00	0.00		0.40	100	1.00	
COD	3.76	100	1.00		63.0	57.5	0.57		–	–	–		21.0	100	1.00	
TN	14.0	0.00	0.00		20.4	0.00	0.00		–	–	–		2.00	100	1.00	
TP	0.02	100	1.00		1.17	0.00	0.00		–	–	–		0.90	80.0	0.80	
BOD_5	2.45	100	0.00		7.70	100	1.00		–	–	–		12.0	20.0	0.20	
Dissolved oxygen	7.44	100	1.00		–	–	–	–	–	–	–		–	–	–	–
Nitrite	–	–	–	–	–	–	–	–	Not detected	100	1.00		–	–	–	–
Fluoride	–	–	–	–	–	–	–	–	0.06	100	1.00		–	–	–	–
Nitrate nitrogen	–	–	–	–	–	–	–	–	10.67	100	1.00		–	–	–	–
Chloride	–	–	–	–	–	–	–	–	3.69	100	1.00		–	–	–	–
The water environmental quality niche suitability index X_{11} for Yanhe Village was 0.68																

Monitor project performance standards: S_{111}, *Environmental Quality Standards for Surface Water* GB 3838-2002 (III class water); S_{112}, *Environmental Quality Standards for Surface Water* GB 3838-2002 (V class water); S_{113}, *Quality Standard for Ground Water* GB/T 14848-2017 (II class water); and S_{114}, *Discharge Standard of Pollutants for Municipal Wastewater Treatment Plant* GB 18918-2002 (Class A standard).

Table 8.21 Calculation of air environmental quality niche suitability index for Yanhe Village (S_{2j} unit: g/m³)

Monitoring content	Monitoring value							
	S_{121}	S_{122}	S_{123}	S_{124}	S_{125}	S_{126}	R_{12i} (%)	X_{12i}
$PM_{2.5}$	27	9	37	11	6	5	100	1
PM_{10}	52	18	80	22	12	9	100	1
SO_2	11.375	15.015	10.135	23.750	21.735	19.025	100	1
NO_x	73.875	65.115	50.036	40.150	42.350	30.175	100	1
The air environmental quality niche suitability index X_{12} for Yanhe Village was 1.00								

Monitor project performance standards: *Ambient Air Quality Standards* GB 3095-2012

Table 8.22 Calculation of solid waste treatment status niche suitability index for Yanhe Village

Monitoring project	Household garbage fixed point clearance rate S_{131}	Household garbage harmless disposal rate S_{132}	R_{13i} (%)	X_{13i}
Monitoring results	100%	93.75%	100	1
The solid waste treatment status niche suitability index X_{13} for Yanhe Village was 1.00				

Monitor project performance standards:S_{131}, *Technique Code for Village Rehabilitation* GB 50445-2008 andS_{132}, *Guidelines for the Construction of Beautiful Village* GB/T 32000-2015

The calculation results of the solid waste disposal status niche suitability indicator for Yanhe Village are shown in Table 8.22.

(4) Acoustic environmental quality niche suitability indicator X_{14}

The key parameter of the evaluation calculation is a sound level detection index for different monitoring points at Yanhe Village. Because the built-up area of Yanhe Village is not close to the trunk road and the sound environment is better at night than in the daytime, this evaluation mainly took the daytime monitoring data of Yanhe Village for calculation. The calculation results of the basic data were obtained through field investigation and testing.

The calculation results of the acoustic environmental quality niche suitability indicator for Yanhe Village are shown in Table 8.23.

2) Calculation of ecological status impact factors niche suitability indicator X_{2i}

(1) Calculation content

The evaluation indicator calculation of the ecological status impact niche suitability of Yanhe Village includes mainly the evaluation and calculation of the ecological status and the landscape.

Table 8.23 Calculation of sound environmental quality niche suitability index for Yanhe Village (Unit: dB (A))

Monitoring value	S_{141}	S_{142}	S_{143}	S_{144}	S_{145}	S_{146}	S_{147}	S_{148}	$R_{14i}(\%)$	X_{14i}
A sound level (daytime)	53.9	46.4	65.0	49.4	68.0	71.5	65.8	40.0	50	0.48
A sound level (nighttime)	–	–	–	–	–	–	–	–	–	–
The sound environmental quality niche suitability index for Yanhe Village X_{14} was 0.48										

Monitor project performance standards: *Environmental Quality Standard for Noise* GB 3096-2008 (Class 1)

Note As Yanhe Village is located in a rural area with shallow hills and in a hilly region with no large industrial enterprises nearby and far from major traffic arteries, the sound environmental quality within the scope of the village was better at night than during the day, so the nighttime sound environmental quality of Yanhe Village was not included in this calculation

(2) Evaluation basis

The basis of the evaluation calculation is mainly *Technical Code for Ecosystem Status Assessment* HJ 192-2015, *Standard for the Assessment of Regional Biodiversity* HJ 623-2011, *Standard for Planning of Town* GB 50188-2007, *Guidelines for the Construction of Beautiful Village* GB/T 32000-2015 and*National Ecological Civilization Construction Demonstration Village and Town Index (Trial).*

(3) Evaluation indexes

① The basic calculation indexes of ecological status niche suitability indicator X_{2i} include the following:

Eco-environmental quality impact factors X_{211}—forest and grassland areas, water and wetland areas, arable land area, construction land area and unused area; and

Biological abundance impact factor X_{212}—wild animal abundance indicator and wild vascular plant abundance indicator.

② Landscape impact factors X_{22}—town appearance impact indicator and village appearance impact indicator.

For the niche suitability evaluation of a "village", the "village and town landscape" indicator needs to be evaluated only as "village appearance". The village appearance impact indicator includes mainly architectural form and courtyards, public spaces and sketches, public green space and environmental health.

(4) Calculation method

① Calculation formula for eco-environmental quality, X_{211}

$$\frac{\begin{array}{c}\mathrm{A_{bio}}(0.35\times \text{woodland}+0.21\times \text{grassland}+0.28\times \text{water wetland}+\\ 0.11\times \text{Arable land}+0.04\times \text{Construction land}+0.01\times \text{unused})\end{array}}{\text{Evaluation area}} \tag{8.14}$$

where $\mathrm{A_{bio}}$ is the normalized coefficient of the habitat quality indicator, and the reference value is 511.2642131067. (The value is based on the *Technical Code for Ecosystem Status Assessment* HJ 192-2015).

② Calculation formula for biological abundance, X_{212}

Because the area of Yanhe Village is small, the calculation of biological abundance in this case takes Gucheng County, where Yanhe Village is located, as the statistical range of biological species. The calculated weights of "wild animal abundance indicator X_{212i}" and "wild vascular plant abundance indicator X_{212j}" were both 0.5.

$$\text{Wild animal abundance indicator}\,(X_{212i})=\frac{\text{Evaluate the number of wild animal species in the area}}{635} \tag{8.15}$$

In the formula, 635—the reference maximum of wild animal species in a county.

$$\text{Wild vascular plant abundance indicator } (X_{212j}) = \frac{\text{species of wild vascular plants in evaluation area}}{3662} \quad (8.16)$$

where 3662 is the maximum number of wild vascular plant species in a county.

(5) Scoring criteria of village appearance, X_{22}

According to the evaluation standard of "village appearance" in Table 8.17, the actual construction level of Yanhe Village is excellent in terms of architectural forms, courtyards, public spaces and sketches, public green space and environmental sanitation, so the overall evaluation level of the village appearance was "excellent".

(6) Calculation of ecological status impact factor niche suitability indicator X_{21} for Yanhe Village

According to the green village and town plan niche suitability model established by this evaluation method, each ecological status impact factor of niche suitability indicator is calculated according to Formula (8.10).

① Niche suitability indicator of eco-environmental quality of Yanhe Village X_{211}

The evaluation indexes of the eco-environmental quality niche suitability indicator for Yanhe Village X_{211} are shown in Table 8.24.

The guarantee rate of eco-environmental quality is shown in Formula (8.10).

The eco-environmental quality niche suitability indicator for Yanhe Village X_{211} can be calculated by Formula (8.14), with the required calculation parameter S_{21i} being obtained through field measurement. All basic data in this case need not be normalized, so A_{bio} is set as 1 when Formula (8.14) is used.

The total land area of Yanhe Village is 1200 hm^2.

The calculation results of the eco-environmental quality niche suitability indicator for Yanhe Village are shown in Table 8.24.

② Biological abundance niche suitability indicator for Yanhe Village X_{212}.

There are approximately 39 wild animal species in Yanhe Village. Among them, birds include pheasant, magpie, and grey magpie. Animals include squirrels, wild boars,

Table 8.24 Calculation of eco-environmental quality niche suitability indicator for Yanhe Village

Measured value	Evaluation project				
	Area of forest and grassland	Wetland area	Arable land	Construction area	Unused land area
S_{211i} (hm^2)	1106.9	18.2	39.7	13.0	24.5
R_{211i} (%)	100	50	100	50	50
X_{211i}	1106.9	9.1	39.7	6.5	12.3
X_{211}	0.52				

Table 8.25 Calculation of biological abundance niche suitability indicator for Yanhe Village

Evaluation project	Measured values and calculated values			
	S	R (%)	X	X_{212}
Wild animal abundance indicator X_{212i}	39	50	0.03	0.068
Wild vascular plant abundance indicator X_{212j}	784	50	0.105	

and weasels. Rare animals include white cranes, tragopan temminckii, pangolins, and giant salamanders [2].

There are approximately 784 wild plant species in Yanhe Village. Among them are 18 rare tree species, such as *Metasequoia*, dove tree, ginkgo and *Eucommia ulmoides* [2].

The calculation results of the biological abundance niche suitability indicator for Yanhe Village are shown in Table 8.25.

The wild animal abundance indicator for Yanhe Village was calculated using Formula (8.15), and the wild vascular plant abundance indicator for Yanhe Village was calculated using Formula (8.16).

③ Ecological status impact factor niche suitability indicator for Yanhe Village X_{21}

$$X_{21} = \frac{X_{211} + X_{212}}{2} = 0.29$$

(7) Village appearance impact factor niche suitability indicator for Yanhe Village X_{22}

According to Table 8.17 and the field survey results, the appearance of Yanhe Village was graded with a 5-point system, and the number of evaluation grades was converted into measured values according to 0–1. Then, the niche suitability indicator of the appearance of Yanhe Village was calculated according to Formula (8.10).

The calculation results of the village appearance impact factor niche suitability indicator for Yanhe Village are shown in Table 8.26.

3. Overall calculation of niche suitability

1) Calculation formula

According to the green village and town plan niche suitability evaluation model established in Sect. 8.3.2, the overall niche suitability evaluation of the Yanhe Village plan was calculated according to Formula (8.13).

Table 8.26 Calculation of village appearance impact factor niche suitability indicator for Yanhe Village

Evaluation value	Evaluation project			
	Architectural forms and courtyards	Public spaces and sketches	Public green spaces	Environmental health
Score grade	5	5	5	4
Reduced value S_{22i}	1	1	1	0.8
R_{22i} (%)	100	100	100	100
X_{22i}	1	1	1	0.8
X_{22}	0.95			

2) Calculation weight

According to *Technical Code for Ecosystem Status Assessment* HJ 192-2015, the weights of the niche suitability evaluation impact factor of the Yanhe Village plan are shown in Table 8.27.

3) Calculation results

According to Formula (8.13), the value of X, the overall calculation result of the niche suitability of the Yanhe Village plan is shown in Table 8.28.

4. Comprehensive evaluation of niche suitability

According to Table 8.19, the comprehensive evaluation conclusions of the niche suitability of the Yanhe Village plan are as follows:

(1) The niche suitability of the Yanhe Ecological Village plan is 0.69, and the evaluation grade is comparatively suitable.
(2) The sub-item evaluation results show that the biodiversity indicator score of Yanhe Village was low, at only 0.29, which is less suitable. The main reasons were that in the Yanhe Village plan, no in-depth and detailed investigations were conducted for this indicator, there was no clear evaluation basis, and there was no practical and feasible measure for "biodiversity protection of Yanhe

Table 8.27 Weights of evaluation impact factors of niche suitability of Yanhe Village plan

Impact factor *X*	Water environment	Air environment	Solid waste treatment status	Acoustic environment	Ecological status	Village appearance
Weight *W*	0.1	0.1	0.1	0.1	0.3	0.3

Table 8.28 Overall calculation of niche suitability of Yanhe Village plan

Evaluation project	Impact factor	Calculation content		
		X_i	W_i	X
Environmental quality factor	Water environmental quality	0.68	0.1	0.69
	Air environmental quality	1.00	0.1	
	Solid waste treatment status	1.00	0.1	
	Acoustic environmental quality	0.48	0.1	
Ecological status impact factor	Ecological status	0.29	0.3	
	Village appearance	0.95	0.3	

Village", which led to the low value of the plan implementation guarantee rate R_i of this index. These factors all affect the overall score.

(3) The indicator score of the acoustic environmental quality niche suitability for Yanhe Village was 0.48, which is basically suitable. The main reason for the low score was that the rapid development of rural tourism projects in Yanhe Village led to a large number of people and cars in the village during the daytime.
(4) When revising plans and formulating new development plans for Yanhe Village, more attention should paid to the protection of biodiversity within the village area.
(5) The number of tourists and cars within the village area should be reasonably controlled, and the quality of the sound environment should be strengthened.

8.4 Evaluation of Green Village and Town Water Resource Utilization and Sewage Disposal Plan

China has an overall water shortage, and water conservation is one of the core green village and town construction and planning tasks. Water resource utilization and sewage treatment are key points of green village and town plan implementation evaluation, including 3 components: evaluation of the water resource utilization plan; evaluation of the town sewage treatment plan; and evaluation of the village sewage treatment plan.

8.4.1 Evaluation Method for Green Village and Town Water Resource Utilization Plan

1. Evaluating contents of water resource utilization plans

1) Review and evaluate the basic plan data

This section examines how to perfect basic water resource utilization plan data and evaluates the authenticity and practicability of plan contents and objectives. The evaluation contents are as follows:

(1) Information about physical geography and resources includes meteorological data; topography, soil, vegetation, and mineral products; fisheries and other biological resources; marine resources; and tourism and the number, distribution, development and utilization of key cultural resources.
(2) Socioeconomic data include demographic or census data; national economic statistics or census data; and land-use and irrigated area statistics or census data.
(3) Hydrological data include survey statistics of water resource development and utilization infrastructure; measurement data of water collection and use of the water supply infrastructure; survey statistics of water supply, water consumption and related areas; survey statistics such as water conservation protection measures and implementation; and survey statistics such as water prices and water resource management measures.
(4) Water ecological environment data include statistical data of point source emissions and river inflow related to life and industry in villages; analysis of surface water quality and groundwater quality monitoring; data of the quantity, distribution and development and protection status of various nature reserves and eco-environmentally sensitive areas; and investigation and analysis of water ecology and water environment.

2) Evaluating water resources development and utilization status

(1) Evaluation of water resource quantity includes water vapor transportation, precipitation, evaporation, surface water resources, groundwater resources, total water resources, and available amount of water resources.

In the evaluation of a green ecological village and town water resource utilization plan, the key point is whether the planning includes "three waters" (precipitation, surface water, groundwater), total water resource calculation and total available amount of water resources.

The calculation method for total water resources is shown in Formula (8.17):

$$W = R_s + U_P = R + U_P - R_g \tag{8.17}$$

where W is total water resources;

R_s is surface runoff (the difference between river runoff and river base);

U_P is precipitation infiltration recharge;

R is river runoff, namely, surface water resources; and

R_g is river base.

The calculation method for total water resources in different regions refers to the current *Code of Practice for Water Resources Bulletin* GB/T 23598-2009.

(2) Evaluation of measuring available water resources includes the amount of available surface water resources, the amount of groundwater that can be exploited, and the total amount of available water resources.

The focus of the above 2 estimates is the calculation of the total amount of local available water resources. The evaluation includes whether the plan includes water supply infrastructure and water supply capacity, amount of water supply, amount of water consumption and other data and analysis, water wastage analysis and sewage discharge survey.

(3) Comprehensive evaluation of endowment conditions and supporting capacity of water resources includes the amount of water resources per capita and the amount of water per mu of cultivated land and comprehensively evaluates the endowment conditions and supporting capacity of water resources on the basis of comparative analyses of similar regions in China and abroad.

3) Evaluation of water supply infrastructure construction status

This evaluation combines water resource conditions and economic and social development patterns to analyse and evaluate the rationale of the water supply infrastructure layout, which includes engineering type (centralized or decentralized), water source selection and water supply range, number and size of decentralized water supply facilities, and operational status.

4) Evaluation of water resource development and utilization analysis

(1) Development and utilization rate of surface water resources can be expressed by the ratio of the average annual water supply (including the amount of transferred water) formed by local surface water resources in recent years to the average annual surface water resources over the years.
(2) Exploitation rate of shallow groundwater resources in plains areas can be expressed by the ratio of the average annual exploitation of shallow groundwater in the recent plains area to the total annual average groundwater resources over many years.
(3) Development and utilization rate of water resources can be expressed by the ratio of the average annual water supply (including the amount of transferred water) formed by local water resources in recent years to the average annual total water resources in local areas.

5) Evaluation of water quality status analysis

The method and content of the evaluation are carried out in accordance with the current national standard *Code of Practice for Water Resources Bulletin* GB/T 23598-2009, including the evaluation of surface water quality and groundwater quality.

(1) Evaluation indicators of surface water quality are implemented according to the current standard *Environmental Quality Standards for Surface Water* GB 3838-2002; and

(2) Evaluation indicators of groundwater quality are implemented in accordance with the current standard *Quality Standard for Ground Water* GB/T 14848-2017.

2. Evaluation of the water resource utilization plan technical route

The evaluation of the green village and town water resource utilization plan technical route is shown in Fig. 8.10.

3. Evaluation of water resource utilization plan

1) Evaluation index

According to *Water Law of the People's Republic of China, Code for Water Resources Planning* GB/T 51051-2014, *A Guide to Water Resources Assessment* SL/T 238-1999, *Code of Practice for Water Resources Bulletin* GB/T 23598-2009, evaluation criteria and scoring methods for green village and town water resources plans can

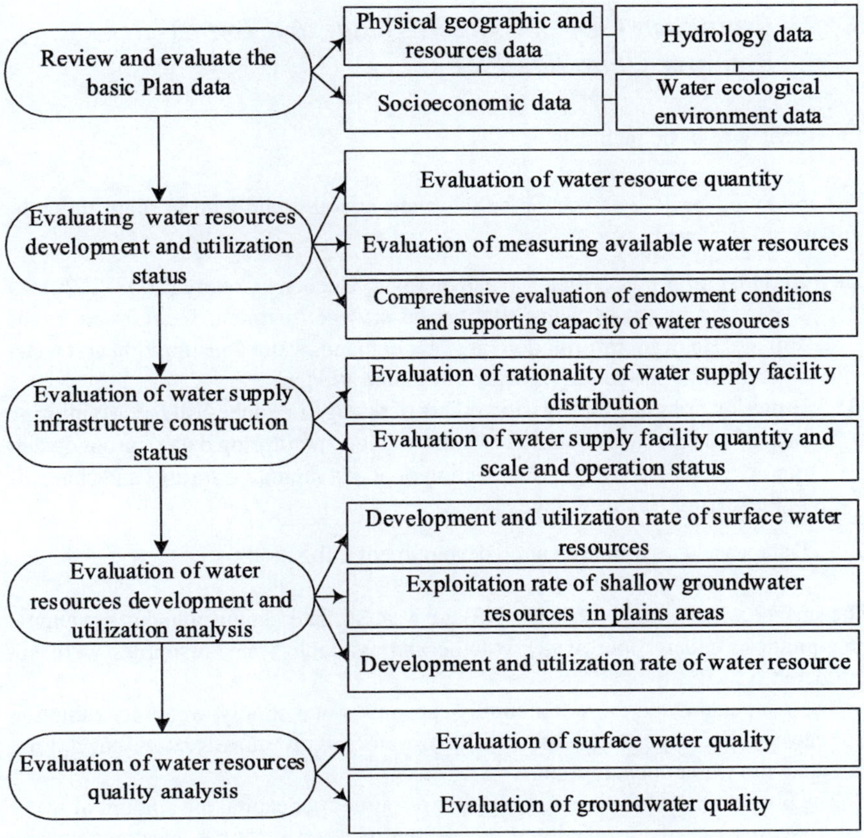

Fig. 8.10 Evaluation of green village and town water resource utilization plan technical route

be established, with evaluation indexes are divided into 4 categories that contain 13 items in total.

2) Evaluation method

Evaluation of implementation of a green village and town water resource utilization plan adopts the scoring method of 0–1. Any evaluation index that meets or reaches the current national, sectoral or industry standards or norms is scored as 1; otherwise, it is scored as 0, and the full score is 13. If the total score is 70% or more, then the plan meets the standard.

The evaluation criteria and scoring methods of green village and town water resource utilization plans are shown in Table 8.29. In the actual evaluation operation, the evaluation results can be obtained directly by entering the actual scores of each sub-item of the evaluated village or town into the corresponding scoring column in the table.

8.4.2 *Evaluation Case of a Green Village and Town Water Resource Utilization Plan*

1. Review of basic plan data

The following basic data related to the water resource utilization plan of Yanhe Village are reviewed:

(1) The Plan of Yanhe Village, Wushan Town, Gucheng County (2008–2020).
(2) Types and layout of water supply and sewage treatment facilities in Yanhe Village. Relevant information on water demand, water consumption and water conservation protection measures in Yanhe Village.
(3) Emission volume and emission points of sewage in Yanhe Village. Monitoring data of surface water quality in Yanhe Village, monitoring data of groundwater quality in Yanhe Village. Distribution of pools, mountain springs and channels in the area of Yanhe Village.

2. Data review of water resource development utilization

The surface water volume of Yanhe Village was estimated in the plan, but the annual precipitation, underground water volume and available water resources were not estimated.

The plan calculated the water supply capacity, water supply, water consumption and sewage discharge of the whole village water supply infrastructure but did not calculate the water consumption.

The plan estimated the amount of water resources per capita, the amount of water resources per mu of farmland and the degree of water resource development and utilization.

Table 8.29 Evaluation standards and methods for green village and town water resource utilization plans

Evaluation items	Evaluation indexes	Evaluation standards	Evaluation results			
			Qualified (1)	Sub-standard (0)	Score	Total score
Review of basic information in planning compilation (4 points)	(1) Information about natural geography and resources: Meteorology; landforms; soil, vegetation and mineral products; fisheries and other living resources; marine resources; tourism and key cultural heritage resources	Information coverage ratio ≥80%; "Marine resources" indicators are only for coastal areas				
	(2) Information about social and economic factors: Demography; national economic statistics; land use and irrigated areas	Information coverage ratio ≥80%				
	(3) Information about hydrology: Statistics on infrastructure for development and utilization of water resources; observation and metreing of water consumption for water supply infrastructure; water supply and water consumption statistics; water resource conservation measures and implementation statistics; analysis and statistics of water prices and water resource management	Information coverage ratio ≥80%				

(continued)

Table 8.29 (continued)

Evaluation items	Evaluation indexes	Evaluation standards	Evaluation results			
			Qualified (1)	Sub-standard (0)	Score	Total score
	(4) Information about water eco-environment: Rural household and industrial point source emissions and river discharge statistics; monitoring and analysis of surface water quality and groundwater quality; information on various nature reserves and ecologically sensitive areas; survey information on water ecology and water environment	Information coverage ratio ≥80%				
Water resource exploitation and utilization current conditions (3 points)	(5) Evaluation of water resource quantity: Precipitation; amount of surface water; amount of groundwater; amount of total water resources; estimation of water resource availability	Item estimation ≥80%				
	(6) Calculation and evaluation of water resource availability: Supply ability of water supply infrastructures; amount of water supply; amount of water usage; loss analysis in water usage; amount of sewage discharge	Item estimation ≥80%				

(continued)

Table 8.29 (continued)

Evaluation items	Evaluation indexes	Evaluation standards	Evaluation results			
			Qualified (1)	Sub-standard (0)	Score	Total score
	(7) Comprehensive evaluation of endowment conditions and supporting capacity of water resources: Water resources per capita; water resource per 1/100 km^2 of cultivated land	To reach the average of similar regions in the country				
Current conditions of water supply infrastructure (1 point)	(8) Evaluation of current conditions of water supply infrastructure: Rationale evaluation of water supply facility layout, amount, scale and working conditions	Meet requirements of current *Design Code for Rural Water Supply Engineering* SL 687-2014				
Water resource degree of development and utilization (3 points)	(9) Exploitation and utilization ratio of surface water: Ratio of the average annual water supply formed by the recent local surface water resources to the average annual surface water resources over the years	Development and utilization of water resources should not be higher than local water resources per capita				

(continued)

Table 8.29 (continued)

Evaluation items	Evaluation indexes	Evaluation standards	Evaluation results			
			Qualified (1)	Sub-standard (0)	Score	Total score
	(10) Amount of shallow groundwater exploitation in plains areas: Ratio of the average annual exploitation of shallow groundwater in plains area to total annual average groundwater resources in recent years					
	(11) Exploitation and utilization ratio of water resources: Ratio of the average annual water supply (including the amount of transferred water) formed by local water resources in recent years to the average annual total water resources in the local area					
Analysis and evaluation of water resource quality status (2 points)	(12) Evaluation of surface water quality: Rural drinking water supply from surface water must meet Class III; water for agriculture and general landscaping must meet Class V	Every test result should meet the requirements of GB 3838-2002				

(continued)

Table 8.29 (continued)

Evaluation items	Evaluation indexes	Evaluation standards	Evaluation results			
			Qualified (1)	Sub-standard (0)	Score	Total score
	(13) Evaluation of groundwater quality: Centralized drinking water sources and water for industry and agriculture must meet Class III	Every test result should meet the requirements of current standard GB/T 14848-2017				

3. Analysis and evaluation of water quality status

The environmental monitoring station of Gucheng County Environmental Protection Bureau was entrusted to carry out several types of monitoring of surface water and groundwater quality in Yanhe Village. The test results showed that the surface water quality and groundwater quality in Yanhe Village met the requirements of *Environmental Quality Standards for Surface Water* GB 3838-2002 and *Quality Standard for Ground Water* GB/T 14848-2017, respectively.

4. Effect evaluation of water resource utilization plan

According to the above investigation and analysis, Table 8.29 was used to evaluate the implementation status of the water resource utilization plan of Yanhe Village item by item. According to the evaluation criteria, Yanhe Village was divided into 10 points, and the evaluation results were qualified. See Table 8.30.

8.4.3 Evaluation Methods for Green Village and Town Sewage Disposal Plan

1. Evaluating the content of a green village and town sewage disposal plan

The evaluation contents of a green village and town sewage disposal plan can be divided into 2 categories: green town evaluation and green village evaluation.

1) Evaluating green town sewage disposal plan

The green town plan sewage disposal plan must meet the “drainage plan” requirements in the current national *Standard for Planning of Town* GB 50188-2007.

(1) Amount of drainage. This includes the amount of domestic sewage and rainwater. The estimation of domestic sewage and rainwater must meet the requirements in the current national *Standard for Planning of Town* GB 50188-2007.
(2) Drainage system. It is advised to choose separate systems, although a combined system can be chosen if conditions for separate systems are not available. Sewage should be pre-treated in a septic tank or domestic sewage digester before being discharged into the pipe network.
(3) Drainage arrangement. Sewage discharge should meet the requirements in the current national standard *Integrated Wastewater Discharge Standard* GB 8978-2002, and sewage used for farmland irrigation should meet the requirements in the current national standard *Standards for Irrigation Water Quality* GB 5084-2005. Rainwater should be drained by making full use of surface runoff and canals, and sewage should be discharged through pipes or culverts. Pipes and canals for rainwater and sewage should be designed to take advantage of gravity flow.

Table 8.30 Evaluation results of water resource utilization plan of Yanhe Village

Evaluation items	Evaluation indexes	Evaluation results			
		Qualified (1)	Sub-standard (0)	Score	Total score
Review of basic information in planning compilation (4 points)	(1) Information about natural geography and resources: Meteorology; landforms; soil, vegetation and mineral products; fisheries and other living resources; marine resources; tourism and key cultural heritage resources	1		3	10
	(2) Information about social and economic factors: Demography; national economic statistics; land use and irrigated areas	1			
	(3) Information about hydrology: Statistics on infrastructure for development and utilization of water resources; observation and metreing of water consumption for water supply infrastructure; water supply and water consumption statistics; water resource conservation measures and implementation statistics; analysis and statistics of water prices and water resource management		0		
	(4) Information about water eco-environment: Rural household and industrial point source emissions and river discharge statistics; monitoring and analysis of surface water quality and groundwater quality; information on various nature reserves and ecologically sensitive areas; survey information on water ecology and water environment	1			

(continued)

Table 8.30 (continued)

Evaluation items	Evaluation indexes	Evaluation results			
		Qualified (1)	Sub-standard (0)	Score	Total score
Water resource exploitation and utilization current conditions (3 points)	(5) Evaluation of water resource quantity: Precipitation; amount of surface water; amount of groundwater; amount of total water resources; estimation of water resource availability		0	1	
	(6) Calculation and evaluation of water resource availability: Supply ability of water supply infrastructures; amount of water supply; amount of water usage; loss analysis in water usage; amount of sewage discharge		0		
	(7) Comprehensive evaluation of endowment conditions and supporting capacity of water resources: Water resources per capita; water resource per 1/100 km^2 of cultivated land	1			
Current conditions of water supply infrastructure (1 point)	(8) Evaluation of current conditions of water supply infrastructure: Rationale evaluation of water supply facility layout, amount, scale and working conditions	1		1	
Water resource degree of development and utilization (3 points)	(9) Exploitation and utilization ratio of surface water: Ratio of the average annual water supply formed by the recent local surface water resources to the average annual surface water resources over the years	1		3	

(continued)

Table 8.30 (continued)

Evaluation items	Evaluation indexes	Evaluation results			
		Qualified (1)	Sub-standard (0)	Score	Total score
	(10) Amount of shallow groundwater exploitation in plains areas: Ratio of the average annual exploitation of shallow groundwater in plains area to total annual average groundwater resources in recent years	1			
	(11) Exploitation and utilization ratio of water resources: Ratio of the average annual water supply (including the amount of transferred water) formed by local water resources in recent years to the average annual total water resources in the local area	1			
Analysis and evaluation of water resource quality status (2 points)	(12) Evaluation of surface water quality: Rural drinking water supply from surface water must meet Class III; water for agriculture and general landscaping must meet Class V	1		2	
	(13) Evaluation of groundwater quality: Centralized drinking water sources and water for industry and agriculture must meet Class III	1			

Note Detailed evaluation standards are shown in Table 8.29

(4) Sewage disposal facilities. When using a centralized disposal system for sewage, sewage disposal facilities should be located downstream from the town near receiving water or farmland irrigation areas. Discharge standards for sewage disposal facilities should meet the requirements in the current national standard *Discharge Standard of Pollutants for Municipal Wastewater Treatment Plant* GB 18918-2002. Reclaimed water utilization should meet the requirements in the current national standards *Design Code for Reclaimed Water System of Building* GB 50336-2018 and *Design Code for Sewage Recycling Project* GB 50335-2016.

2) Evaluating green village sewage disposal plan

(1) Choose appropriate sewage control technology. The main technologies for domestic sewage management in a village are source control, household biogas, and decentralized sewage disposal with low energy consumption. Villages in developed areas can build centralized sewage disposal facilities based on the amount of sewage.
(2) Collection and discharge of rainwater and sewage. Adopt a combined system of rainwater and sewage treatment and then choose side ditches and natural canals for delivery. Treated rainwater should be considered for utilization as a resource, and local natural conditions should be considered in rainwater discharge. First, rainwater enters the local water cycle through pits, depressions, farmland and other methods, and it should not be directly discharged into water in function divisions defined by the state. On-site purification and utilization of rainwater should be encouraged, and it should be disposed of through the natural purification of plants, green spaces or soil.
(3) Sewage disposal facilities. According to the distribution of the village and farms, the sewage disposal system can be constructed and maintained on a regular basis through centralized or decentralized processing or a combination of the two. In a centralized processing system, sewage disposal plants should be located downstream of the town near receiving water or farmland irrigation areas. Discharge standards for sewage disposal facilities should meet the requirements in the current national standard *Discharge Standard of Pollutants for Municipal Wastewater Treatment Plant* GB 18918-2002.

3) Evaluating green village and town sewage disposal plan technical route

It is generally appropriate to evaluate township and village sewage treatment plans according to the current relevant national standards and norms.

The evaluation of the green village and town sewage disposal plan technical route is shown in Fig. 8.11.

2. Evaluation method for green town sewage disposal plan

1) Evaluation index

According to *Standard for Planning of Town* GB 50188-2007, *Discharge Standard of Pollutants for Municipal Wastewater Treatment Plant* GB 18918-2002, *Design Code for Reclaimed Water System of Building* GB 50336-2018 and *Design Code for*

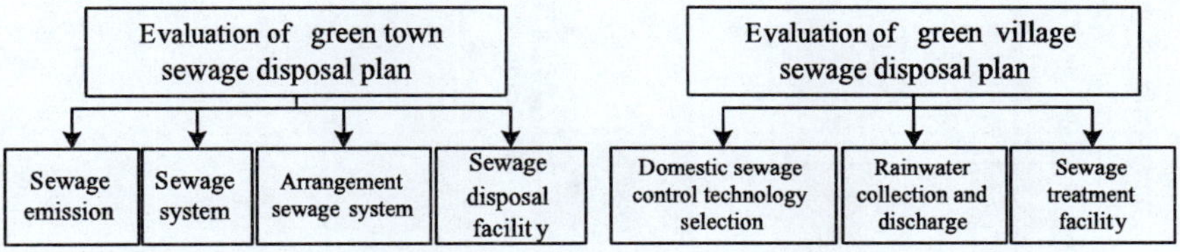

Fig. 8.11 Evaluation of green village and town sewage disposal plan technical route

Sewage Recycling Project GB 50335-2016, the evaluation of a green town sewage disposal plan is divided into 4 categories of indicators containing a total of 8 items.

2) Evaluation method

The evaluation of a green town sewage disposal plan adopts a 0–1 scoring method. An evaluation index that meets or surpasses the current national/department/industry standards or norms is scored as 1. Those that do not meet the relevant standards or specifications are scored as 0. A full mark is 8. If the total score is 70% or more, then the plan meets the qualified standards.

The evaluation standards and methods of a green town sewage disposal plan are shown in Table 8.31.

3. Evaluation method for green village sewage disposal plan

1) Evaluation index

According to *Technical Specifications of Domestic Pollution Control for Town and Village* HJ 574-2010, *Guidelines for the Construction of Beautiful Village* GB/T 32000-2015, *National Ecological Civilization Construction Demonstration Village and Town Indicators (Trial)*, *Standards for the Establishment of National Eco-villages (Trial)*, and *Discharge Standard of Pollutants for Municipal Wastewater Treatment Plant* GB 18918-2002, the evaluation of a green village sewage disposal plan is divided into 3 categories of indicators containing a total of 4 items.

2) Evaluation method

The evaluation of a green village sewage disposal plan adopts a 0–1 scoring method. An evaluation index that meets or exceeds the current national/department/industry standards or norms is scored as 1. Those that do not meet the relevant standards or specifications are scored as 0. A full mark is 4. If the total score reaches 70% or more, then the plan meets the standard.

The evaluation standards and methods of a green village sewage disposal plan are shown in Table 8.32.

Table 8.31 Evaluation standards and methods of a green town sewage disposal plan

Evaluation items	Evaluation indexes	Evaluation standards	Evaluation results			
			Qualified (1)	Sub-standard (0)	Score	Total score
Displacement estimation (1 point)	(1) Amount of sewage and rain	Estimation of sewage and amount of rainwater should meet requirements of GB 50188-2007				
Drainage system (1 point)	(2) System of diversion or combination of rainwater and sewage; sewage disposal before discharging into pipe network	Meet requirements of GB 50188-2007				
Sewage system deployment (3 points)	(3) Quality of sewage emissions	Meets requirements of GB 8978-2002				
	(4) Quality of sewage used for farmland	Meets requirements of GB 5084-2005				

(continued)

Table 8.31 (continued)

Evaluation items	Evaluation indexes	Evaluation standards	Evaluation results			
			Qualified (1)	Sub-standard (0)	Score	Total score
	(5) Discharge method of rainwater	Meets requirements of GB 50188-2007				
Sewage disposal facilities (3 points)	(6) Location of sewage disposal plant	Location of sewage disposal plants meets requirements of GB 50188-2007				
	(7) Quality of emissions from sewage disposal facilities	Meets requirements of GB 18918-2002				
	(8) Sewage recycling project	Usage of reclaimed water meets requirements of GB 50335-2016 and GB 50336-2018				

Table 8.32 Evaluation standards and methods of green village sewage disposal plan

Evaluation items	Evaluation indexes	Evaluation standards	Evaluation results			
			Qualified (1)	Sub-standard (0)	Score	Total score
Select proper control methods of domestic sewage (1 point)	(1) Source control, domestic biogas technology, low energy consumption and decentralized sewage disposal technology; establish centralized sewage disposal facilities	Choose at least one sewage control technology				
Collection and emissions of rainwater and sewage (2 points)	(2) The system of diversion or combination of rainwater and sewage; sewage disposal before discharging into the pipe network	Meet requirements of HJ 574-2010				
	(3) Disposal rate of domestic sewage	Meet *Construction Indicators of National Ecological Civilization Demonstration Villages and Towns (Trial)* Disposal rate of domestic sewage ≥90%				
Evaluation of sewage disposal facilities (1 point)	(4) Quality of emissions from sewage disposal facilities	Meet requirements of GB 18918-2002				

8.4.4 *Evaluation Case of a Green Village and Town Sewage Disposal Plan*

1. Selecting domestic sewage control technology

The development plan clearly shows that the domestic sewage control technologies of Yanhe Village include household biogas technology, constructed wetland technology and minor "integrated" centralized sewage treatment facilities.

2. Rainwater collection and emissions evaluation

According to the development plan, a separate system is used for the current collection and discharge of rainwater from Yanhe Village. Domestic sewage is used to fertilize fields after treatment by decentralized biogas digesters, or it can reach the standard of discharge after treatment by centralized sewage treatment facilities. The sewage treatment rate of Yanhe Village is no less than 90%.

3. Evaluation of sewage disposal plan

The inspection was carried out by the environmental monitoring station of Gucheng County Environmental Protection Bureau. The monitoring results showed that the discharged water quality of the sewage treatment facilities in Yanhe Village met the requirements of *Discharge Standard of Pollutants for Municipal Wastewater Treatment Plant* GB 18918-2002.

The evaluation standards and methods for the Yanhe Village green village sewage disposal plan are shown in Table 8.32. The method was used to evaluate the implementation status of sewage treatment planning in Yanhe Village item by item, and the total score was 4. The evaluation results indicate that the plan met the standards. See Table 8.33.

8.5 Evaluation of Green Village and Town Waste Disposal Plan

8.5.1 *Evaluation Content of Green Village and Town Waste Disposal Plan*

Evaluation of a green village and town waste disposal plan includes the following 7 elements:

(1) Analysis of domestic waste daily output in built-up areas of villages or seats of government of towns;
(2) Analysis of the basic composition of domestic waste;
(3) Number of waste containers (dustbins);

Table 8.33 Evaluation results of sewage disposal plans in Yanhe Village

Evaluation items	Evaluation indexes	Evaluation results			
		Qualified (1)	Sub-standard (0)	Score	Total score
Select proper control methods of domestic sewage (1 point)	(1) Source control, domestic biogas technology, low energy consumption and decentralized sewage disposal technology; establish centralized sewage disposal facilities	1		1	4
Collection and emissions of rainwater and sewage (2 points)	(2) The system of diversion or combination of rainwater and sewage; sewage disposal before discharging into the pipe network	1		2	
	(3) Disposal rate of domestic sewage	1			
Evaluation of sewage disposal facilities (1 point)	(4) Quality of emissions from sewage disposal facilities	1		1	

Note Detailed evaluation standards are shown in Table 8.32

(4) Classification of domestic waste;
(5) Model of waste collection and transportation;
(6) Waste collection sites and transfer stations; and
(7) Disposal of domestic waste.

The evaluation of a green village and town household waste disposal plan mainly includes basic plan data compilation, status analysis of waste disposal, evaluation of household garbage classification, garbage collection and transportation mode, and household garbage treatment.

8.5.2 Evaluation Method for Green Village and Town Waste Disposal Plan

1. Evaluation Index

According to the evaluation content of a waste disposal plan, the evaluation criteria and scoring methods of a green village and town waste disposal plan can be established. The evaluation indexes are divided into 3 categories containing a total of 9 items, as shown in Table 8.34. The third indicator, "household garbage disposal", refers to various treatments or disposal methods as "parallel projects"; that is, any or several combinations of garbage disposal methods are advisable.

2. Evaluation method

The evaluation of a green village and town waste disposal plan adopts a 0–1 scoring method. The evaluation index that meets or exceeds the current national/department/industry standards or norms is scored as 1. Those that do not meet the relevant standards or specifications are score as 0. A full mark is 9. If the total score reaches 70% or more, then the plan meets the standard.

The evaluation standards and methods of green village and town waste disposal plan are shown in Table 8.34.

8.5.3 Evaluation Case of Green Village and Town Waste Disposal Plan

The sorting, collection and treatment of domestic garbage in Yanhe Village began in 2005. In the Yanhe Village plan, the author collected and sorted the data of garbage treatment for the past 10 years and analysed the basic components of household garbage.

According to *Technical Specifications of Domestic Pollution Control for Town and Village* HJ 574-2010, domestic waste is classified simply as organic waste and inorganic waste in the Yanhe Village plan. The garbage collection and transportation in Yanhe Village follow the mode of "house classification, village collection, town transfer, county and city disposal". A classified garbage collection point was built in the village and called a "resource recycling centre". Forty-eight sorting bins were set up in Yanhe Village, and full-time health workers were assigned to collect garbage. Because Yanhe Village is a rural area, constructing domestic garbage disposal equipment and facilities such as landfills was not suitable. Therefore, organic waste is put

Table 8.34 Evaluation standards and methods of green village and town waste disposal plan

Evaluation items	Evaluation indexes	Evaluation standards	Evaluation results			
			Qualified (1)	Sub-standard (0)	Score	Total score
Planning and preparation of basic data and analysis of the status quo of garbage disposal (3 points)	(1) Daily production of domestic garbage in village built-up areas or township	Calculate according to current standard GB 50188-2007 or on-site investigation and forecast				
	(2) Analysis of basic components of domestic garbage	Conduct site investigation and collect complete data				
	(3) Number of garbage collection containers: Service radius of each collection container is 50–80 m	Meet requirements of GB 50188-2007				
Domestic waste classification and garbage collection and transportation methods (3 points)	(4) Classification of domestic waste: Towns can refer to the current industry standard *Urban Domestic Waste Classification and Evaluation Criteria* CJJ/T 102-2004 evaluation; villages are evaluated according to the current norm *Technical Specifications for Rural Living Pollution Control* HJ 574-2010	Meet requirements of CJJ/T 102-2004 or HJ 574-2010				

(continued)

Table 8.34 (continued)

Evaluation items	Evaluation indexes	Evaluation standards	Evaluation results			
			Qualified (1)	Sub-standard (0)	Score	Total score
	(5) Garbage collection mode: According to "house classification, village collection, town transfer, county and city disposal mode" evaluation	Meet requirements of HJ 574-2010				
	(6) Garbage collection points and transfer stations: Planning whether to set up domestic garbage collection points or transfer stations	Villages set up garbage collection points; towns set up garbage transfer stations, and the service radius meets the current HJ 574-2010 requirements				

(continued)

Table 8.34 (continued)

Evaluation items	Evaluation indexes	Evaluation standards	Evaluation results			
			Qualified (1)	Sub-standard (0)	Score	Total score
Domestic waste disposal (3 points)	(7) Domestic garbage disposal or disposal methods: Organic waste and farm animal and poultry manure enter the household biogas tank anaerobic disposal; landfill disposal; composting disposal; incineration disposal, etc.	The anaerobic treatment of biogas digesters meets the requirements of HJ 574-2010; landfill disposal complies with the current GB 16889-2008 and relevant standards; composting is treated according to the current CJJ 52-2014 specification; incineration disposal meets the current GB 18485-2014 standard; farm livestock and poultry manure treatment meet GB 18055-2012 standards				
	(8) Domestic waste resource utilization	Manufacturing biogas or compost or recycling resources				
	(9) Harmless disposal of domestic garbage: Such as sanitary landfill, incineration and resource utilization	Harmless disposal rate: village requirements are 100%, township requirements are ≥95%				

into biogas tanks for anaerobic treatment, and biogas residue and slurry are used to fertilize fields. Recyclable inorganic waste is recycled by the residents themselves or recycled by full-time health workers. Harmful garbage, such as old batteries and pesticide bottles, is sent to the town for unified treatment by full-time health workers.

The harmless treatment rate of domestic garbage in Yanhe Village is 100%.

The evaluation standards and methods in Table 8.34 were used item by item to evaluate the implementation status of the Yanhe Village domestic garbage treatment plan. The total score was 9, and the evaluation results indicated that the plan met the standards. See Table 8.35.

8.6 Evaluation of Green Village and Town Clean Energy Utilization Plan

Energy conservation and the adoption of clean energy are the core of the evaluation of a green village and town plan. The evaluation contents of a green village and town clean energy utilization plan include a survey of village and town clean energy utilization status, village and town living energy accounting, comprehensive assessment of village and town clean energy utilization status, current plan and implementation comparison analysis.

8.6.1 Evaluation Accounting Basis of Green Village and Town Clean Energy Utilization Plan

1. Types of village and town clean energy utilization

The evaluation object of a green village and town clean energy utilization plan is the current situation of village and town clean energy utilization, which means according to the green village and town plan, evaluating the status of using renewable energy in accordance with local conditions such as solar power, biogas, wind power, micro-hydropower and biomass energy or using non-renewable energy that causes very little pollution in production and consumption, such as natural gas, liquefied petroleum gas and other clean energy sources.

The main types of village and town clean energy utilization are shown in Table 8.36.

2. Accounting method for village and town domestic energy consumption

The measurement unit of the accounting method for village domestic energy consumption is shown in Table 8.37.

The standard coal accounting method and equivalent electricity accounting method are used to calculate village and town domestic energy consumption. The

Table 8.35 Evaluation results of the waste disposal plan in Yanhe Village

Evaluation items	Evaluation indexes	Evaluation results			
		Qualified (1)	Sub-standard (0)	Score	Total score
Planning and preparation of basic data and analysis of the status quo of garbage disposal (3 points)	(1) Daily production of domestic garbage in village built-up areas or township	1		3	9
	(2) Analysis of basic components of domestic garbage	1			
	(3) Number of garbage collection containers	1			
Domestic waste classification and garbage collection and transportation methods (3 points)	(4) Classification of domestic waste	1		3	
	(5) Garbage collection mode	1			
	(6) Garbage collection points and transfer stations	1			
Domestic waste disposal (3 points)	(7) Domestic garbage disposal or disposal methods	1		3	
	(8) Domestic waste resource utilization	1			
	(9) Harmless disposal of domestic garbage	1			

Note Detailed evaluation standards are shown in Table 8.34

Table 8.36 Main types of village and town clean energy utilization

Clean energy category	Use patterns of clean energy
Electric energy	Traditional power generation, wind power, hydropower
Gas	Liquefied petroleum gas, natural gas
Biomass energy	Biogas, straw gasification centralized gas supply, straw solidification moulding, straw carbonization
Solar energy	Solar water heater, solar cooker, solar house, solar photovoltaic power generation
Small power utilization	Small wind power, micro-hydropower
Other renewable energy	Geothermal energy, ocean energy

standard coal accounting method is based on the primary energy combustion calorific value as the conversion benchmark, which reflects mainly the quantity difference in village and town energy use, while the equivalent electricity accounting method takes the conversion capacity of secondary energy as the conversion benchmark and focuses on the quality difference in village and town energy use. Different accounting methods can be chosen according to different demands.

1) Standard coal accounting method

Fuels such as coal, gas, oil and solid biomass fuel can be accounted for as standard coal by Formula (8.18).

$$Q = \frac{h}{h_{ce}} \times N \tag{8.18}$$

where Q is the amount in standard coal of energy being accounted for, kgce;

h is the low calorific value of energy being accounted for, kJ/kg or kJ/Nm^3;

h_{ce} is the low calorific value of standard coal, 29270 kJ/kgce; and

N is the total consumption amount of energy being accounted for, kg or Nm^3.

2) Valuing low calorific values of different energies

Low calorific values of common energies can be found in Table 8.38. If the low calorific value of an energy accounted for is not available, then testing should be performed according to *Determination of Calorific Value of Coal* GB/T 213-2008, *Testing Methods for Heat Value of Biomass Fuels* GB 5186-1985 or other current national standards.

The low calorific value of common energies in accounting for village and town domestic energy consumption is shown in Table 8.38.

3) Formula (8.19) should be used for standard coal accounting for electricity.

$$Q = \frac{k_e}{1000} \times N \tag{8.19}$$

Table 8.37 Measurement units of various energy standards

Name	Electric energy	Firewood	Straw	Coal burning	Gas	Fuel	Charcoal	Biogas	Livestock	Hot water from solar energy
Unit	kWh	kg	kg	kg	Nm^3	kg	kg	Nm^3	kg	kJ

Table 8.38 Low calorific value of common energies

Energy name	Average low calorific value	Standard coal coefficient
Raw coal	20908 kJ/kg	0.7143 kgce/kg
Washed coal	26344 kJ/kg	0.900 kgce/kg
Coke	28435 kJ/kg	0.9714 kgce/kg
Fuel oil	41816 kJ/kg	1.4286 kgce/kg
Gasoline	43070 kJ/kg	1.4714 kgce/kg
Kerosene	43070 kJ/kg	1.4714 kgce/kg
Diesel	42652 kJ/kg	1.4571 kgce/kg
Liquefied petroleum gas	50179 kJ/kg	1.7143 kgce/kg
Natural gas	38931 kJ/m^3	1.3300 kgce/m^3
Water gas	10454 kJ/m^3	0.3571 kgce/m^3
Human faeces	18817 kJ/kg	0.643 kgce/kg
Horse manure	13799 kJ/kg	0.471 kgce/kg
Pig manure	12545 kJ/kg	0.429 kgce/kg
Sheep, donkey, horse, mule manure	15472 kJ/kg	0.529 kgce/kg
Chicken manure	18817 kJ/kg	0.643 kgce/kg
Soybean rod, cotton rod	15890 kJ/kg	0.543 kgce/kg
Rice straw	12545 kJ/kg	0.429 kgce/kg
Straw	14635 kJ/kg	0.500 kgce/kg
Corn pole	15472 kJ/kg	0.529 kgce/kg
Weed	13799 kJ/kg	0.471 kgce/kg
Leaves	14635 kJ/kg	0.500 kgce/kg
Firewood	16726 kJ/kg	0.571 kgce/kg
Biogas	20908 kJ/m^3	0.714 kgce/m^3

where Q is the amount in standard coal of energy accounted for, kgce;

k_e is the conversion coefficient, gce/kWh, calculated according to the standard coal consumption of thermal power generation (this evaluation method uses data from the "China Energy Statistics Yearbook (2014)", 122.9 gce/kWh); and

N is the total power usage, kWh.

4) Formula (8.20) should be used for standard coal accounting for solar hot water.

$$Q = \frac{N_{hs}}{\eta \cdot h_{ce}} \tag{8.20}$$

where N_{hs} is heat consumption in solar hot water, kJ;

η is average thermal efficiency of a small boiler, value = 0.65; and

h_{ce} is low calorific value of standard coal, 29270 kJ/kgce.

3. Accounting Method for Equivalent Electricity

Formula (8.21) is used to calculate equivalent electricity.

$$W = \eta \times E \times N \tag{8.21}$$

where W is equivalent electricity accounting of the energy accounted for, kWh;

η is the maximum conversion efficiency of the energy accounted for converted into electrical energy, that is, the equivalent electricity conversion efficiency, %;

E is the energy contained in the accounted-for energy unit usage, kWh/kg or kWh; and

N is the total amount of energy used for accounting, kg or Nm^3.

Referring to *Standard for Reporting and Verifying Domestic Energy Consumption of Rural Dwellings* CECS 309-2012, the conversion coefficient of the equivalent electricity accounting method corresponding to common energy is shown in Table 8.39.

4. Results of energy accounting

The results of village and town domestic energy accounting include the total amount of domestic energy consumption, the total amount of clean energy consumption, and the rate of clean energy utilization. The total amount of domestic energy consumption and the total amount of clean energy consumption are calculated according to Formula (8.22).

Table 8.39 Conversion coefficient of the equivalent electricity accounting method corresponding to common energy

Terminal energy	Equivalent electric conversion rate %	Total energy	Equivalent electro-conversion factor
Electricity	100	1.000 kWh/kWh	1.000 kWh/kWh
Natural gas	66.1	10.814 kWh/Nm^3	7.148 kWh/Nm^3
Gasoline, kerosene	66.1	11.967 kWh/kg	7.908 kWh/kg
Diesel	66.1	11.848 kWh/kg	7.831 kWh/kg
Raw coal	45.4	5.814 kWh/kg	2.640 kWh/kg
Standard coal	45.4	8.131 kWh/kg	3.691 kWh/kg
Municipal hot water	23.6	2.778×10^{-4} kWh/kJ	6.556×10^{-5} kWh/kJ
Municipal steam	34.8	2.778×10^{-4} kWh/kJ	9.667×10^{-5} kWh/kJ
Biomass energy	35.0	Depend on the type	–

$$\varphi_i = \frac{Q_i}{Q} \times 100\% \tag{8.22}$$

where φ_i is the energy consumption ratio of the *i*-type;

Q is the total amount of domestic energy consumption or clean energy consumption, kgce or kWh; and

Q_i is the amount of energy consumption of *i*-type, kgce or kWh.

The amount of domestic energy consumption per unit of building area and the amount of clean energy consumption per unit of building area are calculated according to Formula (8.23).

$$q = \frac{Q}{A} \tag{8.23}$$

where q is the amount of domestic energy consumption per unit of building area and the amount of clean energy consumption per unit of building area;

Q is the total amount of domestic energy consumption or clean energy consumption, kgce or kWh; and

A is the residential building area, m^2.

8.6.2 Main Contents of Green Village and Town Clean Energy Utilization Plan

The evaluation of green village and town clean energy utilization contains 2 parts: the survey on the performance of the clean energy utilization plan and the evaluation of clean energy utilization status.

1. Performance survey of village and town clean energy utilization plan

(1) Obtain the evaluated village or town clean energy plan through data collection;
(2) Through data collection, questionnaire survey, interviews, typical household surveys and other methods, obtain the status of implementing the clean energy plan in the evaluated village or town; and
(3) Compare the current implementation situation with the plan, and analyse the implementation of the clean energy plan in the evaluated village or town.

2. Evaluation index of village and town clean energy utilization plan

The green evaluation index of clean energy utilization status is divided into 3 parts: the control index, general index and preference index.

1) Evaluation index of control item

(1) Village and town annual income per capita

The village and town clean energy level is closely related to the economic level of the residents. Farmers with a high economic level tend to use clean energy and commodity energy. Their energy-consuming equipment is more efficient and more conducive to energy conservation. However, the energy-consuming efficiency of farmers with a low economic levels is low. Only when the village or town economy reaches a certain level do the evaluation results have guiding significance for the energy plan.

This evaluation takes the per capita annual income in the construction index of the national ecological county in 2007 as the recommended standard for economic evaluation, with the county annual income in economically developed areas being ≥6000 yuan/person and the county annual income in less developed areas being ≥4,500 yuan/person.

(2) Satisfaction in human settlements

A good living environment not only is beneficial to the health of farmers but also has great significance for improving the quality of farmers' lives and promoting the harmonious development of energy resources, the environment and the economy. According to the findings of the author's research group, the green village and town living environment should satisfy most residents. Therefore, a satisfaction rate of more than 75% in human settlements is set as the control index of green evaluation.

(3) Penetration of clean energy

According to *Guidelines for the Construction of Beautiful Village* GB/T 32000-2015, clean energy use by 70% of farmers is the evaluation standard.

(4) Village and town grid coverage

The village and town grid coverage reflects the social attributes of energy utilization in that it reflects the modernization of the lifestyle of village residents. Only by guaranteeing a certain grid coverage rate can we increase the proportion of electricity and reduce the rate of extensive direct burning of straw and firewood for cooking and heating. The survey found that electric energy has become commonly used in villages and towns. Considering the feasibility of setting the control index, we chose 80% as the control.

2) Evaluation index of general items

The evaluation index of general items of a green village and town clean energy utilization plan includes energy conservation, environmental protection, social and technical support and others, with 4 primary indicators and 16 secondary indicators.

The evaluation index of general items of a green village and town clean energy utilization plan is shown in Table 8.40.

3) Evaluation index of preference items

Clean energy utilization in rural areas is affected by differences in natural conditions, resources and the environment and social and economic activities in the region. There

Table 8.40 Evaluation index of general items of a green village and town clean energy utilization plan

Primary indexes	Secondary indexes	Connotations of indexes
Energy savings B_1	Proportion of peasants using energy-saving heating equipment C_1	$\frac{\text{Number of households using energy-saving heating equipment}}{\text{Total number of households in village or town}} \times 100\%$
	Proportion of peasants using energy-saving cooking equipment C_2	Same as above
	Proportion of peasants using energy-saving appliances C_3	Same as above
	Proportion of peasants using energy-saving lighting C_4	Same as above
	Proportion of energy-saving buildings C_5	$\frac{\text{Energy-saving building area in village or town}}{\text{Total construction area of village or town}} \times 100\%$
Environmental protection B_2	Indoor environmental satisfaction C_6	$\frac{\text{Number of people who are satisfied with the indoor environment}}{\text{Total number in village or town surveys}} \times 100\%$
	Carbon dioxide emissions reduction rate C_7	$\frac{\text{Reduced carbon dioxide emissions by clean energy}}{\text{Alternative traditional energy carbon dioxide emissions}} \times 100\%$
	Sulfur dioxide emissions reduction rate C_8	Same as above
	Nitrogen oxide emissions reduction rate C_9	Same as above
	TSP emissions reduction rate C_{10}	Same as above
Sociality B_3	Clean energy utilization C_{11}	$\frac{\text{Clean energy use}}{\text{Total energy used in village or town}} \times 100\%$
	Biomass energy-efficient utilization C_{12}	$\frac{\text{Biomass energy-efficient use}}{\text{Total use of biomass energy}} \times 100\%$

(continued)

Table 8.40 (continued)

Primary indexes	Secondary indexes	Connotations of indexes
	Clean energy demand satisfaction C_{13}	$\frac{\text{Number of months that farmers can use clean energy}}{12} \times 100\%$
Technical support B_4	Solar energy utilization ratio of peasants C_{14}	$\frac{\text{Number of households using solar energy}}{\text{Total number of households in village or town}} \times 100\%$
	Biogas utilization ratio of peasants C_{15}	$\frac{\text{Number of households using biogas}}{\text{Total number of households in village or town}} \times 100\%$
	Clean energy use ratio of peasants C_{16}	$\frac{\text{Number of households using clean energy}}{\text{Total number of households in village or town}} \times 100\%$

are obvious regional differences. This evaluation sets the preference index from the perspective of whether village and town clean energy use should be adapted to local conditions and give full play to the advantages of regional resources.

The evaluation index of preference items for green village and town clean energy utilization is shown in Table 8.41.

8.6.3 *Evaluation Method for Green Village and Town Clean Energy Utilization Plan*

1. Judging principles of evaluation results

When evaluating the implementation status of a green village and town clean energy utilization plan, the evaluation results are judged according to the following principles:

(1) For those that do not meet the control evaluation standards, the general item evaluation and preference evaluation will not be carried out, and the direct judgement will be unqualified.
(2) For those that meet the general standard but do not satisfy the preference evaluation, the final evaluation results are only those of the general evaluation.
(3) For those that meet the standards for the bonus point preferences, the final evaluation results are the sum of the general evaluation results and the preference evaluation results.

2. Dimensionless method for general evaluation index value

Table 8.41 Evaluation index of preference items for green village and town clean energy utilization

Preference items	Regional suitability indexes	Criteria	Sources of indexes
Biomass energy utilization	In areas with abundant biomass resources, whether to build a biomass-forming fuel production plant independently or jointly based on fuel demand	Yes	Technique Code for Village Rehabilitation
	In a village with dense population and large and medium-sized farms, whether to build large and medium-sized centralized biogas supply systems	Yes	
Solar energy utilization	Whether new public buildings in the village use sun rooms	Yes	
Wind energy utilization	Whether wind-power generation technology is adopted in areas where the annual average wind speed is greater than 2–3 m/s and where wind turbines are installed	Yes	
Water use	Whether micro-hydropower or small hydropower is used in mountainous areas far from the power system	Yes	
Marine energy utilization	Whether coastal villages far from the power system use small tidal power generation technology to supply electricity	Yes	
Geothermal energy utilization	Whether small-scale geothermal power generation technology is used to supply electricity in villages far from the power system but rich in geothermal resources	Yes	
	Whether villages with power supply and abundant geothermal resources use heat-pump technology to supply cooling in summer or heating in winter	Yes	

The quantitative indicators of the general indicator system for village and town green energy utilization are dimensional. It is not possible to directly compare and calculate them, so the data must be standardized.

When the relative treatment method is adopted, the "positive indicator" and "inverse indicator" are treated differently. For the positive indicator, the larger the actual value is, the greater the role played in the evaluation. The general indicators in this evaluation are positive indicators, that is, calculated by Formula (8.24).

$$C_i = \frac{x_i}{x_{\max}} \tag{8.24}$$

where C_i is the evaluation value of each indicator in index layer C;
x_i is the actual value of each indicator; and
$x_{\max}$ is the maximum value of each indicator.

3. Process of evaluation

First, judge the control item indicators, and if they are satisfied, then evaluate the general item index. The general evaluation index system of the clean energy utilization rate is divided into 2 levels using a weighted summation and evaluation method. These methods are shown in Formula (8.25) and Formula (8.26).

$$l_i = w_i x_i \tag{8.25}$$

$$W = \sum w_i x_i \tag{8.26}$$

where l_i is the evaluation indicator of index i;
x_i is the score of the index;
w_i is the weight of the index; and
W is the comprehensive indicator of the evaluation index of this group and the score of the upper-level evaluation index.

4. Determination of Index Weight

The average weight is used to determine the index weight, and the general index weight of the evaluation of a green village and town clean energy utilization plan is obtained, as shown in Table 8.42.

5. Evaluation criteria of green level of village and town clean energy utilization status

The comprehensive index W reflects the green level of the current situation of village and town clean energy utilization, and the green level is divided into 5 levels: the A level indicates that the green level of the evaluation object is the best, and the E level indicates that it is the worst.

Table 8.42 General index weight of evaluation of green village and town clean energy utilization plan

Primary indexes	Weights	Secondary indexes	Weights
Energy-saving B_1	0.25	Proportion of peasants using energy-saving heating equipment C_1	0.2
		Proportion of peasants using energy-saving cooking equipment C_2	0.2
		Proportion of peasants using energy-saving appliances C_3	0.2
		Proportion of peasants using energy-saving lighting C_4	0.2
		Proportion of energy-saving buildings of village or town C_5	0.2
Environmental protection B_2	0.25	Indoor environmental satisfaction C_6	0.2
		Carbon dioxide emissions reduction rate C_7	0.2
		Sulfur dioxide emissions reduction rate C_8	0.2
		Nitrogen oxide emissions reduction rate C_9	0.2
		TSP emissions reduction rate C_{10}	0.2
Sociality B_3	0.25	Clean energy utilization C_{11}	0.33
		Biomass energy-efficient utilization C_{12}	0.33
		Clean energy demand satisfaction C_{13}	0.33
Technical support B_4	0.25	Solar energy utilization ratio of peasants C_{14}	0.33
		Biogas utilization ratio of peasants C_{15}	0.33
		Clean energy use ratio of peasants C_{16}	0.33

The green level evaluation criteria of village and town clean energy utilization status are shown in Table 8.43.

After completing the general item evaluation, conduct the preference evaluation. When the village or town to be evaluated has more than 1 preference indicator that meets the requirements, the star rating is added to the general item index evaluation level, meaning that the green level of the village and town clean energy utilization status has reached a certain level and has regional suitability.

Table 8.43 Green level evaluation criteria of village and town clean energy utilization status

Level	A	B	C	D	E
Score *W*	1.0–0.80	0.79–0.60	0.59–0.40	0.39–0.20	0.19–0

If the general evaluation level of a village and town is B and meets more than 1 preference index, the current level of clean energy utilization is B+. Other analogies are similar.

8.6.4 Evaluation Case of Green Village and Town Clean Energy Utilization Plan

1. Basic data investigation and household energy consumption accounting

1) Basic data investigation

Through the basic data obtained from the household survey, the basic situation of living energy in Yanhe Village can be obtained.

The household survey and analysis of current energy utilization in Yanhe Village are shown in Table 8.44.

2) Consumption accounting of household energy

According to the above basic data, the relevant formula proposed in Sect. 8.6.3 can be used to calculate the overall situation of the annual living energy of Yanhe Village.

The analysis of the overall situation of living energy in Yanhe Village is shown in Table 8.45.

3) Analysis of the overall situation of household energy consumption

According to the survey, the clean energy types used in Yanhe Village were solar energy and biogas, with 252 households using solar water heaters and 208 households using biogas. The clean energy penetration rate was 83.33%. The energy consumption of Yanhe Village was calculated, and the total annual energy consumption was 487,416 kg_{ce}. The clean energy utilization rate of Yanhe Village was 30.1%, with firewood accounting for 41.42%, followed by solar water heaters at 18.82%, electric energy at 16.10%, liquefied petroleum gas at 12.39%, and biogas at 11.27%.

Table 8.44 Household survey and analysis of current energy utilization in Yanhe Village

Energy types	Status of use	Energy quantity	Ratio of total households
Solar water heaters	Number of users	252	83.33%
Biogas	Number of users	208	68.75%
Liquefied petroleum gas	Number of users	252	83.33%
	Annual usage	2348 cans/year	83.33%
Electric energy	Electricity fee	56212 yuan/year	–
	Annual usage	638333 kWh/year	–
Firewood	Number of users	221	72.92%
	Annual usage	353535 kg/year	–

Table 8.45 Analysis of the overall situation of living energy in Yanhe Village

Energy types	Status of use	Energy quantity	Ratio of total households	Unit standardization		Standard coal/kg_{ce}
Solar water heaters	Number of users	252	83.33%	Annual calorie consumption/kJ	1746×10^6	91771
Biogas	Number of users	208	68.75%	Annual usage/m^3	76960	54949
Liquefied petroleum gas	Number of users	252	83.33%	Annual usage/kg	35220	60377
	Annual usage	2348 cans/year	83.33%			
Electric energy	Electricity fee	56212 yuan/year	–	Annual usage/kWh	638333	78451
	Annual usage	638333 kWh/year	–			
Firewood	Number of users	221	72.92%	Annual usage/kg	353535	201868
	Annual usage	353535 kg/year	–			

The analysis of the overall structure of living energy in Yanhe Village is shown in Fig. 8.12.

2. Implementation effect evaluation of clean energy plan

1) Status evaluation of clean energy utilization plan in Yanhe Village

The main clean energy sources used in Yanhe Village are solar energy and biogas. Solar energy is used mainly for streetlamps, and biogas is used for living energy.

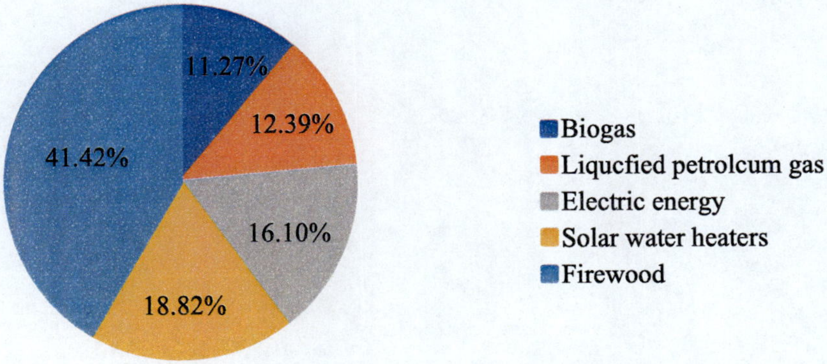

Fig. 8.12 Analysis of the overall structure of living energy in Yanhe Village

(1) Solar streetlamp. According to the road and building layout of Yanhe Village, solar energy collection devices are set on each streetlamp for self-powering.
(2) Biogas. The source of household energy consumption in Yanhe Village is mainly biogas, which is the main fuel for farmers in Yanhe Village. Liquefied petroleum is used as auxiliary energy in the coldest winter periods. The utilization rate of biogas in the whole village is 80%. The first group, second group and third group have biogas digesters for domestic use. Xincun, Tianyi Tea House and Yinxing Mountain Villa do not.

The layout of clean energy sources in Yanhe Village is shown in Fig. 8.13.

2) Implementation effect of the clean energy utilization plan of Yanhe Village

From 2009 to 2016, the implementation effect analysis of the clean energy utilization plan of Yanhe Village is shown in Table 8.46.

According to the analysis of Table 8.46, the implementation of clean energy planning in Yanhe Village is in good condition, and the planning objectives have basically been completed.

(1) Planning formulation. The clean energy utilization plan of Yanhe Village is complete, covering the entire village and each household, as is shown in Table 8.46.

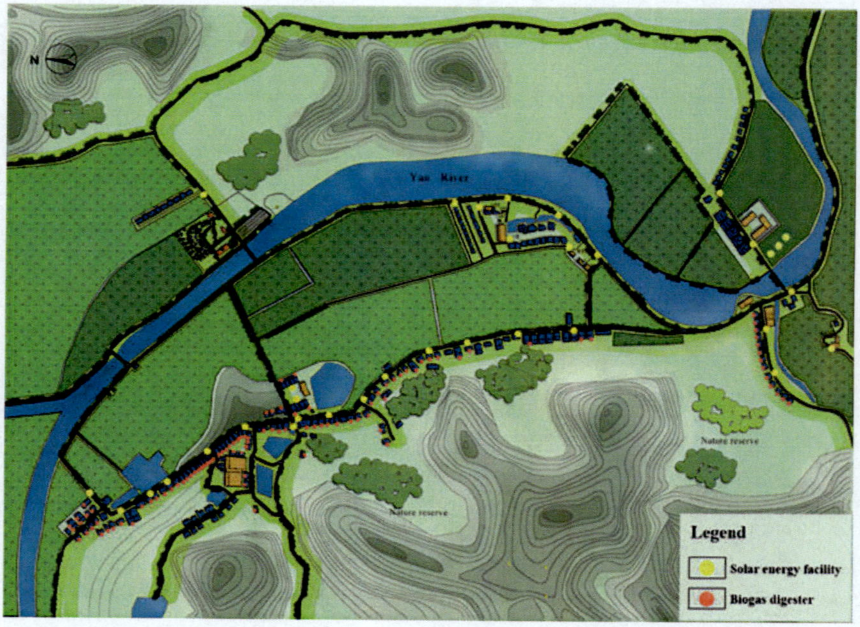

Fig. 8.13 Layout of clean energy sources in Yanhe Village

Table 8.46 Implementation effect analysis of the clean energy utilization plan of Yanhe Village

Clean energy types	Plan compilation situation	Situation of implementation			Comparison results
		In 2009	In 2012	In 2016	
Solar energy	60 solar lights	97 households use solar water heaters; the use rate of solar water heaters reaches 46.63%	37 solar lights; solar water heater usage rate reached 80%	Solar water heater usage rate reached 83.33%	During the implementation of the plan, the plan is adjusted, and the popularity of solar lamps is transferred to the popularity of solar water heaters
Biogas	Biogas utilization rate is 80%	112 household biogas digesters; biogas utilization rate is 53.85%	Building a household biogas tank; biogas utilization rate is 80%	208 household biogas digesters; biogas utilization rate is 83.33%	Number of household biogas tanks constructed and biogas utilization rate meet planning requirements

(2) Planning implementation. The entire village clean energy utilization rate reached 83.33%, meeting the requirement in *Guidelines for the Construction of Beautiful Village*—the proportion of farmers' clean energy use is not less than 70%, which also meets the requirement of the plan. The utilization rate of solar water heaters has reached 83.33%, and the construction rate of household biogas tanks reached the plan requirements for Yanhe Village.

(3) Plan adjustment. During the implementation of the plan, it was found that the solar lighting panels deteriorated quickly and were easily damaged, which was not suitable for village popularization. Therefore, the plan was adjusted, solar water heaters were popularized, and the penetration rate reached 83.33%.

3) Green level evaluation of clean energy utilization in Yanhe Village

(1) Evaluation process

① Evaluation index of control items

According to the survey, the annual per capita income of Yanhe Village is 12,000 yuan, the degree of satisfaction with the living environment is 80%, and the penetration rate of clean energy is 83.3%, all of which meet the requirements of the clean energy utilization status control evaluation index in Sect. 8.6.2. Thus, the assessment can enter the general index evaluation stage.

② Evaluation index of general items

Through interviews, household surveys and other field research methods, as well as the accounting results of living energy, the measured value of the secondary index of clean energy utilization evaluation in Yanhe Village was obtained, which was transformed into the index evaluation value in the evaluation. Using formula $X = \sum WW_i X_i$, the clean energy utilization comprehensive evaluation result in Yanhe Village is 0.693, which is shown in Table 8.47.

③ Evaluation index of preference items

Yanhe Village does not meet the preferred evaluation criteria, so the output level of the evaluation results is the general evaluation result.

(2) Analysis of evaluation results

① The comprehensive evaluation result of clean energy utilization in Yanhe Village is 0.693, and the evaluation grade is B.

② The sub-item evaluation results show that the "Sociality" index of clean energy utilization in Yanhe Village is lower, only 0.384, which needs to be further improved.

8.7 Evaluation of Green Village and Town Plan Implementation and Eco-environmental Restoration

The protection and restoration of the ecological environment is the premise and foundation for the survival and development of green villages and towns, the long-term task of green village and town planning and construction, and the key evaluation content of green village and town plan implementation.

Through quantitative analysis and evaluation of the rural landscape pattern at the green ecological level in the plan implementation process, as well as effective judgement due to natural factors and human activities, such as the construction of a rural ecological environment with adverse effects or damage, the implementation of green village and town ecological landscape construction and repair projects can be used to provide technical support and to determine the needs and how to revise the eco-environmental plan of to provide a decision-making basis.

8.7.1 Related Concepts and Evaluation Content

1. Related concepts

(1) Landscape refers to the specific regional features formed by the natural environment and the human environment in the region as perceived by human beings. It is the result of activities or interactions between humans and nature and is a complex concept that can reflect ecology, economy, culture and other social functions and values.

Table 8.47 Evaluation index and calculation of clean energy utilization in Yanhe Village

Primary index B_i	Weight of primary index W	Secondary index C_i	Actual value of secondary index /%	Weight of secondary index W_i	Evaluation value of index X_i	W_iX_i	Item evaluation results
Energy-saving B_1	0.25	Proportion of peasants using energy-saving heating equipment C_1	–	–	–	–	–
		Proportion of peasants using energy-saving cooking equipment C_2	83.33	0.25	0.833	0.208	0.702
		Proportion of households using energy-efficient home appliances C_3	98	0.25	0.98	0.245	
		Proportion of peasants using energy-saving lighting equipment C_4	98	0.25	0.98	0.245	
		Village and town energy-saving building proportion C_5	1.51	0.25	0.015	0.004	
Environmental protection B_2	0.25	Indoor environmental satisfaction C_6	80	0.2	0.8	0.16	0.911
		Carbon dioxide emissions reduction rate C_7	82.5	0.2	0.825	0.165	
		Sulfur dioxide emissions reduction rate C_8	98	0.2	0.98	0.196	
		Nitrogen oxide emissions reduction rate C_9	98	0.2	0.98	0.196	

(continued)

Table 8.47 (continued)

Primary index B_i	Weight of primary index W	Secondary index C_i	Actual value of secondary index /%	Weight of secondary index W_i	Evaluation value of index X_i	W_iX_i	Item evaluation results
		TSP emissions reduction rate C_{10}	97	0.2	0.97	0.194	
Sociality B_3	0.25	Clean energy utilization rate C_{11}	30.1	0.33	0.301	0.099	0.384
		Biomass energy efficient utilization rate C_{12}	11.27	0.33	0.113	0.037	
		Clean energy demand satisfaction rate C_{13}	75.00	0.33	0.75	0.248	
Technical support B_4	0.25	Solar energy utilization ratio of peasants C_{14}	68.75	0.33	0.6875	0.2269	0.777
		Biogas utilization ratio of peasants C_{15}	83.33	0.33	0.833	0.275	
		Clean energy use ratio of peasants C_{16}	83.33	0.33	0.833	0.275	
Comprehensive evaluation results			0.693				

Note Yanhe Village has no heating facilities in winter, so index C_1 is vacant

(2) Landscape pattern, also known as landscape spatial pattern, usually refers to the spatial configuration and distribution of various landscapes, including the number of landscapes and landscape types.

(3) Patch refers to a spatial unit that is different in appearance or nature from the surrounding environment (background), is nonlinear and has internal homogeneity. It is the most basic element in the spatial pattern.

(4) Corridor is a linear or strip landscape feature that differs in appearance or nature from adjacent areas in the landscape. The structure of the corridor has a great influence on the landscape ecological process. The landscape is divided by the corridor and connected by the corridor, so the corridor has the dual function of connection and barrier.

(5) Basement refers to the background structure with the widest distribution, the greatest continuity, and controlling landscape dynamics.

(6) Village and town basement is the most widely distributed and continuous background structure in the village or town, including the natural base composed of

soil, water, and biology; the agricultural base composed mainly of agricultural production; and the cultural base composed of local historical and cultural elements.

(7) Public space refers to a public place of activity composed of physical space and its environmental elements.

(8) Landscape ornaments refers to small facilities in the public environment that provide conditions for spatial management and for villagers (residents) to rest, decorate, light, and display.

2. Village and town landscape types

In green village and town plan implementation and eco-environmental restoration activities, the key point of evaluation is the impact of village construction on the ecological environment. Therefore, village and town landscape types are divided into 2 levels from the perspective of land use.

(1) First-level landscape type

There are 3 first-level landscape types: natural landscape; semi-natural, semi-artificial landscape; and artificial landscape. The natural landscape is less affected by human activities and is basically in a natural state. The semi-natural, semi-artificial landscape has traces of human activities and contains some natural ecological features. The artificial landscape is formed by human activities and has lost its original natural features.

(2) Second-level landscape type

There are 6 second-level landscape types: forestland, water body, bare land, cultivated land, construction land and road. Road land also belongs to construction land. However, because roads also have the nature of corridors in the analysis and measurement of landscape patterns, this evaluation also uses "road" as a second-level evaluation index of landscape analysis.

The classification of village and town landscape types is shown in Table 8.48.

3. P-C-L-E correlation analysis

The village and town landscape will be affected and disturbed by various construction activities determined and implemented by the village and town plan, thus causing

Table 8.48 Classification of village and town landscape types

First-level classification	Natural landscapes	Semi-natural, semi-artificial landscapes	Artificial landscapes
Second-level classification	Woodland	Arable land	Construction land
	Water body		Road
	Bare land		

changes in landscape patterns, such as changes in type, composition and function. In the construction activities implemented within the scope of the village or town plan, it is necessary to analyse and evaluate the correlation and interaction among the 4 elements of planning, construction, landscape pattern change and eco-environmental change (P-C-L-E).

The green village and town P-C-L-E correlation analysis path is shown in Fig. 8.14.

In applying P-C-L-E and landscape pattern change analysis, correlation analysis technology can be an objective analysis and evaluation at different points of the village and town plan implementation period, the influence of construction activities on the ecological environment, and the status of the environment in coping with the interruption. It can also be used for dynamic monitoring, adjustment and prediction of construction activities determined by the village and town plan, to evaluate the implementation effect of the rural eco-environmental plan, and to determine whether there is a need for eco-environmental restoration or eco-environmental plan revision to provide a decision-making basis.

4. Content of evaluation

The evaluation of green village and town plan implementation and eco-environmental restoration includes remote sensing data processing; calculation of patch-type index, landscape index, diversity, fragmentation and stability of landscape pattern and other landscape indexes; dynamic analysis of landscape pattern, landscape level, diversity, fragmentation and stability of landscape pattern; and correlation evaluation of P-C-L-E.

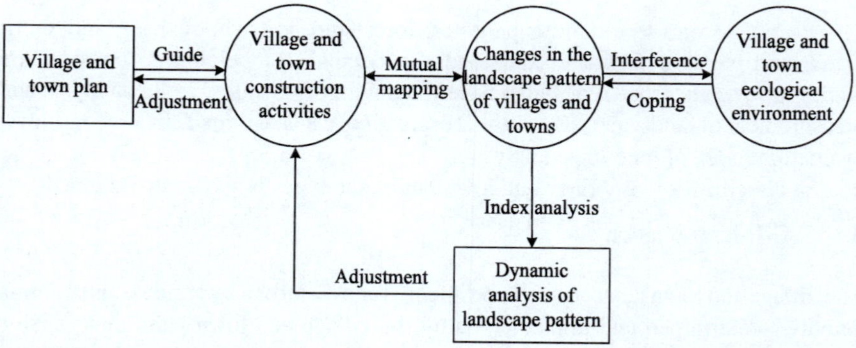

Fig. 8.14 Green village and town P-C-L-E correlation analysis path

8.7.2 Selection and Calculation of Green Village and Town Landscape Pattern Index

The analysis of landscape pattern characteristics is usually carried out through 3 levels: patch level, class level and landscape level, and the scale of the 3 levels is gradually expanded.

The landscape pattern index is a quantitative indicator that reflects the landscape structure and its composition and spatial configuration. Common landscape pattern indexes include the patch level, class level, and landscape level. According to the characteristics of the landscape types, this study selects 13 indexes, such as the patch shape index and patch area, to evaluate village and town landscape patterns.

1. Dynamic analysis index of green village and town landscape pattern changes

There are 3 main types of dynamic indexes for analysing of village and town landscape pattern changes: landscape diversity, landscape fragmentation and landscape stability.

(1) Landscape diversity index

This index can be used to analyse the diversity and variability of different types of landscape in terms of spatial structure and temporal dynamics. It can reflect the complexity of the landscape.

The indicators of landscape diversity include the number of patches, patch richness, landscape diversity (such as the Shannon diversity index), and fractal dimension.

(2) Landscape fragmentation index

This index reflects the interference intensity of village and town construction activities on the ecological environment by the fragmentation degree of the ecological landscape caused by nature and humans. The higher the degree of landscape fragmentation is, the stronger the interference of village and town construction activities in the ecological environment.

The indicators describing the degree of landscape fragmentation include the number of patches, patch density, and mean patch area.

(3) Landscape stability index

This index reflects the ability of the system to resist interference and recover its original state after being disturbed. The ability of the system to resist interference in this evaluation is expressed by landscape heterogeneity. Due to the inhomogeneity and complexity of a certain patch class distributed in time or space, landscape heterogeneity increases, and ecosystem stability is enhanced.

The indicators of landscape stability include patch density and landscape evenness.

2. Selection of village and town landscape index based on dynamic analysis

According to the above analysis, a total of 26 landscape indexes, including 13 "patch areas", were selected in this evaluation to carry out dynamic analysis and calculation of village and town landscape pattern changes caused by construction activities in the planning and implementation process.

The selection of village and town landscape indexes based on dynamic analysis shown in Table 8.49, with "Δ" for each selected evaluation index.

Table 8.49 Selection of village and town landscape indexes based on dynamic analysis

Number	Name of landscape indexes	Horizontal dynamic analysis of patch type	Dynamic analysis at landscape level	Analysis of diversity, fragmentation and stability of landscape patterns
1	Patch area *CA*	Δ	Δ	
2	Patch area percentage *PLAND*	Δ		
3	Number of patches *NP*	Δ	Δ	Δ
4	Patch density *PD*	Δ	Δ	Δ
5	Largest patch index *LPI*	Δ	Δ	
6	Total edge *TE*	Δ	Δ	
7	Edge density *ED*	Δ	Δ	
8	Landscape shape index *LSI*		Δ	
9	Mean patch area *MPA*	Δ	Δ	Δ
10	Average shape index *MSI*	Δ	Δ	
11	Contagion index *CONT*		Δ	
12	Shannon Diversity Index *SHDI*		Δ	Δ
13	Landscape evenness index *SHEI*		Δ	Δ

8.7.3 Evaluation of Village and Town Landscape Pattern and Eco-Environmental Restoration

1. Dynamic analysis method for landscape pattern changes

This evaluation method uses satellite remote sensing data, ArcGIS software and ENVI software to correct and process the data, classify the landscape, and convert it into raster data. Fragstats software, which is commonly used for landscape pattern analysis, was used to dynamically calculate the patch class level, landscape level and landscape pattern index. On this basis, we analysed and studied the characteristics of landscape patterns in different time periods in the area and then combined the landscape patterns at different time points in the region to complete the dynamic analysis of village and town landscape pattern changes.

The dynamic analysis path of the village and town landscape pattern construction is shown in Fig. 8.15.

2. Evaluation results of landscape and eco-environmental restoration

According to the above evaluation method, the evaluation of village and town landscape patterns and eco-environmental restoration is expressed in the form of level evaluation; that is, the evaluation results are divided into 5 levels: excellent, good, fair, relatively poor, and poor.

The higher the level of the evaluation results is, the better the effect of eco-environmental protection in the implementation process of the evaluated plan.

When the evaluation results of P-C-L-E are all positive, the evaluation grade of village and town landscape pattern and eco-environmental restoration is “excellent”. For a village or town with “excellent” evaluation results, there is no need to carry out environmental restoration projects or revise the plan.

When the P-C-L-E evaluation results have 1 weak positive effect, the evaluation grade is “good”. For a village or town with “good” evaluation results, there is no

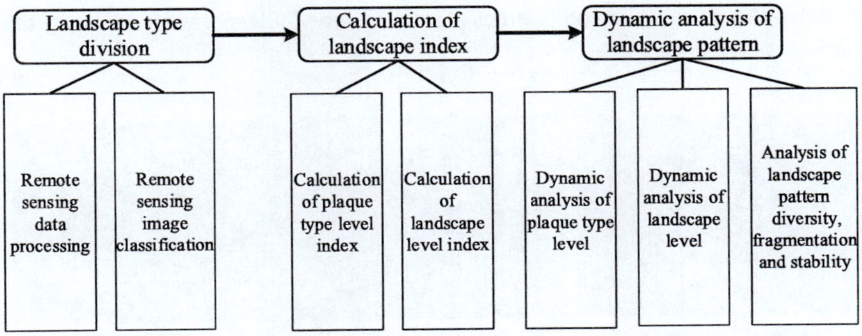

Fig. 8.15 Dynamic analysis path of village and town landscape pattern construction

need to carry out environmental restoration projects or revise the village and town plan, but the plan implementation should be strictly supervised and controlled.

When the P-C-L-E evaluation results have 2 weak positive effects or 1 negative effect, the evaluation grade is "fair". For a village or town with "fair" evaluation results, it is necessary to carry out partial environmental restoration projects or revise the plan according to the situation.

When the P-C-L-E evaluation results have 3 weak positive effects or 2 negative effects, the evaluation grade is "relatively poor". For a village or town with "relatively poor" evaluation results, it is necessary to carry out environmental restoration projects and revise the plan.

When the P-C-L-E evaluation results are all negative, the evaluation grade is "poor". For a village or town with "poor" evaluation results, the damaged ecological environment must be rectified, environmental restoration projects must be carried out immediately, and the plan must be revised.

8.7.4 Evaluation Case of Green Village and Town Plan Implementation and Eco-Environmental Restoration

1. Processing of remote sensing data and classification of remote sensing images in Yanhe Village

1) Remote sensing data processing

Using remote sensing image data for Yanhe Village in 2005, 2011 and 2016, ArcGIS vector editing tools were separately used to pre-process boundary rules and other data and to mark the area of research in Yanhe Village. Remote sensing images of Yanhe Village and the surrounding areas were obtained for 2005, 2011 and 2016.

The comparison of remote sensing images of Yanhe Village from 2005 to 2016 is shown in Fig. 8.16.

A comparison of the three remote sensing images clearly shows that from 2005 to 2016, the village construction activities in Yanhe Village greatly changed the land-use pattern.

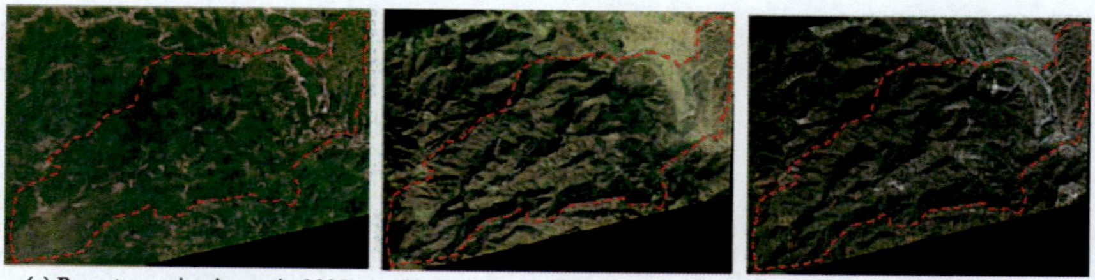

(a) Remote sensing image in 2005 (b) Remote sensing image in 2011 (c) Remote sensing image in 2016

Fig. 8.16 Comparison of remote sensing images of Yanhe Village from 2005 to 2016

2) Classification of remote sensing images

This evaluation used ENVI5.1 software for landscape type classification processing, including mainly visual solution and supervised classification.

(1) Visual solution

This is also called visual interpretation. The evaluation area of this case is Yanhe Village, which is a small area. Based on many field investigations and measurements and combined with high-definition image data, using the classification method of visual interpretation can bring high precision to the research results.

(2) Supervised classification

This is also called the "training classification method". The main steps of supervised classification are category definition, typical sample selection, classifier selection, execution supervision classification, evaluation classification result and post-classification processing.

After the sample selection is finished, the training sample evaluation is performed, the sample separability is calculated, the sample is appropriately adjusted according to the size of the separability value, and the supervised classifier selected when performing the supervised classification is the maximum likelihood classifier.

(3) Landscape classification

This evaluation classifies the landscape by the human landscape impact on the natural environment and the landscape patch classification of land use. The landscape was divided into 6 categories to construct the landscape classification system of Yanhe Village.

The landscape classification system of Yanhe Village is shown in Table 8.50.

Table 8.50 Landscape classification system of Yanhe Village

Landscape types	Description	Partial typical sample
Woodland	Land covered by natural forests, secondary forests, and plantations	
Arable land	Land on which crops are grown, including dry land, paddy fields and field ridges	
Water	Rivers, ponds, ditches, etc. in the area	
Construction land	Land where buildings and structures have been built in the countryside	
Road	County roads, township roads and village roads for pedestrians and various means of transportation	
Bare land	Bare ground with no plant growth	

The ROI tool in ENVI was used to define training samples. First, the sample name was confirmed and a region of interest created for each type of sample.

The operational interface of ENVI5.1 software supervision and classification is shown in Fig. 8.17.

The ROI separability tool was used to measure the degree of difference in the training samples. The sample was appropriately adjusted so that the separability value was greater than 1.9, which is a qualified sample, and the maximum likelihood classifier was then selected to perform supervised classification. The landscape classification maps of Yanhe Village in 2005, 2011 and 2016 were obtained.

The landscape classification images of Yanhe Village from 2005 to 2016 are shown in Fig. 8.18.

2. Indicator calculation of landscape in Yanhe Village

1) Import into the landscape pattern analysis software

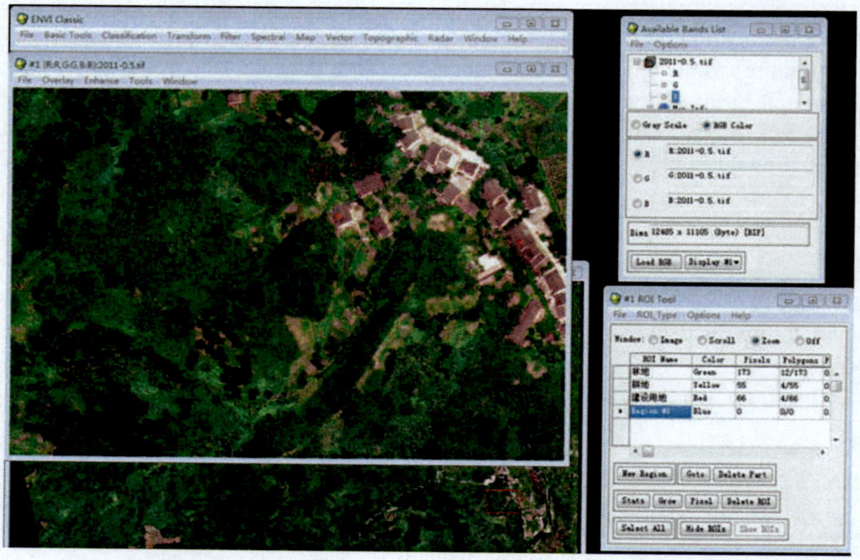

Fig. 8.17 Operational interface of ENVI5.1 software supervision and classification

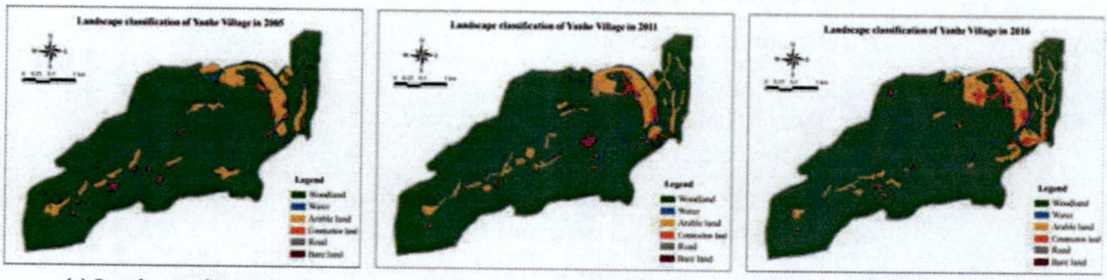

(a) Landscape image in 2005 (b) Landscape image in 2011 (c) Landscape image in 2016

Fig. 8.18 Landscape classification images of Yanhe Village from 2005 to 2016

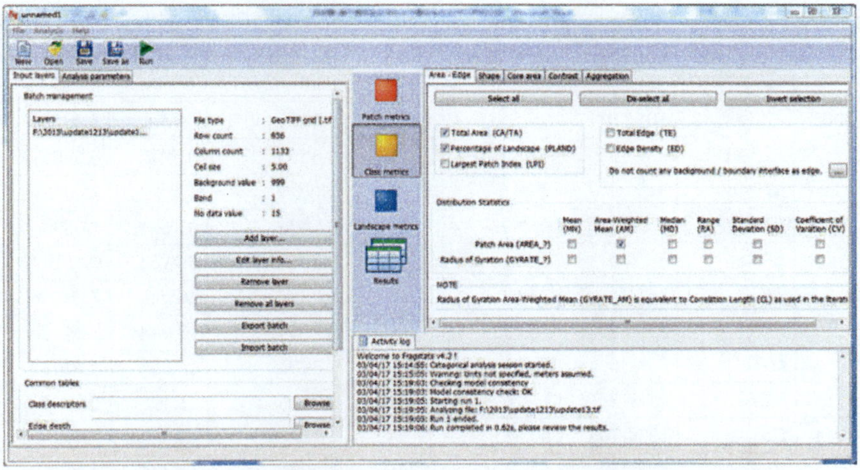

Fig. 8.19 Indicator calculation process of Fragstats 4.2 software pattern

The classification results are transferred in ArcGIS into raster data and imported into landscape pattern analysis software to obtain the desired landscape indicator results. Fragstats 4.2 software was chosen for this review, as it is the most commonly used landscape pattern analysis software. It can calculate a series of landscape pattern indicators at three levels: patch level, patch type level and landscape level. The landscape structure and spatial pattern can be quantitatively analysed.

In this evaluation, 21 indicators were selected at the level of patch type and landscape indicator. Basic data for 2005, 2011 and 2016 in the planning implementation process of Yanhe Village were selected to analyse the dynamic pattern of the landscape pattern of Yanhe Village.

The indicator calculation process of the Fragstats 4.2 pattern is shown in Fig. 8.19.

2) Calculation of patch type indicator

In this evaluation, 9 landscape pattern indicators were selected and calculated at the patch type level: patch area (*CA*), percentage of patch area (*PLAND*), number of patches (*NP*), patch density (*PD*), largest patch index (*LPI*), total edge (*TE*), edge density (*ED*), mean patch area (*MPA*), and mean shape index (*MSI*).

The landscape indicator calculation of the patch type level of Yanhe Village is shown in Table 8.51.

3) Calculation of landscape level indicator

Twelve landscape pattern indicators were selected and calculated for the landscape level analysis: total patch area (*CA*), total number of patches (NP), patch density (PD), largest patch index (*LPI*), total edge (*TE*), edge density (*ED*), landscape shape index (*LSI*), mean patch area (*MPA*), mean shape index (*MSI*), contagion index (*CONT*), Shannon diversity index (*SHDI*), and landscape uniformity index (*SHEI*).

Table 8.51 Calculation of patch type level and landscape indicator of Yanhe Village

Year	Patch type	Forestland	Cultivated land	Water	Construction land	Road	Bare land
2005	*CA*	1052.018	76.565	11.165	5.2225	6.74	8.0075
	PLAND	90.7133	6.602	0.9627	0.4503	0.5812	0.6905
	NP	31	19	31	18	4	25
	PD	1.2876	0.7892	1.2876	0.7476	0.1661	1.0384
	LPI	38.7451	0.5975	0.1174	0.0567	0.1963	0.0617
	TE	86110	24880	14710	6900	26350	7330
	ED	35.7655	10.3338	6.1098	2.8659	10.9444	3.0445
	MPA	33.936	4.0297	0.3602	0.2901	1.685	0.3203
	MSI	2.0632	1.7267	1.6815	1.7906	11.1981	1.3884
2011	*CA*	1034.795	80.885	11.2125	11.6275	15.5825	5.645
	PLAND	89.2259	6.9744	0.9668	1.0026	1.3436	0.4867
	NP	37	34	30	26	4	9
	PD	1.5368	1.4122	1.246	1.0799	0.1661	0.3738
	LPI	37.9045	0.4185	0.0814	0.148	0.3788	0.1052
	TE	88990	37270	15400	11820	39530	3870
	ED	36.9617	15.48	6.3963	4.9094	16.4187	1.6074
	MPA	27.9674	2.379	0.3738	0.4472	3.8956	0.6272
	MSI	1.9699	1.9786	1.877	1.7646	11.4191	1.4411
2016	*CA*	1018.288	84.505	10.49	21.8875	21.2925	3.285
	PLAND	87.8025	7.2865	0.9045	1.8873	1.836	0.2833
	NP	38	38	31	33	4	11
	PD	1.5783	1.5783	1.2876	1.3706	0.1661	0.4569
	LPI	15.1825	0.4698	0.0679	0.4279	0.4599	0.0308
	TE	125830	41010	14700	16460	77770	3550
	ED	52.2631	17.0334	5.9749	6.8366	32.3015	1.4745
	MPA	26.797	2.2238	0.3384	0.6633	5.3231	0.2986
	MSI	1.9881	1.998	1.6421	1.5549	19.1184	1.5772

The calculation of landscape level and landscape pattern indicators of Yanhe Village is shown in Table 8.52.

3. Dynamic analysis of landscape pattern in Yanhe Village

1) Dynamic analysis of patch type

According to the landscape indicator calculation results at the patch type level in Table 8.51, the dynamic changes in 6 landscape types in Yanhe Village in 2005, 2011 and 2016 were analysed.

Table 8.52 Calculation of landscape level and landscape pattern indicators of Yanhe Village

Landscape indexes	2005	2011	2016
CA	1159.718	1159.748	1159.748
NP	128	140	155
PD	11.0372	12.0716	13.365
LPI	80.4366	78.6893	60.7442
TE	83140	98440	138660
ED	71.6898	84.8805	119.5605
LSI	5.8138	6.5965	8.6518
MPA	9.0607	8.2839	7.4822
MSI	2.0435	2.1559	2.2469
CONT	75.8221	74.8615	73.8708
SHDI	0.8857	0.9151	0.9394
SHEI	0.4552	0.4703	0.4827

(1) Landscape pattern basement of Yanhe Village

According to the calculation results in Table 8.51, the Yanhe Village forestland patch area occupies a total area of the village domain of 90.7–87.8%, and the largest patch index, total length, and average patch border area were significantly greater than the other types. The forest landscape in Yanhe Village has much better integrity and connectivity than other landscape types, and the basement is the Yanhe Village landscape pattern analysis.

(2) Changes in patch area of various types

The forestland area decreased by 2.9107%; the bare land area decreased by 0.4072%; the cultivated land, construction land and road area increased significantly; the cultivated land area increased by 0.6845%; the construction land area increased by 1.4370%; and the road area increased by 1.2548%. The above analysis shows that the total area of natural landscape patches in Yanhe Village is gradually decreasing, and the total area of artificial landscape patches is gradually increasing.

(3) Changes in the number of patches of each type

The number of patches in forestland, cultivated land and construction land has gradually increased. The water and road patches have remained basically unchanged. The number of bare land patches has generally declined.

(4) Change in maximum patch indicator

The largest patch index of forestland decreased from 38.7451 to 15.1825. The largest water patch decreased from 0.1174 to 0.0679, the area of the largest construction land and road type patch increased gradually, the largest construction land patch increased from 0.0567 to 0.4279, and the largest road patch increased from 0.1963 to 0.4599.

(5) Changes in the total boundary length of different types of patches

The total edge length of the four patch types of forestland, cultivated land, construction land and road showed a continuous growth trend. The total edge length of the bare patch type gradually decreased, and the total edge length of the water patch type changed little. The change rule of the edge density index of the six landscape types is the same as that of the total edge length.

(6) Changes in the average patch area of each type

The mean patch area of the forestland and cultivated land types gradually decreased. The mean patch area of the construction land and road types gradually increased, and the mean patch area of the water type changed little.

(7) Change in the average shape index of each type of patch

The mean shape index of the cultivated land, roads and bare land types gradually increased. The mean shape index of the construction land type gradually decreased, and the mean shape index of the forestland type changed little.

2) Dynamic analysis of landscape level

According to the calculation results of the landscape level of Yanhe Village in Table 8.52, the landscape level of Yanhe Village in 2005, 2011 and 2016 was compared and analysed, and an analysis chart of the landscape pattern indicator change trend of Yanhe Village was drawn.

The changing trend analysis of the landscape pattern indicator of Yanhe Village is shown in Fig. 8.20.

Dynamic analysis of landscape level:

(1) Figure 8.20A shows that the number of patches, the density of patches, the total length of the boundary and the boundary density gradually increased. This reflects an increase in the degree of fragmentation of the Yanhe Village landscape and the degree of isolation between patch types. With the continuous fragmentation of the landscape, the maximum patch index and average patch area of Yanhe Village gradually decreased.
(2) Figure 8.20B shows that the landscape shape index and average shape index of Yanhe Village are obviously larger. The landscape structural complexity reflects greater complexity in the patch shape in the Yanhe Village landscape.
(3) Figure 8.20C shows that the landscape aggregation index decreased with the construction and development of Yanhe Village. The aggregation index is small, indicating that there are a large number of small patches in the landscape, which reflects decreased connectivity and increased fragmentation of the overall landscape of Yanhe Village. This is consistent with the index results of patch density and boundary density.

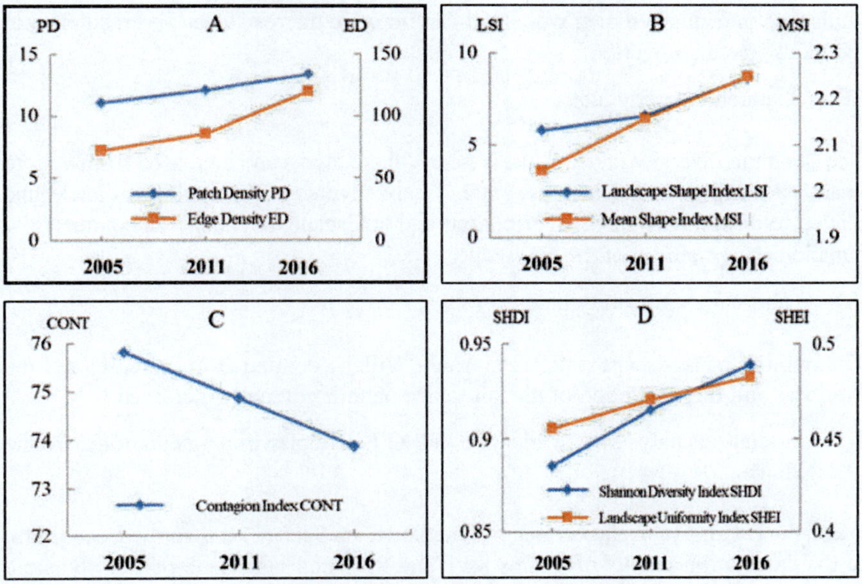

Fig. 8.20 Changing trend analysis of landscape pattern indicators of Yanhe Village

(4) Figure 8.20D shows that the Shannon diversity index increased gradually. The land use of Yanhe Village tends to be rich, and all kinds of patches present a balanced distribution trend in the landscape.

(5) Figure 8.20D shows that the uniformity index of the landscape gradually increased. The evenness of landscape patch distribution in Yanhe Village increased.

3) Analysis of the diversity, fragmentation and stability of landscape patterns

(1) Patch number and average patch density

The horizontal analysis of patch types in Yanhe Village shows that the number of patches in forestland, cultivated land and construction land increased gradually, with the number cultivated land type patches being the largest, but the average patch area decreased gradually, indicating that the spatial shape tended to be complex and strongly affected by human activities.

The area of construction land and road increased obviously, and the maximum patch index and average patch area gradually increased, showing increasingly higher connectivity.

As corridor structures, water bodies and roads connected and separated the whole landscape, and bare land showed characteristics of atrophy.

The forestland type of Yanhe Village landscape had the most extensive distribution, the greatest continuity, and absolute superiority. As the base structure of the landscape of Yanhe Village, woodland can improve the resistance and resilience of the landscape organization.

(2) Shannon diversity index

The Shannon diversity index of the Yanhe Village landscape increased from year to year, reflecting the rapid increase in landscape diversity. The ecological landscape tended to have a reasonable distribution, and landscape diversity was conducive to enhancing the stability of the landscape.

(3) Landscape uniformity index

The number of landscape patches in Yanhe Village continued to increase, and the evenness and fragmentation of the landscape pattern gradually increased.

4. Correlation analysis and evaluation of P-C-L-E of plan implementation in Yanhe Village

When the P-C-L-E correlation analysis method from Sect. 8.7.2 is applied, according to the calculation results of Tables 8.51 and 8.52 and landscape pattern dynamic analysis, the results of eco-environmental protection, the possibility of environmental remediation and the evaluation of whether to implement plan revision in Yanhe Village are obtained.

1) Correlation analysis of the main landscape pattern change index and eco-environmental change

The analysis of the changes in the main types of patch-level indicators and ecological environment of Yanhe Village is shown in Table 8.53.

The analysis of changes in the main types of landscape level indicators and ecological environment of Yanhe Village is shown in Table 8.54.

2) Correlation Evaluation of P-C-L-E of plan implementation in Yanhe Village

During the plan implementation in Yanhe Village, the construction of farmhouses, infrastructure and public facilities and the ecological tourism industry exerted great pressure on the ecological environment, leading to changes in the landscape pattern and causing a certain impact on the ecological environment of Yanhe Village.

According to the correlation analysis of landscape pattern change and eco-environmental change in Tables 8.53 and 8.54, the correlation evaluation conclusion of P-C-L-E in the implementation of the Yanhe Village plan can be drawn:

(1) Impact of patch type-level change on the ecological environment

The change in landscape pattern is as follows: at the patch type level, construction activity led to a decreasing trend of forestland from year to year; the cultivated land area increased, and the cultivated land form tended to be complex, indicating that

Table 8.53 Analysis of the changes in the main types of patch-level indicators and ecological environment of Yanhe Village

Analysis items	Change in landscape pattern (%)						Construction activities
	Forestland	Cultivated land	Water	Construction land	Road	Bare land	Interference/Response Eco-environment
PLAND	Falling 2.91	Rising 0.68	Falling 0.06	Rising 1.44	Rising 1.25	Falling 0.41	Interference: total proportion of forestland decreased
							Response: Woodland as landscape base has little influence at present
NP	Rising 0.23	Rising 1.00	Same	Rising 0.83	Same	Falling 0.56	Interference: farmland and construction land increase
							Response: Control the increase in construction projects according to the plan requirements
LPI	Rising 23.6	Falling 0.13	Falling 0.05	Rising 0.37	Rising 0.26	Falling 0.03	Interference: Arable land and water have a fragmentation tendency
							Response: Protect farmland and water integrity
TE	Rising 0.46	Rising 0.65	Falling 0.0007	Rising 1.39	Rising 1.95	Falling 0.52	Interference: The total boundary length of road and construction land grows most rapidly
							Response: Control the increase in construction projects according to the plan requirements

Table 8.54 Analysis of changes in the main types of landscape level indicators and ecological environment of Yanhe Village

Analysis items	Change in landscape level indicator (%)	Change in eco-environment
CONT	Falling 0.03	Total degree of aggregation decreases and degree of fragmentation increases
SHDI	Rising 0.06	Land use tends to be rich, and diversity increases
SHEI	Rising 0.07	The uniformity of landscape increases, and the anti-interference ability of the environment is enhanced

construction activity tended to squeeze the cultivated land. Since the forestland area of Yanhe Village is large and widely distributed, constituting the basement of the landscape pattern of Yanhe Village, the above changes had a positive effect on the current ecological environment of Yanhe Village in general.

The implementation of a road traffic construction project in Yanhe Village increased the connectivity of all villager groups and tourist attractions and increased the convenience of traffic, which had a positive effect on the change in the ecological environment in Yanhe Village.

(2) Impact of landscape-level change on the ecological environment

The increasing degree of landscape fragmentation in Yanhe Village is reflected in the fact that during the planning and implementation process of Yanhe Village, with the large increase in construction projects, the land-use forms in Yanhe Village were under pressure, which led to more abundant land-use forms in Yanhe Village. However, there was no negative impact on the change in the ecological environment.

(3) Impact of diversity, fragmentation and stability of landscape patterns on the ecological environment

The increases in the Shannon diversity index and landscape uniformity index reflect that with the increase in land-use richness and stability in Yanhe Village, the impact on eco-environmental change was positive, but the increase in landscape fragmentation should be controlled in future construction activities.

5. Comprehensive evaluation conclusion

(1) The landscape pattern change in Yanhe Village at the patch type level had a positive effect on the ecological environment as a whole;
(2) The horizontal landscape pattern change in Yanhe Village had a weak positive effect on the ecological environment; and
(3) The diversity, fragmentation and stability of the landscape pattern in Yanhe Village had positive effects on the ecological environment as a whole.

Therefore, the overall ecological environment of Yanhe Village is good. During the plan implementation in Yanhe Village, the interference of various construction activities on the ecological environment of the village was enhanced, but the overall degree was controllable. At the present stage, there is no need to implement an environmental restoration project in Yanhe Village or to revise the eco-environmental plan.

References

1. Liu YF, Ming DP, Yang JY (2002) Optimization of land use structure based on ecological green equivalent. J Wuhan Univ (Inf Sci Ed) (05):493–498 + 515
2. Gucheng County People's Government (2020) Natural resources. https://www.hbgucheng.gov.cn/gcgk/zrdl/201906/t20190626_1698811.shtml. Accessed 1 Sept 2020

Chapter 9
Evaluation of Green Village and Town Environmental Infrastructure Construction

Abstract Evaluating the implementation effect of green village and town environmental infrastructure construction plays an important role in protecting and improving the green ecological environment and enhancing village and town green living quality and green production capacity. In this chapter, six evaluation methods of green village and town environmental infrastructure construction are established. Taking Yanhe Village of Hubei Province, a national green ecological village, as a case study, evaluation methods for green village and town environmental infrastructure construction are introduced and applied.

Environmental infrastructure construction is an important foundation for green village and town environmental protection and sustainable improvement. The implementation evaluation of environmental infrastructure construction is based on the current corresponding standards and specifications, including evaluation of the following six aspects: water source protection and water-saving facilities; sewage reduction and reuse facilities; garbage classification, collection and harmless treatment facilities; clean energy facilities; greening and landscape facilities; and disaster prevention and mitigation and early-warning facilities.

The original version of this chapter was revised: Table 9.1 has been corrected. The correction to this chapter is available at https://doi.org/10.1007/978-981-16-2098-0_12

X. Wang and H. Hua, *Green Village and Town Construction in China*,
https://doi.org/10.1007/978-981-16-2098-0_9

9.1 Evaluation of Green Village and Town Drinking Water Source Protection and Water-Saving Facility Construction

9.1.1 Evaluation Method for Green Village and Town Drinking Water Source Protection and Water-Saving Facility Construction

1. Evaluation requirements

There is a close relationship between the construction of drinking water source protection facilities and the construction of water conservation facilities, which is the basic guarantee for green village and town construction and development. They should be promoted simultaneously in the construction process and combined in the evaluation method. The evaluation of green village and town drinking water source protection and water-saving facility construction should be based on relevant national standards and specifications.

2. Evaluation index and method

According to the implementation evaluation content of drinking water source protection and water-saving facility construction, the evaluation standard and scoring method of green village and town drinking water source protection and water-saving facility construction can be established with a total of 9 evaluation indexes.

The evaluation criteria and scoring methods for the protection of green village and town drinking water sources and water-saving facility construction are shown in Table 9.1.

The evaluation of green village and town drinking water source protection and water-saving facility construction adopts the 0–1 scoring method. If all the selected evaluation indexes meet or exceed the current national, department and industry standards or specifications, the score is 1, and "√" is marked for the selected evaluation index item. If all the selected evaluation indexes do not meet the relevant standards or specifications, the score is 0, and there are no points for non-participating items. Item (9) in the table is a bonus point to encourage all regions to use reclaimed water, and a bonus point is awarded if the evaluated region uses reclaimed water. A score of more than 70% of the possible total score is considered eligible.

The sources of domestic water are different in villages or towns in different regions and with different natural conditions, and the ways of obtaining domestic water and facility construction are not necessarily the same. Therefore, the full marks in Table 9.1 also differ, and the participating items and full marks should be reasonably selected according to the actual situation of the village or town being evaluated.

In the application of Table 9.1 to an actual evaluation, the actual scores of each sub-item of the evaluated village or town should be filled in as the corresponding

Table 9.1 Evaluation standards and scoring methods for the protection of green village and town drinking water sources and water-saving facility construction

Evaluation items	Evaluation indexes	Evaluation standards		Results of evaluation			
				Qualified (1)	Sub-standard (0)	Score	Total score
Drinking water conservation facility construction	Decentralized drinking water source protection and facility construction						
	(1) Water source protection range (choose one of the three categories, 1 point per column)	Surface water	Meets the Environmental Protection Guidelines for Dispersed Drinking Water Sources issued by the Ministry of Environmental Protection				
		Water cellar					
		Groundwater					
	(2) Water source construction and protection (choose one of the four categories, 1 point per column, full score 2–4 points)	Lake reservoir water source	Take the water intake point as close as possible to the middle of the river, the centre of the lake, or as far as possible from the riverbank or lake shore				
			The water source was preliminarily treated by sedimentation and basic disinfection				
			Built a signboard				
			Water quality meets requirements of GB 3838-2002				
		Water cellar	Built a special rainwater collection pool				

(continued)

Table 9.1 (continued)

Evaluation items	Evaluation indexes	Evaluation standards		Results of evaluation			
				Qualified (1)	Sub-standard (0)	Score	Total score
			Simple sedimentation and other purification disposal facilities, drainage channels to prevent surface runoff pollution				
			Protective net or fence and sign				
			Water quality meets requirements of GB 3838-2002				
		Well	Have built wells, kerbs and manhole covers, signs				
			The water quality meets requirements of GB/T 14848-2017				
		Mountain spring water	Built protective pools, fences and other protective facilities and signs				

(continued)

Table 9.1 (continued)

Evaluation items	Evaluation indexes	Evaluation standards		Results of evaluation			
				Qualified (1)	Sub-standard (0)	Score	Total score
			Water quality meets requirements of GB/T 14848-2017				
	(3) Water source protection requirements without water purification disposal (1 point)	Water source must meet requirements of GB 5749-2006					
	Centralized drinking water source protection and facility construction						
	(4) Water source protection scope (applicable only to towns,1 point)	The scope of the protected area is divided into primary and secondary areas, and the quasi-protected areas are classified when necessary.	Scope of the protected area meets requirements of HJ/T 338-2018				
	(5) Water quality requirements for centralized drinking water sources (applicable only to towns, choose one of three categories, 1 point per column, full score 3–4 points)	River type	Primary reserve meets the GB 3838-2002 water quality standard II				
			Secondary reserve meets the GB 3838-2002 water quality standard III				

(continued)

Table 9.1 (continued)

Evaluation items	Evaluation indexes	Evaluation standards		Results of evaluation			
				Qualified (1)	Sub-standard (0)	Score	Total score
			The quasi-protected areas shall refer to the water quality requirements of the second-class protected areas				
			Built water intake facilities, isolation facilities, signs				
		Lake water	Primary reserve meets the GB 3838-2002 water quality standard II				
			Secondary reserve meets the GB 3838-2002 water quality standard III				
			Built water intake facilities, isolation facilities and signs				

(continued)

Table 9.1 (continued)

Evaluation items	Evaluation indexes	Evaluation standards		Results of evaluation			
				Qualified (1)	Sub-standard (0)	Score	Total score
		Groundwater	Primary reserve meets the GB 3838-2002 water quality standard II				
			Secondary reserve meets the GB 3838-2002 water quality standard III				
			The quasi-protected areas shall refer to the water quality requirements of the second-class protected areas				
			Built water intake facilities, isolation facilities, signs				
Water-saving facility construction	(6) Water-saving domestic appliances (1 point per column, full score 5 points)	Water nozzle	All test data meet requirements of CJ/T 164-2014				
		Toilet and toilet system	All test data meet requirements of CJ/T 164-2014				

(continued)

Table 9.1 (continued)

Evaluation items	Evaluation indexes	Evaluation standards		Results of evaluation			
				Qualified (1)	Sub-standard (0)	Score	Total score
		Shower	All test data meet requirements of CJ/T 164-2014				
		Washing machine	All test data meet requirements of CJ/T 164-2014				
		Water-saving appliance penetration rate	≥80% (calculation method can be seen in GB/T 26928-2011)				
	(7) Village water supply network (1 point)	Water supply network leakage rate	All survey data meet requirements of GB/T 26928-2011, and the leakage rate of pipe network shall not be greater than 1% of the correction stipulated in CJJ 92-2016				
	(8) Domestic sewage disposal (1 point)	Coverage rate of rural households in domestic sewage disposal	Not less than 70%				
	(9) Use water-saving facilities such as reclaimed water (+1 point)	Pluses	Encourage the use of water-saving facilities such as rainwater and reclaimed water				

(continued)

Table 9.1 (continued)

Evaluation items	Evaluation indexes	Evaluation standards	Results of evaluation			
			Qualified (1)	Sub-standard (0)	Score	Total score

Form-filling explanation:

(1) The types of water sources evaluated in the village (town) are mainly: ____________

(2) The main methods of water supply are: ____________

(3) The full score of evaluation is ____________ points, and the actual score is ____________ points

(4) The result of the evaluation is ____________ (qualified)/(sub-standard)

score column of the "evaluation score" in the table, and the evaluation results should be obtained by filling out the "illustrations".

9.1.2 Evaluation Case of Green Village and Town Drinking Water Source Protection and Water-Saving Facility Construction

1. Investigation of drinking water source protection and water-saving facility construction

1) Condition of domestic water source

The water source for domestic water use at Yanhe Village is a decentralized domestic water source, mainly shallow groundwater, which is obtained by drilling wells. A small number of households use mountain spring water as a source of domestic water. The wells at Yanhe Village are motor-pumped wells, which are built mainly by farmers. This water is directly used by farmers after purification by simple sedimentation.

Figure 9.1 shows the actual situation of well construction for farmers in Yanhe Village.

Farmers using mountain spring water have built high water tanks, and the water flows through the farmer's self-built pipe network. This water is directly used by farmers after purification by simple sedimentation.

A real scene of a domestic high water tank in Yanhe Village is shown in Fig. 9.2.

Fig. 9.1 Actual situation of well construction for farmers in Yanhe Village

Fig. 9.2 Real scene of a high water tank for domestic water in Yanhe Village

2) Quality detection of drinking water and surface water

In 2016, all the indexes of drinking water and surface water quality in Yanhe Village met the requirements of the current national standards after being tested by professional testing units.

3) Construction of water-using facility

A single-control stainless-steel or copper-alloy single-handle ceramic plate sealing nozzle is commonly used indoors in buildings of Yanhe Village; showers adopt double-controlled stainless-steel or copper-alloy single-handle ceramic plates to seal the nozzles; toilets generally use a squatting pan and simply adopt a one-handle ball valve to control the amount of water used to flush the bowl.

In Yanhe Village, 80% of the households use solar showers, and all indicators reach *Domestic Water Saving Devices* CJ/T 164-2014 standards. Villagers generally use automatic washing machines, and the indicators reach the CJ/T 164-2014 standards.

According to the investigation, no leakage was found in the water supply pipe network at Yanhe Village.

Domestic sewage of Yanhe Village is treated by household biogas digesters or centralized sewage treatment facilities, and the coverage rate of sewage treatment for farmers surpasses 80%.

2. Evaluation results of drinking water source protection and water-saving facility construction

In Table 9.1, items (1), (2) and (5) are multiple options set according to the drinking water source protection type and the use of water-saving facilities in villages and towns of different regional types; these options need to be reasonably selected based on the actual situation of the town or village to be evaluated. The "√" column is the evaluation item selected from the multiple options; item (9) is the bonus item. The full score of the evaluation is determined according to the evaluation item.

When the evaluation criteria and scoring method in Table 9.1 were applied item by item to evaluate the implementation of drinking water source protection and water-saving facilities in Yanhe Village, item (4) and item (5) did not apply to the evaluation, so the full score of the evaluation of Yanhe Village was 11 + 1 points.

The actual evaluation score of Yanhe Village is 10 + 1 points, and the overall evaluation result is qualified. See Table 9.2.

9.2 Evaluation of Green Village and Town Domestic Sewage Treatment Facility Construction

9.2.1 Evaluation Method of Green Village and Town Domestic Sewage Treatment Facility Construction

1. Evaluation basis

Domestic sewage treatment is an important part of green village and town construction. According to the requirements of *Guidelines for the Construction of Beautiful Village* GB/T 32000-2015, villages and towns should adopt centralized or decentralized methods to treat domestic sewage according to the distribution of villages and farmers.

Green village and town domestic sewage treatment facilities can be constructed according to actual local conditions, such as the construction of sewage treatment plants and small sewage treatment stations or a combination of those facilities with the construction of biogas digesters for domestic sewage treatment.

2. Evaluation method

The comprehensive evaluation method for green village and town domestic sewage treatment facility construction is as follows:

(1) Conduct investigation and water quality testing.
(2) According to the investigation results, evaluate whether local sewage treatment facilities have been built or whether local sewage treatment measures have been taken (biogas digester, sewage treatment station, etc.). If the domestic sewage

Table 9.2 Evaluation results of drinking water source protection and water-saving facility construction in Yanhe Village

Evaluation items	Evaluation indexes			Results of evaluation			
				Qualified (1)	Sub-standard (0)	Score	Total score
Drinking water conservation facility construction	Decentralized drinking water source protection and facility construction						
	(1) Water source protection range	Meet the Environmental Protection Guidelines for Dispersed Drinking Water Sources issued by the Ministry of Environmental Protection		1			10 + 1
	(2) Water source construction and protection	Well	Have built wells, kerbs and manhole covers, signs√	1		2	
			The water quality meets requirements of GB/T 14848-2017	1			
	(3) Water source protection requirements without water purification disposal	Water source must meet requirements of GB 5749-2006		1		1	
	Centralized drinking water source protection and facility construction						
Water-saving facility construction	(6) Water-saving domestic appliances	Water nozzle	All test data meet requirements of CJ/T 164-2014	1		4	
		Toilet and toilet system	All test data meet requirements of CJ/T 164-2014	1			

(continued)

Table 9.2 (continued)

Evaluation items	Evaluation indexes			Results of evaluation			
				Qualified (1)	Sub-standard (0)	Score	Total score
		Shower	All test data meet requirements of CJ/T 164-2014	1			
		Washing machine	All test data meet requirements of CJ/T 164-2014	1			
		Water-saving appliance penetration rate	≥80%				
	(7) Village water supply network	Water supply network leakage rate	All survey data meet requirements of GB/T 26928-2011, and the leakage rate of pipe network shall not be greater than 1% of the correction stipulated in CJJ 92-2016	1	0	1	
	(8) Domestic sewage disposal	Coverage rate of rural households in domestic sewage disposal	Not less than 70%.√	1		1	
	(9) Use water-saving facilities such as reclaimed water	Pluses	Encourage the use of water-saving facilities such as rainwater and reclaimed water.√	+1		+1	

(continued)

Table 9.2 (continued)

Evaluation items	Evaluation indexes	Results of evaluation			
		Qualified (1)	Sub-standard (0)	Score	Total score
Form-filling explanation: (1) The types of water sources evaluated in the village (town) are mainly: groundwater (2) The main methods of water supply are: distributed water supply (3) The full score of evaluation is 11 points, and the actual score is 10 + 1 points (4) The result of the evaluation is qualified (qualified)/(sub-standard)					

treatment rate exceeds 70%, the answer is yes; then, enter the evaluation stage. If the answer is no, the evaluation is suspended.

(3) According to the test results, evaluate whether the indexes of sewage discharged after local treatment meet the current national standards. If yes, the score is 1; if no, the score is 0.

9.2.2 Evaluation Case of Green Village and Town Domestic Sewage Treatment Facility Construction

1. Investigation of domestic sewage treatment facility construction

1) Sewage treatment plan of Yanhe Village

The sewage of Yanhe Village consists mainly of domestic sewage, production sewage and farmhouse catering sewage. According to the current plan of Yanhe Village, rain and sewage diversion systems are adopted for village drainage, and rain and sewage are separated and discharged nearby in a systematic manner. A sewage treatment facility is located south of Yanhe Village to treat commercial and domestic sewage. The third villager group has a constructed wetland treatment tank for domestic sewage.

2) Current status of sewage treatment in Yanhe Village

At present, a sewage treatment station has been built and put into operation in Yanhe Village. Most peasant households in Yanhe Village have biogas digesters of 8 m^3 and combined biogas digesters of 80 m^3. There are 99 biogas digesters in the village, and the biogas utilization rate is 80%. Domestic sewage treatment is carried out throughout the year. Biogas is used for domestic energy, and biogas slurry and biogas residue are treated by the users themselves as fertilizers.

3) Condition of sewage discharge compliance of the Yanhe Village sewage treatment station

According to the test results of the professional testing company, the effluent indicators of the Yanhe Village sewage treatment station met the requirements of *Discharge Standard of Pollutants for Municipal Wastewater Treatment Plant* GB 18918-2002.

2. Comprehensive evaluation of domestic sewage treatment facility construction

When the comprehensive evaluation methods for village and town sewage treatment facility construction listed in Sect. 9.2.1 are applied, the comprehensive evaluation results are as follows:

(1) Yanhe Village has built sewage treatment facilities such as biogas digesters and sewage treatment stations. The domestic sewage treatment rate exceeds 70%; this, the village can enter the comprehensive evaluation programme for sewage reduction and reuse facility construction.

(2) Biogas slurry and biogas residue of Yanhe Village are treated as farmyard fertilizer, which is in line with the requirements of *Guidelines for the Construction of Beautiful Village* GB/T 32000-2015.
(3) Sewage discharged from the sewage treatment station of Yanhe Village was tested by a professional testing company, and all the indicators met the current national standards.

According to the comprehensive evaluation method of village and town domestic sewage treatment facility construction mentioned above, the comprehensive evaluation of the domestic sewage treatment facility construction in Yanhe Village scores 1 point.

9.3 Evaluation of Green Village and Town Waste Classified Collection and Harmless Treatment Facility Construction

9.3.1 Evaluation Method for Green Village and Town Waste Classified Collection and Harmless Treatment Facility Construction

1. Evaluation technical route

The evaluation of village and town waste classified collection and harmless treatment facility construction should be based on relevant national industrial standards and technical regulations.

The green village and town waste classified collection and harmless treatment facility construction evaluation technical route can be seen in Fig. 9.3.

2. Evaluation index and method

According to the implementation evaluation content, the implementation evaluation standard and scoring method of green village and town waste classified collection and harmless treatment facility construction can be established. The evaluation indexes are divided into two categories with a total of 7 items.

The implementation evaluation of green village and town waste classified collection and harmless treatment facility construction adopts a 0–1 scoring method. If the evaluation index meets the current national/department/industry standards or specifications, the score is 1. If the evaluation index does not meet the relevant standards or specifications, the score is 0. A full score is 7. A score of more than 70% of the total score is considered eligible.

Refer to Table 9.3 for the evaluation criteria and scoring methods of green village and town waste classified collection and harmless treatment facility construction.

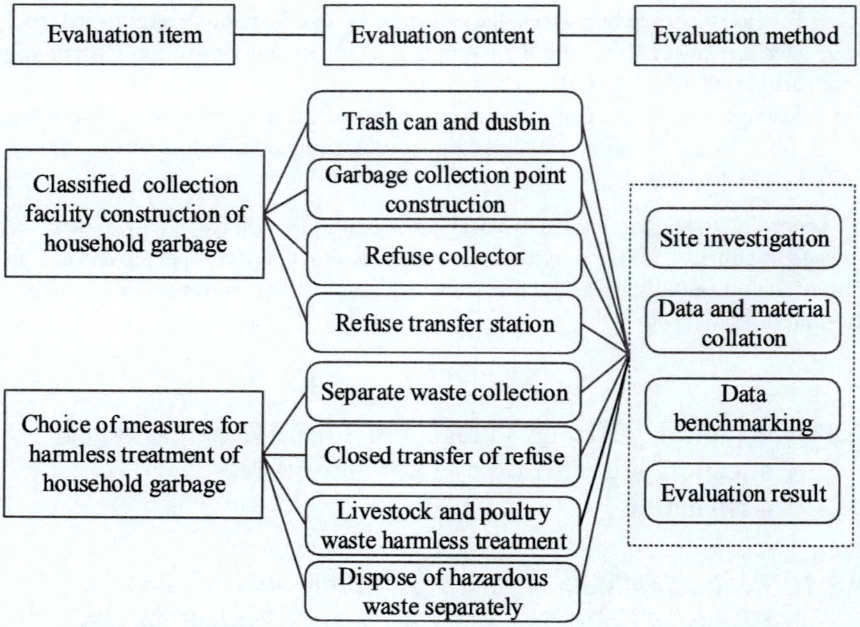

Fig. 9.3 Green village and town waste classified collection and harmless treatment facility construction Evaluation technical route

During the actual evaluation operation, the actual scores of each sub-item of the village or town to be evaluated are filled into the corresponding score column in the table, and the evaluation results can be obtained directly.

9.3.2 Evaluation Case of Green Village and Town Waste Classification Collection and Harmless Treatment Facility Construction

1. Investigation of garbage classification collection and harmless treatment facilities

In 2004, Yanhe Village began to carry out the construction of domestic waste classified collection and treatment-transfer facilities. The analysis of the basic components of domestic waste in Yanhe Village is shown in Table 9.4.

Table 9.3 Evaluation standards and methods of green village and town waste classified collection and harmless treatment facility construction

Evaluation items	Evaluation indexes	Evaluation standards	Results of evaluation			
			Qualified (1)	Sub-standard (0)	Score	Total score
Construction of domestic waste separation and collection facilities (4 points)	(1) Setting of trash cans and dustbins	Each household is equipped with 2 trash cans with shields, anti-leakage protection and clear markings				
	(2) Construction of waste collection points	Waste collection point service radius: Town ≤ 100 m, village ≤ 200 m				
	(3) Setting of garbage collection vehicle	Waste collection vehicle type: Van type human collection vehicle, effective volume 0.5–1 m^3; mechanical collection vehicle, effective volume 2–5 m^3				
	(4) Construction of garbage transfer station	The town sets a waste transfer station. The coverage of the waste transfer station is generally within 5 km, the area is not less than 100 m^2, and it can store all the waste generated in a day				

(continued)

2. Evaluation of waste classified collection and harmless treatment facility construction

The evaluation standards and methods shown in Table 9.3 are used to evaluate the construction status of domestic waste classified collection and harmless treatment

Table 9.3 (continued)

Evaluation items	Evaluation indexes	Evaluation standards	Results of evaluation			
			Qualified (1)	Sub-standard (0)	Score	Total score
Selection of harmless disposal measures for domestic garbage (3 points)	(5) Waste classified collection	Waste sorting must be carried out; it can be classified into organic waste, recyclable waste, non-recyclable waste and harmful waste				
	(6) Garbage closed transport	Equipped with a dedicated waste collection vehicle				
	(7) Separate disposal of harmful waste	Harmful waste individual handling rate reaches 100%				

facilities in Yanhe Village item by item. The total score is 7, and the overall evaluation result is qualified, as shown in Table 9.5.

9.4 Evaluation of Green Village and Town Clean Energy Facility Construction

9.4.1 Evaluation Method of Green Village and Town Clean Energy Facility Construction

1. Evaluation technical route

The construction of clean energy facilities is regional and should be adapted to the local climate, natural conditions, resource environment, economic level and villagers' living habits. Due to the different types and sizes of clean energy facilities in different villages and towns, it is not suitable to simply delimit unified evaluation criteria. Therefore, we propose an evaluation method based on environmental benefits.

When evaluating the implementation effect of village and town clean energy facility construction, six indicators are obtained first: CO_2 emissions reduction benefit, SO_2 emissions reduction benefit, NO_x emissions reduction benefit, TSP emissions reduction benefit, solid waste emissions reduction benefit, and health benefits.

Table 9.4 Analysis of basic components of domestic waste of Yanhe Village

Category	Organic matter		Inorganic matter		Recyclable						Harmful garbage
	Animal	Plant	Lime soil	Brick ceramic	Paper	Plastic and rubber	Textiles	Glass	Metal	Wood and bamboo	
Mass/kg	0.35	2.8	0.25	0.1	1.65	1.1	0.2	1.65	0.2	0.05	0.3

Table 9.5 Evaluation results of construction for waste classified collection and harmless treatment facilities in Yanhe Village

Evaluation items	Evaluation indexes	Results of evaluation			
		Qualified (1)	Sub-standard (0)	Score	Total score
Construction of domestic waste separation and collection facilities (4 points)	(1) Setting of trash cans and dustbins	1		4	7
	(2) Construction of waste collection points	1			
	(3) Setting of garbage collection vehicle	1			
	(4) Construction of garbage transfer station	1			
Selection of harmless disposal measures for domestic garbage (3 points)	(5) Waste classified collection	1		3	
	(6) Garbage closed transport	1			
	(7) Separate disposal of harmful waste	1			

Note Detailed evaluation standards are shown in Table 9.3

Additionally, there are comprehensive environmental benefits of green village and town clean energy facility construction.

See Fig. 9.4 for the green village and town clean energy facility evaluation technical route.

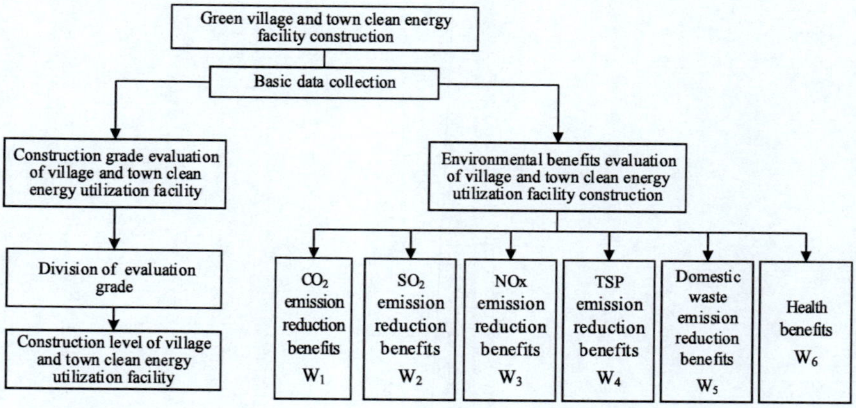

Fig. 9.4 Green village and town clean energy facility evaluation technical route

2. Grade evaluation of village and town clean energy utilization facility construction

According to *Guidelines for the Construction of Beautiful Village*, "The countryside should scientifically use and gradually reduce the direct use of traditional fuels such as wood, grass, straw and bamboo, and promote the use of clean energy such as electricity, solar energy, wind energy, biogas and natural gas. The proportion of peasant households which use clean energy should be more than 70%." The classification criteria for village and town clean energy utilization facility construction can be formulated. The proportions of 5 types of clean energy that rural households use, electricity, solar energy, biogas, natural gas, and wind energy, are mainly evaluated. The average value of the proportion of rural households using all types of clean energy was taken as the proportion of rural households using clean energy. The average value of the proportion of farmers using all kinds of clean energy was taken as the proportion of farmers using clean energy.

If the proportion of peasant households using clean energy in a village or town is $\geq 70\%$, then the village or town meets the standard and receives 1 point. The full score is 1 point.

The score of village or town clean energy facility construction evaluation is divided into five levels: A, B, C, D, and E, ranging from "Excellent" to "Poor", and the corresponding score is 1–0.2. For example, if the proportion of households using clean energy in a village or town is $\geq 70\%$, 1 point will be scored; if the proportion is $\geq 50\%$, 0.8 points will be scored; and so on, no households using clean energy will be scored 0 points as shown in Table 9.6. For practical application, please refer to the evaluation method in Table 9.9 and Sect. 9.4.2.

3. Environmental benefit evaluation of village and town clean energy utilization facility construction

1) CO_2, SO_2, NOx, and TSP Emissions Reduction Benefits

Comparing specific energy supply schemes with traditional energy (firewood, straw, and coal) in villages and towns, we then calculate the emissions reduction of CO_2, SO_2, NO_x, and TSP. According to the estimation of the environmental value of pollutants in the thermal power industry, CO_2 is 0.023 yuan/kg, SO_2 is 6.00 yuan/kg, NO_x is 8.00 yuan/kg, and TSP is 2.20 yuan/kg.

The pollutant production, calorific value and heat conversion efficiencies of household energy are shown in Table 9.7.

Table 9.6 Evaluation grade of village and town clean energy facility construction

Evaluation grade	A	B	C	D	E
Proportion of peasant households using clean energy/%	70 –100	50 –69	30 –49	20 –29	1 –19
Score of evaluation	1	0.8	0.6	0.4	0.2

Table 9.7 Contaminant production, calorific value, and heat conversion efficiencies of household energy

Household energy	Straw	Firewood	Coal	Biogas
CO_2 (m^3/kg)	0.78	0.91	0.85	1.17
SO_2 (g/kg)	0.53	0.63	35.2	0.63
NOx (g/kg)	1.29	0.70	1.88	0.10
TSP (g/kg)	7.50	6.00	1.30	0.73
Calorific value (kJ/kg)	14,635	16,711	23,325	20,934
Heat conversion efficiency (%)	15	18	22	60

Traditional rural household energy (cooking and hot water) is provided mainly by straw, coal and firewood, of which the proportion reaches 57%, followed by coal at 31%.

The formula is as follows:

CO_2 emissions reduction = CO_2 emissions of traditional energy – CO_2 emissions of clean energy, i.e.,

$$CO_2 \text{ emissions reduction} = \sum_{i=1}^{n} \alpha_i M_i - \sum_{i=1}^{n} \beta_i N_i \quad (9.1)$$

where α_i, β_i is units of CO_2 emissions reduction of traditional and clean energy; and M_i and N_i are the amounts of traditional energy and clean energy, respectively.

The calculation formula for CO_2 emissions reduction benefits is as follows:

$$CO_2 \text{ emissions reduction benefit} = CO_2 \text{ emissions reduction} \times 0.023 \text{ yuan/kg} \quad (9.2)$$

In the same way, the emissions reduction benefits of SO_2, NO_x and TSP can be obtained.

2) Organic waste biogas waste reduction benefits

When biogas digesters and returning waste to the fields are used to treat some solid waste, the market value method is used to calculate the solid waste reduction benefits. Generally, the landfill disposal cost of 1 ton of waste is 50–70 yuan. For this evaluation, a value of 60 yuan was used.

$$\text{Domestic waste emissions reduction benefits} = \text{Domestic garbage recycling (t)} \times 60 \text{ (yuan)} \quad (9.3)$$

3) Health Benefits

The deterioration of environmental quality may cause economic losses related to health, such as premature death, illness, and increased medical expenses. According to the disease cost method, biogas digesters improve environmental hygiene, resulting in a reduction in annual medical expenses of 15.9 yuan [1]. For this evaluation, biogas digesters were chosen as the base data for health benefit evaluation calculations. The average annual medical expense of the villagers was reduced by 15.9 yuan by correctly using biogas digesters. This value can be adjusted according to the actual situation of different areas and evaluation times.

The health benefit formula is as follows:

$$\text{Annual Health Benefit} = \text{Number of biogas digester users} \times 15.9\ \text{yuan/person} \tag{9.4}$$

9.4.2 *Evaluation Case of Green Village and Town Clean Energy Facility Construction*

1. Data acquisition

As of the investigation period from April 20 to April 22, 2016, the total number of households in Yanhe Village was 303, with a total population of 1050 people and 3.47 persons per family.

According to the grade evaluation method of village and town clean energy utilization facility construction proposed in Sect. 9.4.1, the clean energy used in Yanhe Village in 2016 included electric energy, solar energy, biogas and natural gas. Some farmers also used a small amount of firewood to cook their family meals in combination with carrying out farmhouse tourism, which was not included in the assessment of clean energy use.

The use of clean energy and the proportion of farmers of Yanhe Village in 2016 are shown in Table 9.8.

Table 9.8 Overall situation of energy consumption of Yanhe Village in 2016

Energy type	Usage status	Amount of energy consumed	Proportion of total peasant households (%)
Electricity	303	638,333 kWh/year	100.00
Solar water heaters	252	–	83.33
Biogas	208	–	68.75
Liquefied petroleum gas	252	2348 cans/year	83.33
Wind and other clean energy sources	–	–	–
Other supplementary energy sources	–	–	(Not included)
Firewood	221	353,535 kg/year	72.92

Table 9.9 Comprehensive evaluation of clean energy utilization facility construction in Yanhe Village

Evaluation grade	A	B	C	D	E
Proportion of peasant households using clean energy (%)	70–100	50–69	30–49	20–29	1–19
Evaluation score	1	0.8	0.6	0.4	0.2
Yanhe Village score	1	–	–	–	–

2. Evaluation process

1) Comprehensive evaluation of clean energy utilization facility construction in Yanhe Village

According to the basic data of clean energy utilization facility construction in Yanhe Village obtained from the site investigation and the evaluation method proposed in Sect. 9.4.1 and Table 9.6, the comprehensive level of clean energy utilization facility construction in Yanhe Village can be evaluated.

A comprehensive evaluation of the clean energy utilization facility construction of Yanhe Village is shown in Table 9.9.

Evaluation conclusion: According to the investigation results of 2016 and the evaluation level in Table 9.9, the comprehensive level of clean energy facility construction in Yanhe Village received the full score of 1 point. The evaluation grade was A.

2) Environmental benefit evaluation of clean energy facility construction in Yanhe Village

The environmental benefit evaluation indicators for the construction of clean energy facilities in Yanhe Village included CO_2 emissions reduction benefit W_1; SO_2 emissions reduction benefit W_2; NO_x emissions reduction benefit W_3; TSP emissions reduction benefit W_4; solid waste reduction benefit W_5; and health benefit W_6. According to the investigation data on the actual situation; Formulas (9.1), (9.2), and (9.3); and Table 9.9, the CO_2, SO_2, NOx and TSP annual emissions reduction benefits, household garbage emissions reduction benefits and health benefits of Yanhe Village can be calculated.

See Table 9.10 for the environmental benefit evaluation results of clean energy facility construction in Yanhe Village.

3. Evaluation results

According to available data in 2016, the construction status of clean energy facilities in Yanhe Village was obtained. In terms of renewable energy utilization, the main biomass energy used in Yanhe Village were biogas and solar energy.

In the village, 83.33% of the peasant households had solar water heaters, 37 solar streetlamps had been installed, the utilization rate of biogas had reached 68.75%, and 208 households had built biogas digesters, with several households building them by

Table 9.10 Environmental benefit evaluation results of clean energy facility construction in Yanhe Village

Evaluation project	Annual emissions reduction (kg)	Annual emissions reduction benefits (yuan)
CO_2	3.97×10^5	9131
SO_2	2311	13,866
NO_x	348	2784
TSP	1591	3500
Domestic waste	2076,789	124,607
Health	–	11,476

themselves and others building them together. These were of different sizes: 8 m^3 and 80 m^3. The clean energy penetration rate of Yanhe Village had reached 83.33%.

In summary, according to the comprehensive calculation of Formulas (9.1) to (9.4), the total annual benefits of the comprehensive environmental benefits of clean energy facility construction in Yanhe Village are:

$$9131 + 13866 + 2784 + 3500 + 124607 + 11476 = 165364 \text{ (yuan/year)}$$

9.5 Evaluation of Village and Town Greening and Landscape Facility Construction

9.5.1 Evaluation Method of Village and Town Greening and Landscape Facility Construction

1. Technical process of evaluation

The evaluation of green village greening and landscape construction can be carried out according to the following 3 steps.

(1) On-site Investigation. The investigation focuses on greening plans and landscape construction plans, benchmarking inspections of green land construction and landscape construction projects, willingness surveys of residents, and field surveys of human landscape construction.
(2) Field Survey. The field survey focuses on measuring the spatial layout of greening projects, measuring the protection status of ancient trees, and measuring physical survey construction and benchmarking.
(3) Comprehensive Evaluation. According to the current relevant standards, norms and construction regulations, a comprehensive evaluation of village greening and landscape facility construction is carried out.

2. Establishment of evaluation indexes

The evaluation of green village and town greening and landscape facility construction is based mainly on *Standard for Planning of Town* GB 50188-2007; *Technique Code for Village Rehabilitation* GB 50445-2008; *Guidelines for the Construction of Beautiful Village* GB/T 32000-2015; *Code for the Design of Urban Green Space* GB 50420-2007; and *Technical Specification for Village Landscape Environmental Engineering* CECS 285-2011.

The evaluation indexes of green village and town greening and landscape facilities include 4 first-level indexes: greening and beautification, natural landscape protection, artificial landscape construction and human landscape protection, and 16 secondary indexes, such as greening and planting. See Table 9.11.

3. Evaluation method

The green village and town greening and landscape facility construction evaluation adopts the 0–1 scoring method. A project that has completed the construction content of the second-level index and met the scoring standard will receive 1 point, and a project that has not will receive 0 points. A full mark is 16 points. If the score is 70% or more of the total score, then the project is qualified.

See Table 9.11 for the evaluation criteria and scoring methods of green village and town greening and landscape facility construction. During the actual evaluation operation, the actual scores of each sub-item of the village or town to be evaluated are filled into the corresponding score column in the table, and the evaluation results can be obtained directly.

9.5.2 Evaluation Case of Village and Town Greening and Landscape Facility Construction

1. Investigation of greening and landscape facility construction

1) Overview of greening and landscape of Yanhe Village

The area where Yanhe Village is located is in the north sub-tropical evergreen deciduous broad-leaved mixed forest zone with more than 1,500 species of vegetation. The main types of vegetation in this area include subtropical evergreen coniferous forest, subtropical deciduous coniferous forest, subtropical deciduous broad-leaved forest, evergreen broad-leaved forest, evergreen broad-leaved mixed forest, and bamboo forest shrub plant, with nearly 200 species of common trees.

A river runs east-west through Yanhe Village. The water area is nearly 0.67 hm^2. There are five ponds in the village, and the aquaculture water area is 0.67 hm^2, with abundant water resources. Since the demonstration project was launched, Yanhe Village has carried out rectification work on the banks of the Yanhe River and achieved good ecological and environmental effects.

Table 9.11 Evaluation standards and methods of green village and town greening and landscape facility construction

Evaluation items	Evaluation indexes	Evaluation standards	Evaluation results			
			Qualified (1)	Sub-standard (0)	Score	Total score
Greening and beautification (5 points)	(1) Green planting	Proportion of native fruit trees, wood flowers and plants is $\geq 70\%$				
	(2) Green project spatial layout	Planting of green plants is coordinated with local topography and land; achieve all, get 1 point				
	(3) Public green space construction	Construction of public green space in the township and the proportion of greenbelt land meet requirements of current standards and specifications; achieve all, get 1 point				
	(4) Protective green space construction	Township: Construction of protective green space or greenbelt for water source, industrial and mining enterprises, aquaculture, railways and roads, high-voltage power-line corridors, windbreaks, etc.; achieve all, get 1 point Village: Greenbelts such as roads, rivers, and farmland were built; achieve all, get 1 point				
	(5) Ancient tree protection	According to the current standards and specifications, protective measures were taken, signs were fully established; 1 point for fully erecting signag				

(continued)

Table 9.11 (continued)

Evaluation items	Evaluation indexes	Evaluation standards	Evaluation results			
			Qualified (1)	Sub-standard (0)	Score	Total score
Natural landscape protection (4 points)	(6) Topographical protection	Ecological conservation of the mountain; no major changes are found in the topography after the completion of the green space construction project; 1 point for implementing 1 item				
	(7) Vegetation protection	Ecological conservation of forests, vegetation, etc.; after the completion of the construction project, no local vegetation has been destroyed; 1 point for implementing 1 item				
	(8) Water system protection	Keep the natural state of the river, do not loosen or straighten the river; unwind the natural state of the pit, do not overuse random landfill; the stone should be selected from local stone, the ecological slope should be planted with bags (cages) or ecological bricks, etc.; achieve all, get 1 point				

(continued)

Table 9.11 (continued)

Evaluation items	Evaluation indexes	Evaluation standards	Evaluation results			
			Qualified (1)	Sub-standard (0)	Score	Total score
	(9) Original form and style of the village and town continue to be preserved	Results of the on-site investigation after the village was rebuilt showed that ≥ 80% of the original villagers believed that the renovation project protected and continued the original form of the village; get 1 point				
Artificial landscape construction (4 points)	(10) Buildings	Village and town architectural design plan refers to the local traditional architectural style; the detailed design of the building has the local traditional style; using the environmental-friendly economic building materials with the local traditional style; 1 point for implementing 1 item				
	(11) Courtyard	Spatial structure conforms to local traditional customs; gardens are planted with native plants; courtyard pavilion uses environmentally friendly materials that are permeable; 1 point for performing 2 tasks				

(continued)

Table 9.11 (continued)

Evaluation items	Evaluation indexes	Evaluation standards	Evaluation results			
			Qualified (1)	Sub-standard (0)	Score	Total score
	(12) Public space	Town area: the construction of park chairs, park lights, waste bins, toilets and other facilities meets requirements of current standards and specifications. Village and town: the facilities and space construction of the children and adults in the green space should meet safety and health requirements, and avoid interference with the surrounding environment; the facility signage shall comply with current national standards and be installed in a location that is compatible with the environment and does not interfere with pedestrian traffic. Achieve all, get 1 point				
	(13) Landscape sketch	Theme, form, materials and volume of the sculptures and landscape pieces in the village green space are in harmony with the environment; encourage the villagers to build small landscape pieces and create resting spaces near their own courtyards or houses; 1 point for implementing 1 item				

(continued)

Table 9.11 (continued)

Evaluation items	Evaluation indexes	Evaluation standards	Evaluation results			
			Qualified (1)	Sub-standard (0)	Score	Total score
Human landscape protection (3 points)	(14) Protection and utilization of local folk customs	Protection and utilization of representative life and production landscape elements that embody local folk customs; achieve all, get 1 point				
	(15) Renovation and protection of traditional houses in ancient town or village	Renovation and protection of ancient towns, ancient villages, traditional dwellings, ancient wells, etc.; get 1 point				
	(16) Historical and cultural heritage and protection of local characteristics	For the protection of historical and cultural heritages and local features, one point will be awarded if the Law of the *People's Republic of China on the Protection of Cultural Relics and the Regulations on the Protection of Historical Villages and Towns* is strictly implemented; get 1 point				

Yanhe Village vigorously developed tea production and developed a variety of operations, forestry production and aquaculture according to local conditions. Yanhe Village plans to implement an agricultural landscape upgrading project featuring ecological farmland, forest orchards and courtyard vegetable gardens, forming a good agricultural ecosystem and characteristic landscape.

2) Construction of greening and landscape facilities in Yanhe Village

In 2005, Yanhe Village began to implement an ecological landscape construction plan. Afforestation and road greening were carried out on the mountain in the village, a 400-metre-long farm melon and fruit ecological corridor was built, ditches in the village and 5 weir ponds were renovated, and a landscape pool was built in the new village of the Weir River.

The 14 newly built farmhouses in Yanhe New Village have different shapes and constitute a village landscape with the architectural style of mountainous areas in western Hubei. The whole village implemented a beautification project in front of the houses and restored a century-old ancient well. A 2-km-long riverbank comprehensive renovation project was completed. The ecological tourism landscape facilities of the Bairi Mountain Scenic Area were built.

The cultural square, village office, village reception centre, village history education centre, stone museum, Tianyi Tea House, Phoenix Stockade Village and other community public leisure places and landscape supporting facilities were built. The village has established a regional rural culture, economic and trade exchange platform and rural training base featuring the famous tea village and "Tea King" competition and won the honour of a national ecological village and green well-off village.

Real scenes of greening and landscape facility construction in Yanhe Village are shown in Fig. 9.5.

2. Evaluation of greening and landscape facility construction

Applying the method in Table 9.11, the implementation status of greening and landscape facility construction in Yanhe Village is evaluated item by item. The total score is 16 points, and the evaluation result is qualified. See Table 9.12.

9.6 Evaluation of Green Village and Town Disaster Prevention, Mitigation and Early-Warning Facility Construction

9.6.1 Evaluation Method of Green Village and Town Disaster Prevention, Mitigation and Early-Warning Facility Construction

1. Evaluation content

(a) Farmhouse greening

(b) Adobe house and outdoor sketches

(c) Tea house outdoor greening

(d) Rural ecological corridor

Fig. 9.5 Real scenes of greening and landscape facility construction in Yanhe Village

The evaluation contents of green village and town disaster prevention, reduction, and early-warning facility construction can be divided into four categories:

(1) Fire control facility construction and early-warning measures;
(2) Flood control facility construction and early-warning measures;
(3) Anti-seismic facility construction and early-warning measures; and
(4) Wind prevention and disaster reduction facility construction and early-warning measures.

See Fig. 9.6 for the green village and town disaster prevention, mitigation and early-warning facility construction evaluation technical route.

2. Evaluation method

According to the evaluation content, the evaluation standard and scoring method of green village and town disaster prevention, mitigation and early-warning facility construction implementation can be established. The evaluation indexes are divided into 4 categories with 17 items in total.

The comprehensive evaluation of green village and town disaster prevention, reduction, and early-warning facility construction adopts a 0–1 scoring method. If the evaluation index meets the current national/sector/industry standard or norm, it

Table 9.12 Evaluation results of greening and landscape facility construction in Yanhe Village

Evaluation items	Evaluation indexes	Evaluation results			
		Qualified (1)	Sub-standard (0)	Score	Total score
Greening and beautification (5 points)	(1) Green planting	1		5	16
	(2) Green project spatial layout	1			
	(3) Public green space construction	1			
	(4) Protective green space construction	1			
	(5) Ancient tree protection	1			
Natural landscape protection (4 points)	(6) Topographical protection	1		4	
	(7) Vegetation protection	1			
	(8) Water system protection	1			
	(9) Original form and style of the village and town continue to be preserved	1			
Artificial landscape construction (4 points)	(10) Buildings	1		4	
	(11) Courtyard	1			
	(12) Public space	1			
	(13) Landscape sketch	1			
Human landscape protection (3 points)	(14) Protection and utilization of local folk customs	1		3	
	(15) Renovation and protection of traditional houses in ancient villages and towns	1			
	(16) Historical and cultural heritage and protection of local characteristics	1			

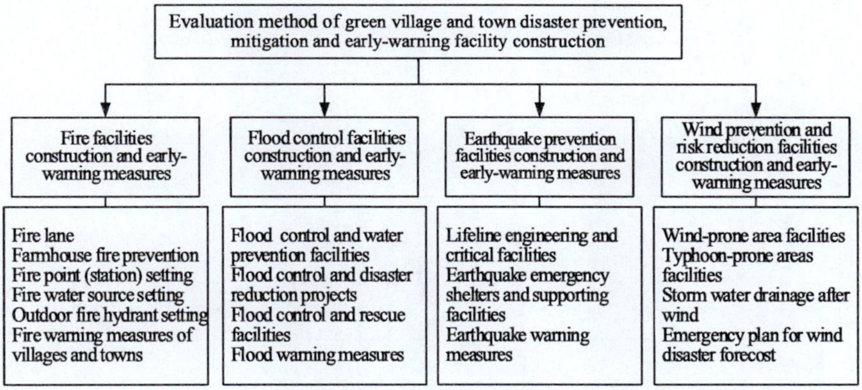

Fig. 9.6 Green village and town disaster prevention, mitigation and early-warning facility construction evaluation technical route

scores 1 point; if not, it scores 0 points. The full score is 17, and if the score reaches or exceeds 70% of the total score, the project is qualified. See Table 9.13.

If an index in Table 9.13, such as item 15, "Typhoon-prone area facilities", is not applicable to a village or town in a non-seaside area, the index is omitted, and the full score of the evaluation is adjusted from 17 to 16. If a project receives 11 points, it reaches more than 70% of the eligibility criteria, and so on. When filling out the form, the evaluator can indicate whether an index is "not participating".

In an actual evaluation, if there are multiple requirements in a certain scoring standard, as long as one of them is satisfied, 1 point will be awarded. In the evaluation content of the four types of indexes, in principle, a certain scoring standard should be only used once to avoid repeated evaluation.

The evaluation standards and methods of green village and town disaster prevention, reduction and early-warning facility construction is shown in Table 9.13.

During the actual evaluation operation, the actual scores of each sub-item of the village or town to be evaluated are filled in the corresponding score column in the table, and the evaluation result can be directly obtained.

9.6.2 Evaluation Case of Green Village and Town Disaster Prevention, Reduction and Early-Warning Facility Construction

1. Investigation of disaster prevention, reduction and early-warning facility construction

1) Disaster prevention and emergency plan of Yanhe Village

Table 9.13 Evaluation standards and methods of green village and town disaster prevention, mitigation and early-warning facility construction

Evaluation items	Evaluation indexes	Evaluation standards	Evaluation results			
			Qualified (1)	Sub-standard (0)	Score	Total score
Fire facility construction and early-warning measures (6 points)	(1) Fire lane	The net width and clearance height of the lane are not less than 4.0 m; the turning radius meets the turning requirements of the fire truck; there are no obstacles such as trees and overhead pipelines that hinder the operation of the fire truck between the fire lane and the building				
	(2) Fire protection of rural buildings	Warehouses for storing firewood and agricultural implements and agricultural materials should be constructed independently; no fewer than 2 fire safety exits for public buildings				
	(3) Fire point (station) setting	There are fixed-location houses and obvious signs; firefighting equipment such as fire trucks, hand-lifted motorized pumps, water guns, hoses, fire extinguishers, and rescue tools; heat preservation measures in cold areas				

(continued)

Table 9.13 (continued)

Evaluation items	Evaluation indexes	Evaluation standards	Evaluation results			
			Qualified (1)	Sub-standard (0)	Score	Total score
	(4) Fire water source setting	Set up an outdoor fire water supply system; build a fire pool or use natural water sources				
	(5) Outdoor fire hydrant setting	The pressure of the fire hydrant port should not be less than 0.1 MP, or the pressure and flow requirements of the most unfavourable point of the building within the protection radius should be met; The diameter of the fire water supply pipe should not be less than 100 mm, the buried depth should meet the requirements, and there are anti-freezing measures in cold areas; fire hydrants are arranged along the road, the spacing is not more than 120 m, and the distance from the exterior wall of the building is not less than 2 m				
	(6) Village and town fire warning measures	Fix fire alarm calls and establish reliable contact with city firefighting centres; have full-time and voluntary firefighting teams; regularly conduct firefighting knowledge publicity and fire drills; formulate fire prevention and emergency response plans				

(continued)

Table 9.13 (continued)

Evaluation items	Evaluation indexes	Evaluation standards	Evaluation results			
			Qualified (1)	Sub-standard (0)	Score	Total score
Flood control facility construction and early-warning measures (4 points)	(7) Flood control and water protection facilities	Flood prevention safety facilities such as cofferdams, safety stations, water shelters, etc.				
	(8) Flood control and disaster reduction project	Riverbank remediation, river dredging, dam construction, flood control, stagnation, etc.				
	(9) Flood prevention and rescue facilities	Emergency evacuation points, medical ambulance, material reserves and alarm devices, etc.				
	(10) Flood prevention warning measures	There are flood control early-warning mechanisms, flood control and risk prevention drills, flood prevention science popularization education, etc.				
Anti-seismic and shockproof facility construction and early-warning measures (3 points)	(11) Lifeline engineering and important facilities	Safeguard transportation, communication, water supply, power supply, energy, fire protection, medical facilities, etc., set up multiple entrances and exits in different directions, emergency command organization to set up backup power, set up suspension and evacuation sites				

(continued)

Table 9.13 (continued)

Evaluation items	Evaluation indexes	Evaluation standards	Evaluation results			
			Qualified (1)	Sub-standard (0)	Score	Total score
	(12) Earthquake emergency shelters and supporting facilities	Choose parks, green spaces, squares, stadiums, etc. as emergency shelters and configure basic facilities such as emergency shelters, medical assistance and sanitation, emergency water supply and power supply, emergency sewage systems and toilets, emergency garbage storage and transportation				
	(13) Early-warning measures for earthquake disasters	There are earthquake prediction and early-warning mechanisms, earthquake prevention and risk prevention drills, earthquake and earthquake prevention science popularized education, etc.				
Wind and disaster reduction facility construction and early-warning measures (4 points)	(14) Facilities in the wind-prone area	Select intensive shelterbelts on the edge of the windward direction, build emergency shelters, and regularly check the wind and risk avoidance of buildings (structures)				

(continued)

Table 9.13 (continued)

Evaluation items	Evaluation indexes	Evaluation standards	Evaluation results			
			Qualified (1)	Sub-standard (0)	Score	Total score
	(15) Facilities in typhoon-prone area	Construction of dams to withstand the impact of storm surges, strengthening the construction of seawalls, dikes, shelters, windbreaks, sheltered anchorages, emergency shelters, and personnel transfer plans				
	(16) After-storm drainage system	Construction of drainage system according to maximum precipitation, timely elimination of post-windstorm				
	(17) Emergency plan for wind disaster forecast	Built a typhoon wind forecast information network, with wind disaster emergency plan, equipped with medical and rescue facilities				

Based on the disaster prevention and emergency plan system of Yanhe Village, we conducted an investigation and review of the construction and implementation of the disaster prevention and emergency plan.

2) Construction of fire protection facilities in Yanhe Village

There is a fire station in Yanhe Village that is equipped with fire protection and rescue equipment and facilities, and CO_2 fire extinguisher equipment is also provided for every 5 households. In the whole village, a fire protection system for the residential area is provided by a village centre fire station-fire truck-5 household fire extinguisher and household fire extinguishing water source protection; the fire lane meets the specified requirements, the village has formulated the "Yanhe Village Forest Fire Prevention Emergency Plan", and a disaster prevention emergency self-help team composed of members of the village committee and four village groups has been established.

3) Construction of flood control facilities in Yanhe Village

Yanhe Village carried out planned repair and maintenance and comprehensive river improvements on the riverbank along the Yanhe River. In places where floods, landslides and other natural disasters are prone to occur in the village, flood prevention and reduction facilities such as flood control levees, intercepting ditches, masonry slope protection and retaining walls have been built. The once-in-50-year flood control requirements have been met.

4) Construction of earthquake-resistant, windproof and disaster-reducing facilities in Yanhe Village

Yanhe Village has adopted the earthquake-resistant design for farmhouses and low-cost earthquake-resistant construction technology to complete the earthquake-resistant and energy-saving farmhouse demonstration project, which has led to the development of the earthquake-resistant and reinforced construction of farmhouses in the whole village and surrounding villages. It was appraised as a demonstration project for the earthquake resistance of rural houses in Hubei Province.

Nine disaster prevention and emergency evacuation sites were built in Yanhe Village, and temporary refuge areas and refuge sites were set up uniformly using landscape land and tea gardens equipped with emergency backup lighting facilities. Additionally, village-level highways around the village were constructed to meet the requirements of earthquake disaster prevention facilities.

Because the village is located in a hilly area, there is no threat of strong winds or typhoons, so this item is not included.

5) Facilities and publicity for flood prevention, disaster reduction and early warning in Yanhe Village

Yanhe Village used a variety of methods, such as propaganda columns, window displays, and signage, to carry out regular and varied publicity and education on

(a) Roads meet fire protection requirements (b) Riverbank remediation project

(c) Yanhe safety emergency square (d) Meteorological information service station

Fig. 9.7 Real scenes of disaster prevention, mitigation and early-warning facilities in Yanhe Village

disaster prevention and reduction. It used television, radio, LED displays, etc. to introduce disaster prevention and emergency knowledge to the villagers and broadcast videos of disaster prevention and emergency knowledge. A village-level weather forecast release information system was established, and an emergency plan for disaster prevention and emergencies in Yanhe Village was formulated.

Real scenes of disaster prevention, mitigation and early-warning facilities in Yanhe Village are shown in Fig. 9.7.

2. Evaluation of disaster prevention, mitigation and early-warning facility construction

The evaluation criteria and scoring methods of Table 9.13 were used to evaluate the construction status of disaster prevention, reduction and early-warning facilities in Yanhe Village. According to the natural conditions of a village located in a central hilly and mountainous area, items 14, 15 and 17 in the table did not participate in the evaluation; thus, 14 items were included in the evaluation. The total score of Yanhe Village is 12, accounting for 85.7% of the full score of 14, more than 70%, and the evaluation result is qualified. See Table 9.14.

Table 9.14 Evaluation results of disaster prevention, mitigation and early-warning facility construction in Yanhe Village

Evaluation items	Evaluation indexes	Evaluation results			
		Qualified (1)	Sub-standard (0)	Score	Total score
Fire facility construction and early-warning measures (6 points)	(1) Fire lane	1		6	12
	(2) Fire protection of rural buildings	1			
	(3) Fire point (station) setting	1			
	(4) Fire water source setting	1			
	(5) Outdoor fire hydrant setting	1			
	(6) Village and town fire warning measures	1			
Flood control facility construction and early-warning measures (4 points)	(7) Flood control and water protection facilities	1		3	
	(8) Flood control and disaster reduction project	1			
	(9) Flood prevention and rescue facilities	1			
	(10) Flood prevention warning measures		0		
Anti-seismic and shockproof facility construction and early-warning measures (3 points)	(11) Lifeline engineering and important facilities	1		2	
	(12) Earthquake emergency shelters and supporting facilities	1			
	(13) Early-warning measures for earthquake disasters		0		
Wind and disaster reduction facility construction and early-warning measures (4 points)	(14) Facilities in the wind-prone area	/		1	
	(15) Facilities in typhoon-prone area	/			
	(16) After-storm drainage system	1			

(continued)

Table 9.14 (continued)

Evaluation items	Evaluation indexes	Evaluation results			
		Qualified (1)	Sub-standard (0)	Score	Total score
	(17) Emergency plan for wind disaster forecast	/			

Reference

1. Li P (2007) Research on the economic and environmental benefits of building Biogas digester for Rural households. Dissertation, Nanjing Agricultural University

Chapter 10
Evaluation of Village and Town Green Building Construction

Abstract Evaluation of the implementation effect of green building construction plays an important role in improving the quality of village and town green building construction projects, guiding the green transformation of existing rural buildings, and improving the technology and management level of village and town green buildings. In this chapter, six specific evaluation methods of village and town green building construction are established, and the application of these methods to demonstration projects is discussed.

Green building is the development direction and key evaluation content of green rural housing construction. The evaluation of environmental infrastructure construction is based on the current corresponding standards and specifications, including the following six aspects: evaluation of construction selection, evaluation of green construction, evaluation of energy-saving efficiency, evaluation of water-saving facility efficiency, evaluation of indoor environmental quality, and evaluation of low-carbon operational management of green buildings in villages and towns.

10.1 Evaluation of Village and Town Green Building Construction Selection

10.1.1 Evaluation Method of Village and Town Green Building Construction Selection

1. Evaluation of technical route

The construction of green building in village and town is guided by the principle of "land savings, energy savings, water savings, material savings and improving indoor environmental quality".

The evaluation of the village and town green building construction selection technical route is shown in Fig. 10.1.

X. Wang and H. Hua, *Green Village and Town Construction in China*,
https://doi.org/10.1007/978-981-16-2098-0_10

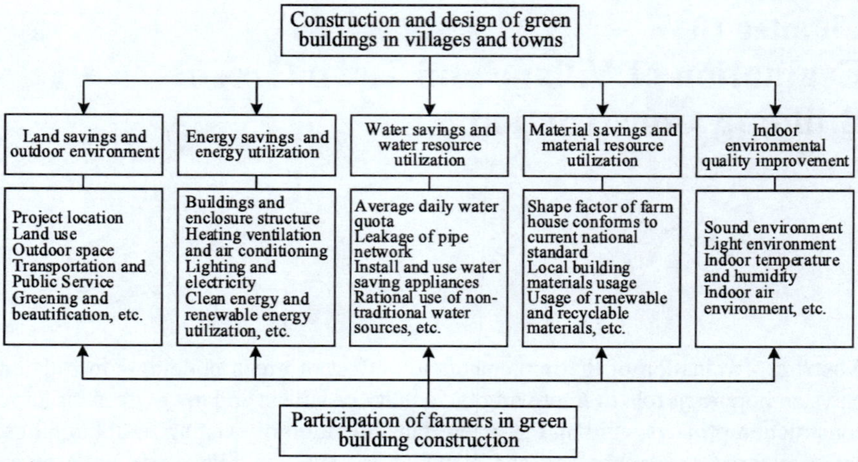

Fig. 10.1 Evaluation of the village and town green building construction selection technical route

2. Evaluation indexes and method

The selection and evaluation of green buildings in villages and towns is based on a scoring method from 0 to 1 point. If the evaluation index meets the current national/sector/industry standards or norms, the score is 1 point; if not, the score is 0 points. The full score is 21, and if the score reaches or exceeds 70%, it is qualified.

The optimization of green building construction design is an important technical aspect and evaluation element of building construction selection. The evaluation standards and methods for village and town green building construction selection are shown in Table 10.1. During the actual evaluation operation, the actual scores of each sub-item of the evaluated village or town is filled into the corresponding column of the "evaluation score" in the table to obtain the evaluation results directly.

10.1.2 Evaluation Case of Village and Town Green Building Construction Selection

1. General construction situation of evaluated building

This evaluation is based on the construction of a typical farm building (evaluated building) in Yanhe Village, and other buildings can be evaluated by reference to this method.

The building being evaluated is a low-cost, energy-saving demonstration farmhouse. The building site is Yanhe Village, which was built by intensively using riverbank land without occupying agricultural land. The building is combined with the development of rural tourism facilities, reflecting the principle of intensive and

Table 10.1 Evaluation standards and methods for village and town green building construction selection

Evaluation items	Evaluation indexes	Evaluation standards	Evaluation results			
			Qualified (1)	Sub-standard (0)	Score	Total score
Land savings and outdoor environment (6 points)	(1) Site selected on the basis of local urban and rural plans and meets the construction control requirements for all kinds of protected areas, cultural relics and historic sites	Design scheme inspection and field survey results meet the requirements of evaluation indexes				
	(2) Sites should not be subject to natural disasters such as floods, landslides, and debris flows and should be far from any pollution sources	Design scheme inspection and field survey results meet the requirements of evaluation indexes				
	(3) Rural homestead and land plan of village building groups should show intensive land-use savings	Residential buildings meet the requirements of GB/T 50,378-2019 "per capita residential land index"; The plot ratio of public buildings meets the requirements of GB/T 50,378-2019 "plot ratio of public buildings"				

(continued)

Table 10.1 (continued)

Evaluation items	Evaluation indexes	Evaluation standards	Evaluation results			
			Qualified (1)	Sub-standard (0)	Score	Total score
	(4) The best efforts should be made to keep original landforms, reduce high fill and deep digging, avoid taking up any local woodland and vegetation, and protect surface water	Field survey results meet the requirements of evaluation indexes				
	(5) Attach importance to the links between public traffic and services	The walking distance to the bus stop is no more than 500 m; The walking distance from the site to the kindergarten and primary school is no more than 500 m				
	(6) Rural courtyards could make full use of natural conditions and artificial environmental factors to beautify yards. Advocate the vertical planting of rural buildings	Landscaping of courtyards, roofs and walls; Ancient and famous trees should be protected by means of guardrails or masonry, etc.				

(continued)

Table 10.1 (continued)

Evaluation items	Evaluation indexes	Evaluation standards	Evaluation results			
			Qualified (1)	Sub-standard (0)	Score	Total score
Energy conservation and utilization of energy (4 points)	(7) Buildings and their envelopes	The site selection, orientation and elevation design of farm buildings meet current standards and specifications				
	(8) Heating, ventilation and air conditioning	Farmhouses adopt high-efficiency and energy-saving heating and ventilation methods in line with local characteristics; Indoor temperature and humidity meet the requirements of the *Code for Thermal Design of Civil Building* GB 50176-2016; Green farmhouses use energy-saving household air conditioners				
	(9) Lighting and electrical appliances	Choose energy-saving lamps and lanterns and energy-saving household appliances				

(continued)

Table 10.1 (continued)

Evaluation items	Evaluation indexes	Evaluation standards	Evaluation results			
			Qualified (1)	Sub-standard (0)	Score	Total score
	(10) Clean energy applications and renewable energy utilization	Use of clean biomass stoves, gas cookers, firedamp stoves, etc.; Use of clean energy such as liquefied petroleum gas and natural gas; Use of solar energy, geothermal energy, biomass and other renewable energy sources; If met any condition mentioned above, get 1 point				
Evaluation of wastewater treatment facility (4 points)	(11) Average daily water consumption quota	Meets the current standards and norms of water-saving quota requirements in the *Standard for water saving design in civil building* GB 50,555-2010				
	(12) Leakage of pipe network	No network leakage				

(continued)

Table 10.1 (continued)

Evaluation items	Evaluation indexes	Evaluation standards	Evaluation results			
			Qualified (1)	Sub-standard (0)	Score	Total score
	(13) Installation and use of water-saving instruments	Water-saving appliances are installed and used in farm buildings				
	(14) Reasonable use of non-traditional water resources	Design and use of reclaimed water systems in farm buildings; Scientific treatment and rational use of farm building domestic wastewater				
Material savings and utilization of local material sources (3 points)	(15) Village building shape has priority	The shape coefficient of rural buildings conforms to the current national standards, and the regular building shape is selected				
	(16) Use local construction materials	Use locally produced building materials				
	(17) Use reusable and recyclable materials	Practical use in farmhouse construction				
Indoor environmental quality (3 points)	(18) Acoustic environment	Indoor noise level of main function room should meet current standards and specifications in the *Code for design of sound insulation of civil buildings* GB 50,118-2010				

(continued)

Table 10.1 (continued)

Evaluation items	Evaluation indexes	Evaluation standards	Evaluation results			
			Qualified (1)	Sub-standard (0)	Score	Total score
	(19) Luminous environment	Main function rooms should have good outdoor views; Daylighting coefficient and quality of architectural lighting should meet current standards and specifications in the *Standard for daylighting design of buildings* GB 50,033-2013 and the *Standard for lighting design of buildings* GB 50,034-2013				
	(20) Indoor air quality	Concentrations of indoor pollutants such as formaldehyde, benzene, ammonia, volatile organic compounds, and inhalable particulate matter should meet the requirements of the *Indoor Air Quality Standard* GB/T 18,883-2002				
Public participation in rural building design (1 point)	(21) Farmers' participation in rural building design	Farmers should be fully involved in the design process of rural housing construction				

efficient land use. It was completed and put into use in December 2008. In 2009, this building was rated as a demonstration building of anti-seismic and energy-saving rural buildings in Hubei Province by the Department of Housing and Urban-Rural Development of Hubei Province.

The complete design of the evaluated building was tested by Professional Engineering Structure Testing Co., Ltd., and met the current national standards. The domestic water of the evaluated building meets the water-saving quota requirements of Standard for Water Saving Design in Civil Building GB 50,555-2010. Domestic sewage enters the sewage treatment plant and is discharged after reaching the standard, and the indoor environmental quality of the evaluated building meets the current national standards.

2. Evaluation of evaluated building selection

According to Table 10.1, the evaluation standards and methods of village and town green building construction selection, the construction selection of the building in Yanhe Village can be evaluated. The evaluation results are shown in Table 10.2. The evaluation results show that there are 21 scoring items for building construction selection, and the evaluation score is a full score—21 points; thus, the construction selection of the evaluated building is qualified.

10.2 Evaluation of Village and Town Green Building Construction

10.2.1 Evaluation Method of Village and Town Green Building Construction

1. Evaluation technical route

The evaluation of the village and town green building construction technical route is shown in Fig. 10.2.

2. Evaluation indexes and method

The evaluation of village and town green building construction adopts a scoring method from 0 to 1 point. If the evaluation indexes meet the current national/sector/industry standards or norms, 1 point is given; if not, 0 points are given. The full score is 19 points, and if the score reaches or exceeds 70%, the project is qualified.

In the actual evaluation, if there are multiple requirements in a certain scoring standard, one point can be given as long as one of them is satisfied. For the five types of evaluation content, one of the scoring standards is selected only once in principle to avoid repeated evaluation.

Table 10.2 Evaluation results of green building construction selection in Yanhe Village

Evaluation items	Evaluation indexes	Evaluation results			
		Qualified (1)	Sub-standard (0)	Score	Total score
Land savings and outdoor environment (6 points)	(1) Site selected on the basis of local urban and rural plans and meets the construction control requirements for all kinds of protected areas, cultural relics and historic sites	1		6	21
	(2) Sites should not be subject to natural disasters such as floods, landslides, and debris flows and should be far from any pollution sources	1			
	(3) Rural homestead and land plan of village building groups should show intensive land-use savings	1			
	(4) The best efforts should be made to keep original landforms, reduce high fill and deep digging, avoid taking up any local woodland and vegetation, and protect surface water	1			
	(5) Attach importance to the links between public traffic and services	1			

(continued)

Table 10.2 (continued)

Evaluation items	Evaluation indexes	Evaluation results			
		Qualified (1)	Sub-standard (0)	Score	Total score
	(6) Rural courtyards could make full use of natural conditions and artificial environmental factors to beautify yards. Advocate the vertical planting of rural buildings	1			
Energy conservation and utilization of energy (4 points)	(7) Buildings and their envelopes	1		4	
	(8) Heating, ventilation and air conditioning	1			
	(9) Lighting and electrical appliances	1			
	(10) Clean energy applications and renewable energy utilization	1			
Evaluation of wastewater treatment facility (4 points)	(11) Average daily water consumption quota	1		4	
	(12) Leakage of pipe network	1			
	(13) Installation and use of water-saving instruments	1			
	(14) Reasonable use of non-traditional water resources	1			
Material savings and utilization of material sources (3 points)	(15) Village building shape has priority	1		3	
	(16) Use local construction materials	1			
	(17) Use reusable and recyclable materials	1			
Indoor environmental quality (3 points)	(18) Acoustic environment	1		3	
	(19) Luminous environment	1			
	(20) Indoor air quality	1			
Public participation in rural building design (1 point)	(21) Farmers' participation in rural building design	1		1	

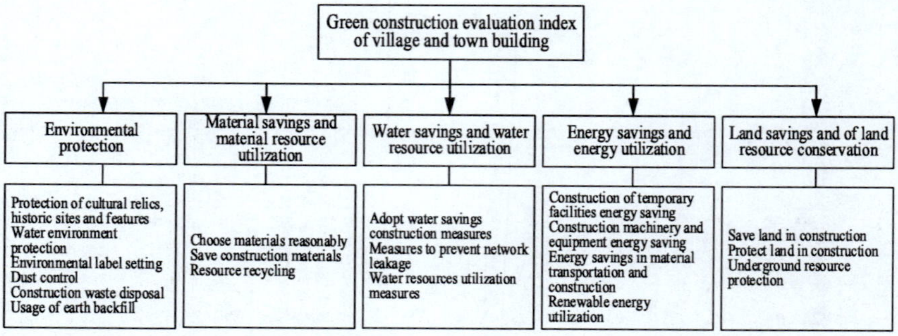

Fig. 10.2 Evaluation of village and town green building construction technical route

The evaluation standards and methods for village and town green building construction are shown in Table 10.3. During the actual evaluation operation, the actual scores of each sub-item of the evaluated village or town is filled into the corresponding "evaluation score" column in the table, and the evaluation results can then be directly obtained.

10.2.2 Evaluation Case of Village and Town Green Building Construction

1. General construction situation of evaluated building

For convenience of comparison, the same demonstration residential building used in Sect. 10.1.2 is selected for green construction evaluation. In 2008, the author provided design and energy-saving technology and guided the construction of the building. The farmers organized local craftsmen to build the building according to green building standards. More than 20 low-cost green construction technologies were integrated and adopted, achieving ideal results. After the building was completed, it became an energy-saving demonstration farm building in Hubei Province.

The information needed to evaluate the green construction of the evaluated building includes basic information and green construction measures.

The basic building information for construction evaluation in Yanhe Village is shown in Table 10.4.

2. Green construction evaluation of evaluated building

According to the above investigation and analysis, the green construction of the evaluated building in Yanhe Village is evaluated item by item in Table 10.3, with a score of 16 points, accounting for 84.2% of the total score of 19 points; as the score is higher than 70%, the green construction of the building is evaluated as qualified.

Table 10.3 Evaluation standards and methods of village and town green building construction

Evaluation items	Evaluation indexes	Evaluation standards	Evaluation result			
			Qualified (1)	Sub-standard (0)	Score	Total score
Environmental protection (6 points)	(1) Protection of cultural relics and features	Take protective measures for cultural relics and ancient trees on site, and choose construction techniques for local rural features				
	(2) Water environmental protection	Protect the original groundwater around the site, reduce groundwater extraction, improve site road and material storage site drainage, establish site toilet septic tanks and clean regularly				
	(3) Green mark setting	Construction signs should include environmental protection; dangerous goods, chemicals storage places and sewage discharge for construction should be isolated and marked for safety				

(continued)

Table 10.3 (continued)

Evaluation items	Evaluation indexes	Evaluation standards	Evaluation result			
			Qualified (1)	Sub-standard (0)	Score	Total score
	(4) Dust control	Closed covering measures should be taken for vehicles transporting removed earth, and shielding and dust suppression measures should be taken for construction operations				
	(5) Building waste disposal	Recycling measures should be taken for waste materials generated during the construction of new, rebuilt and demolished farmhouses				
	(6) Utilization of earthwork backfilling	Reasonable backfill and utilization of excavated earth during construction				
Material savings and local material resource utilization (3 points)	(7) Choose materials reasonably	Green environmental protection materials are used in construction, waste residue is used for energy-saving environmental protection blocks, temporary construction facilities are made of relocatable and recyclable materials, and building materials produced within 50 km of the site are selected				

(continued)

Table 10.3 (continued)

Evaluation items	Evaluation indexes	Evaluation standards	Evaluation result			
			Qualified (1)	Sub-standard (0)	Score	Total score
	(8) Save building materials	Use renewable and reusable material templates and scaffolding, such as aluminium alloy, plastic, glass, steel, and bamboo				
	(9) Resource recycling and utilization	Temporary facilities use existing buildings and municipal facilities and surrounding roads, and the building materials packaging recovery rate is 100%				
Water savings and water resource utilization (3 points)	(10) Take measures for water-saving construction	Reasonably arrange the water supply and drainage system, use water-saving appliances and water metres for on-site domestic water, and use water-saving construction technology				
	(11) Measures for preventing leakage of pipe network	No leakage should be found in pipe networks and water appliances				

(continued)

Table 10.3 (continued)

Evaluation items	Evaluation indexes	Evaluation standards	Evaluation result			
			Qualified (1)	Sub-standard (0)	Score	Total score
	(12) Measures of utilization of water resources	Use recycled water to flush machines, equipment and vehicles. Do not use tap water to spray road surfaces or for green irrigation. Use non-traditional water sources on site				
Energy savings and energy utilization (4 points)	(13) Energy savings of temporary facilities in construction	Energy-saving equipment and automatic control devices are adopted for temporary electricity consumption. The number of energy-saving lighting fixtures is more than 80% on the construction site, and the temporary facilities adopt natural lighting and ventilation and shading facilities for external windows				
	(14) Energy savings of construction machinery	Adopt construction machinery and equipment with high energy utilization efficiency, and do not use construction equipment and tools that the state and local government have explicitly prohibited				

(continued)

Table 10.3 (continued)

Evaluation items	Evaluation indexes	Evaluation standards	Evaluation result			
			Qualified (1)	Sub-standard (0)	Score	Total score
	(15) Material transportation and energy savings of construction	The choice of building materials should shorten the transportation distance and minimize night work and winter construction time				
	(16) Utilization of renewable resources	Rational use of solar, geothermal, wind, biomass or other renewable resources				
Land savings and land resource protection (3 points)	(17) Land savings in construction	Overall layout of compact construction, use of proposed roads for construction services, as far as possible reducing the area				
	(18) Protection of construction site	Protect the surrounding cultural landscape, take measures to prevent soil erosion, and use mountain wasteland as land for removing and abandoning soil; After construction, restore vegetation and landforms				
	(19) Protection of underground resources	Underground cultural relic resources discovered in the construction should be effectively protected, and disposal measures should be appropriate				

Table 10.4 Basic building information for construction evaluation in Yanhe Village

Evaluation type	Evaluation items	Basic information
Basic building information	Site selection	Flooded land and tidal flat of Yanhe Village
	Covered area (m^2)	251
	Building height (m)	9
	Building storeys	2
	Building orientation	East by south
	Structure type	Brick-concrete structure
	Building roof	Precast slab, partly poured on site
	Building foundation	Pebble base and reinforced ring-beam strip foundation
	Building wall	Slag brick, extruded polystyrene board, cement mortar plaster
	Outside door and windows	Heat insulation wooden door, plastic steel double-glass energy-saving windows
Green construction measures	Environmental protection	West Hubei mountainous area countryside-style construction craft and construction waste recycling were selected
	Material savings and resource utilization	The pebbles at the beach 100 m away were used as the rubble foundation of the building, and waste residues were used for brick masonry
	Water savings and water resource utilization	Leakproof water supply and drainage pipes were selected, and local spring water was used for construction
	Energy savings and energy utilization	All building materials were from within a radius of 30 km, artificial hoist cranes and other mechanical devices were used to save electricity, solar street lighting was applied
	Land savings and land resource utilization	Riverbank land was used in order to not occupy farmland lawfully; the curtilage in front of and behind the building was afforested after completion

The green construction evaluation results of the evaluated building in Yanhe Village are shown in Table 10.5.

10.3 Evaluation of Village and Town Green Building Energy-Saving Efficiency

10.3.1 Evaluation Method of Village and Town Building Energy-Saving Efficiency

The energy-saving efficiency evaluation of newly built villages and towns can be performed by using life-cycle assessment or analysis software. The energy-saving assessment of existing buildings should be performed with on-site testing and evaluation by relevant instruments or equipment.

1. Evaluation technology of village and town building energy-saving efficiency based on life cycle

1) Life-cycle division of village and town buildings

Buildings are a construction product with a life cycle. The construction activities of the whole life cycle include the following 4 types: mining, processing and transportation of raw materials; architectural design and construction; building use and maintenance; and building demolition. For convenience of calculation, this evaluation refers to "mining, processing and transportation of raw materials" as "building material production". Therefore, the life-cycle assessment (LCA) of village and town buildings can be divided into 4 stages: building material production, architectural design and construction, building use and maintenance, and building demolition.

The life-cycle division of village and town buildings is shown in Fig. 10.3.

2) Analysis of the energy consumption inventory of village and town buildings

According to the LCA concept, environmental impact analysis and evaluation are carried out on all construction, consumption, recycling and disposal activities throughout the life cycle of village and town buildings. This evaluation focuses on energy-saving efficiency, so the core content of the evaluation is the energy demand and consumption of the building life cycle.

The analysis of the energy consumption inventory of village and town buildings is shown in Fig. 10.4.

3) Life-cycle assessment of the energy-saving efficiency of village and town buildings

The life-cycle assessment of the energy-saving efficiency of village and town buildings includes four stages:

Table 10.5 Green construction evaluation results of evaluated building in Yanhe Village

Evaluation items	Evaluation indexes	Evaluation result			
		Qualified (1)	Sub-standard (0)	Score	Total score
Environmental protection (6 points)	(1) Protection of cultural relics and features	1		4	16
	(2) Water environmental protection	1			
	(3) Green mark setting		0		
	(4) Dust control		0		
	(5) Building waste disposal	1			
	(6) Utilization of earthwork backfilling	1			
Material savings and material resource utilization (3 points)	(7) Choose materials reasonably	1		3	
	(8) Save building materials	1			
	(9) Resource recycling and utilization	1			
Water savings and water resource utilization (3 points)	(10) Take measures of water-saving construction	1		3	
	(11) Measures of preventing leakage of pipe network	1			
	(12) Measures of utilization of water resources	1			
Energy savings and energy utilization (4 points)	(13) Energy savings of temporary facilities in construction	1		4	

(continued)

Table 10.5 (continued)

Evaluation items	Evaluation indexes	Evaluation result			
		Qualified (1)	Sub-standard (0)	Score	Total score
	(14) Energy savings of construction machinery	1			
	(15) Material transportation and energy savings of construction	1			
	(16) Utilization of renewable resources	1			
Land savings and land resource protection (3 points)	(17) Land savings in construction	1		2	
	(18) Protection of construction site	1			
	(19) Protection of underground resources		0		

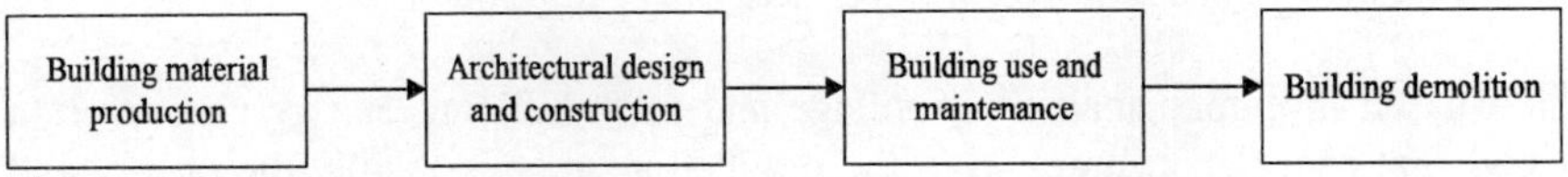

Fig. 10.3 Life-cycle division of village and town buildings

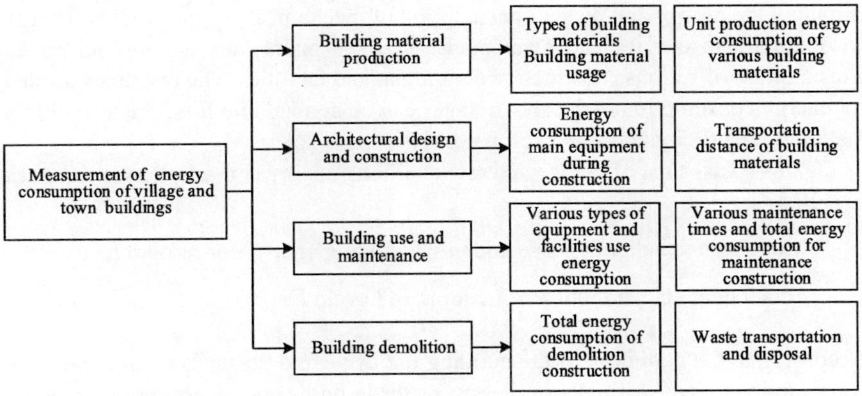

Fig. 10.4 Analysis of the energy consumption inventory of village and town buildings

(1) Building material production stage. The main energy consumption in the production stage includes power consumption in the whole process from raw material mining and processing to manufacturing and gasoline and diesel oil consumption in the transportation of raw materials.
(2) Design and construction stage. Power consumption is the main energy consumption in the design stage. Compared with the energy consumption of the building life cycle, the energy consumption in the design stage is small. The energy consumption at this stage is not counted in the life-cycle assessment of building energy conservation. The energy consumption in the construction stage is mainly the energy consumption of transforming building materials into building structural components and masonry, such as the manual, material, and mechanical use and transportation of building materials. It is an important stage of the life-cycle assessment of building energy consumption.
(3) Use and maintenance stage. The energy consumption in the use and maintenance stage mainly includes the energy consumption of the construction equipment and the energy required for construction maintenance equipment.
(4) Demolition stage. The energy consumption in this stage includes two parts: first, energy consumption from blasting and machinery construction in demolition, and second, energy consumption from the disposal of waste building materials. For convenience of calculation, the energy consumption in the demolition stage is 10% of the total energy consumption in the construction stage [1].

4) List of village and town building energy consumption

LCA-based inventory analysis of village and town building energy consumption shows that energy consumption in the production stage is mainly the energy and power consumption in the production and processing of various building materials; energy consumption in the construction, installation, demolition and disposal stage is mainly the energy and power consumption of mechanical equipment; and energy consumption in the use and maintenance stage is mainly the energy and power consumption of various construction equipment and facilities. The resources needed for energy consumption in the above stages can be divided into fossil fuels, building materials, electricity and water resources.

On this basis, a list of village and town building energy consumption is shown in Fig. 10.5.

5) Metreing formula of village and town building energy consumption

(1) Total energy consumption of building life cycle E_T

According to the definition of the building life cycle and the analysis of the energy consumption list, the formula for measuring the building energy consumption of the building life cycle is

$$E_T = E_m + E_c + E_r + E_d \tag{10.1}$$

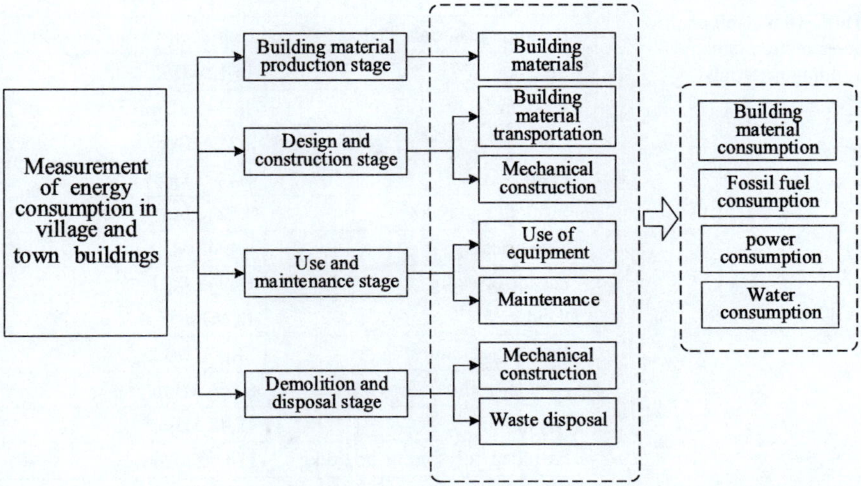

Fig. 10.5 List of the measurement of village and town building energy consumption based on LCA

where E_T is total energy consumption of the building life cycle, MJ;

E_m is total energy consumption in the building material production stage, MJ;

E_c is total energy consumption in the design and construction stage, MJ;

E_r is total energy consumption in the operation and maintenance stage, MJ; and

E_d is total energy consumption in the demolition and disposal stage, MJ.

(2) Total energy consumption in building material production stage E_m

The energy consumption in the building material production stage consists of three parts: raw material mining, building material processing, and transportation of building materials. To simplify the calculation, the evaluation uses the unit energy consumption of building material production to measure consumption [2]; thus, the total energy consumption in the building material production stage is measured as

$$E_m = \sum_{i=1}^{n} J_i m_i \tag{10.2}$$

where E_m is total energy consumption in the building material production stage, MJ;

J_i is the unit energy consumption of building materials i, MJ;

m_i is The usage amount of building materials i, kg, m^2 or m^3; and

i is the number of types of building materials used.

The unit energy consumption of specific building materials is shown in Table 10.6.

Table 10.6 Unit energy consumption of some building materials

Building materials	Unit energy consumption
Steel/kg	15.12 MJ·kg^{-1}
Wood/m^3	48.345 MJ·m^{-3}
Cement/kg	3.28 MJ·kg^{-1}
Plate glass/kg	6.4 MJ·kg^{-1}
Insulating glass/m^2	2.52 MJ·m^{-2}
Slag brick/kg	1.4943 MJ·kg^{-1}
Facade brick/kg	1.4943 MJ·kg^{-1}
Sand/m^3	13 MJ·m^{-3}
Building ceramics/kg	21.096 MJ·kg^{-1}
Asphalt coil/m^2	6.446 MJ·m^{-2}
Drywall/m^2	46.88 MJ·m^{-2}
Extruded polystyrene board/kg	115.65 MJ·kg^{-1}

(3) Total energy consumption in design and construction stage E_c

As mentioned above, the total energy consumption in the design and construction phase is mainly the energy consumption of construction, with emphasis on the total energy consumption of construction and building material transportation; the energy consumption of the design stage is negligible. The formula for measuring consumption in the design and construction stage is

$$E_c = E_{pro} + E_{trans} \tag{10.3}$$

where E_{pro} is total energy consumption of construction, MJ, and

E_{trans} is total energy consumption of building material transportation, MJ.

$$E_{pro} = \sum_{i=1}^{n} E_i \tag{10.4}$$

where E_i is the energy consumption of the main equipment in the construction process, MJ.

$$E_{trans} = \sum_{i=1}^{n} D_i \times m_i \tag{10.5}$$

where D_i is the average distance from the source of building materials i to the construction site, km, and

m_i is the usage amount of building materials i, kg, m^2 or m^3.

(4) Total energy consumption in operation and maintenance stage E_r

The energy consumption in the operation and maintenance stage is composed mainly of the equipment and facilities use of buildings and the maintenance and repair of buildings. The measurement formula is

$$E_r = E_{use} + E_{mat} \tag{10.6}$$

where E_{use} is energy consumption of construction equipment, MJ, and

E_{mat} is energy consumption of building maintenance and repair, MJ.

Building energy consumption comes mainly from air conditioning, lighting, ventilation and other equipment, and the measurement formula is

$$E_{use} = \sum_{i=1}^{n} R_i \tag{10.7}$$

where R_i is the amount of i energy or resource, kg, m^2 or m^3.

The energy consumption of building maintenance and repair construction generally includes two types of energy consumption: overhaul and minor repair (daily maintenance). Overhaul repair includes structural reinforcement and renovation, and its energy consumption E_{mat} is calculated by referring to the energy consumption method in the construction stage to obtain "E_c" for the construction energy consumption of maintenance and repair. It can also be calculated as approximately 50–30% of the energy consumption in the construction of similar new projects according to the repair work quantities.

In overhaul repair, building materials need to be replaced or added, and the energy consumption of the actual use of building materials is calculated by referring to the energy consumption in the building material production stage.

Since the maintenance and repair of farm buildings is not included in national housing repair regulations, farmers usually perform such tasks themselves; therefore, the data are difficult to collect. For that reason, the minor repair (daily maintenance) construction energy consumption is not calculated.

(5) Total energy consumption in demolition and disposal stage E_d

The total energy consumption in the demolition and disposal stage of a building consists of the energy consumption from the demolition and disposal of waste building materials. The total energy consumption during the demolition and disposal phase is calculated according to Formula (10.8). It can also be calculated as 10% of the energy consumption in similar new construction projects[1].

$$E_d = E_{demol} + E_{trans} \tag{10.8}$$

where E_{demol} is total energy consumption of demolition, MJ, and

E_{trans} is total energy consumption of waste building material transportation, MJ.

$$E_{demol} = \sum_{i=1}^{n} r_i \tag{10.9}$$

where r_i is the construction amount of the demolition of building materials *i*, *t*, m^2 or m^3.

$$E_{reans} = \sum_{i=1}^{n} D_i m_i \tag{10.10}$$

where D_i is the average distance from the source of building materials *i* to the disposal site, km, and

m_i is the disposal amount of building materials *i*, kg, m^2 or m^3.

2. Evaluation technology of building energy-saving efficiency based on analysis software

1) Analysis software

E_{cotect} software is an auxiliary ecological design software developed by Square One Company that has good adaptability to the evaluation of village and town building energy efficiency.

2) Evaluation steps

Using E_{cotect} analysis software, the evaluation of building energy-saving efficiency is divided into four steps.

Step 1: Obtain relevant data and basic data

Basic data related to the evaluation should be obtained, including design drawings; the building energy-saving design scheme; basic project cost data; local meteorological, geographical, and hydrological data; and other relevant data. To evaluate the efficiency of an existing building energy-saving renovation project without the corresponding data, the data needed for the evaluation should be supplemented and improved through field investigation and field surveys.

Step 2: Select the appropriate evaluation software

To obtain the required data, appropriate building energy efficiency evaluation software should be selected. Ecotect analysis software was used in this evaluation.

Step 3: Apply analysis software to evaluate building energy efficiency

The evaluation contents include mainly the following:

(1) The determination of the evaluation building and compared building;
(2) The collection, sorting and induction of basic building information; and

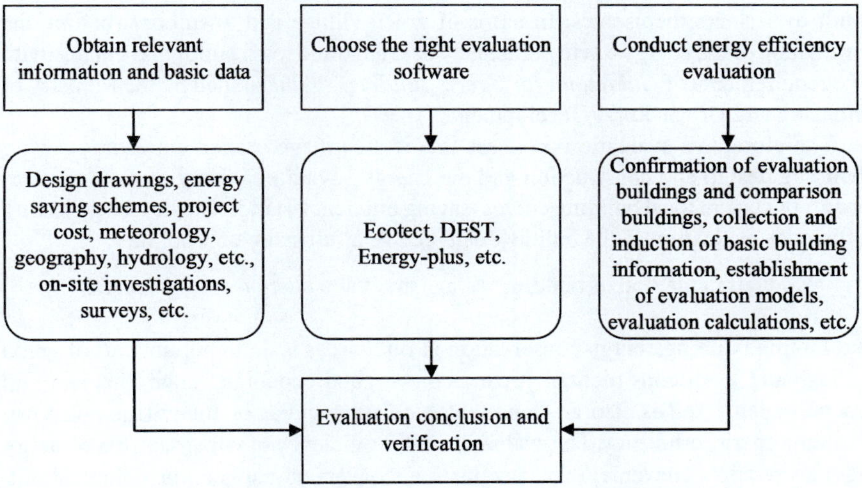

Fig. 10.6 Evaluation process of village and town building energy efficiency

(3) The establishment of the evaluation model and evaluation calculation.

Step 4: Verification of evaluation conclusion and suggestions for improvement

The evaluation process of village and town building energy efficiency using the analysis software is shown in Fig. 10.6.

3. Evaluation method of village and town building energy-saving efficiency

1) Evaluation of the energy-saving efficiency of a single building

According to the basic concept of building energy-saving efficiency, the energy-saving efficiency of an evaluated building is the annual energy consumption of the benchmark building minus the annual energy consumption of the evaluated building and then divided by the annual energy consumption of the benchmark building.

$$\text{Energy - saving efficiency of the evaluated building}$$
$$= \frac{\text{Annual energy consumption of benchmark building} - \text{annual energy consumption of the evaluated building}}{\text{Annual energy consumption of benchmark building}} \times 100 \tag{10.11}$$

2) Evaluation of the energy-saving efficiency of village and town buildings

According to the relevant regulations of the Ministry of Housing and Urban-Rural Development on energy-saving building construction, all new buildings in cities must be energy saving. However, there is no clear requirement for low-rise buildings built by farmers themselves. In terms of green village and town construction, the evaluation of the energy-saving efficiency of village and town buildings is in line with the requirements of *Guidelines for Green Buildings (Trial)* issued by the Ministry of Housing and Urban-Rural Development.

Therefore, this evaluation proposes that if the village carried out energy-saving building design and construction and the energy-saving efficiency met the requirements of current local building energy-saving efficiency standards, the energy-saving efficiency evaluation of the building can receive a full score of 1 point.

4. Evaluation method of building village and town energy-saving benefits

Promoting building energy conservation in rural areas is an important part of green village and town construction. It can achieve good economic, environmental and social benefits and is also an important evaluation indicator for village and town building energy efficiency. The evaluation of the efficiency of village and town energy savings requires convenient data acquisition, convenient application, comparability and combination with the evaluation methods of energy efficiency.

The village and town building energy-saving benefits are reflected in the energy-saving economic value (*V*) achieved through energy-saving measures in the life cycle of a building. The evaluation steps are as follows:

1) Calculate the total energy consumption of evaluated buildings and comparative buildings E_T

The same method as that in Formula (10.1) is used to calculate the total energy consumption of the life cycle of the energy-saving building (E_{B1}) and the total energy consumption of the life cycle of the local non-energy-saving building (E_{B2}) separately, as shown in Formula (10.12):

$$E_T = E_m + E_c + E_r + E_d \tag{10.12}$$

2) Calculate the total energy savings of evaluated buildings and compared buildings E_B

The calculation formula is

$$E_B = E_{B2} - E_{B1} \tag{10.13}$$

where E_B is total energy savings of village and town buildings, kg_{ce};

E_{B1} is total energy consumption of the evaluated building, kg_{ce}; and

E_{B2} is total energy consumption of the compared building, kg_{ce}.

3) Calculate the economic benefits of evaluated buildings V

The calculation formula for evaluating the energy-saving economic benefits of the building life cycle is

$$V_1 = E_B \times P_1 + P_2 \tag{10.14}$$

where V_1 is the economic benefits of the evaluated building, yuan;
E_B is total energy savings of the building, kg_{ce};
P_1 is unit price per kilogram of standard coal, kg/yuan; and
P_2 is government subsidies for farmers to build energy-efficient buildings, yuan.

10.3.2 Evaluation Case of Village and Town Building Energy-Saving Efficiency

1. Evaluation of new building energy-saving efficiency

In this case, a new residential building in Yanhe Village was selected for an LCA energy-efficiency evaluation. The new energy-saving building was called the "evaluated building", and the evaluated "comparison building" was an imaginary building without an energy-saving design according to the local general construction method. To simplify the calculation, the service time of both the evaluated building and the compared building is set at 50 years.

1) Basic information comparison of evaluated building and compared building

The comparisons of the basic information and energy-saving scheme between the evaluated building and compared building are shown in Tables 10.7 and 10.8.

2) Building energy consumption evaluation based on LCA

Table 10.7 Basic information comparison between evaluated building and compared building

Basic information	Evaluated building	Compared building
Structure type	Brick structure	
Age of construction	2009	
Covered area/m^2	251	
Building height/m	9	
Building storeys	2	
Roof form	Pre-fabricated plate, partially cast in place	
Foundation	Cobble, broken brick stone, concrete and steel floor girders	
Wall materials	Slag brick, extruded polystyrene board, cement mortar plaster	Slag brick, cement mortar plaster
Door	Heat insulation wooden door	Ordinary wooden door
Windows	Energy-saving windows of plastic steel double glass	Plastic steel single-glass windows

Table 10.8 Energy-saving scheme comparison between evaluated building and compared building

Comparison project	Evaluated building	Compared building
Covered area/m^2	251	
Foundation	Pebble base with reinforced foundation ring-beam strip foundation	
Main structure	Reinforced concrete ring-beam and constructional column through	
Roof construction	Cement mortar and asphalt felt waterproof layer	
Masonry materials	Cement mortar	
	Slag brick masonry	
	Extruded polystyrene plate	No energy-saving measures
Major energy-saving measures	Insulation treatment outside the wall	
	Roof insulation treatment	
	Energy-saving windows of double insulating glass	

According to the evaluation method of village and town building energy-saving efficiency established in Sect. 10.3.1, we can evaluate the total life-cycle energy consumption of the evaluated building and compared building.

(1) Measurement of energy consumption in the production of building materials stage E_m

According to the actual consumption of building materials in the evaluated building and the estimated amount in the compared building, the energy consumption in the building material production stage is calculated and compared by referring to the unit energy consumption of relevant building materials in Table 10.6.

The comparison of energy consumption between the evaluated building and compared building in the material production stage is shown in Table 10.9.

According to Formula (10.2), the energy consumption of the evaluated building and compared building in the building material production stage are

$$E_{m(\text{evaluated building})} = 495368.96 \text{ MJ}$$

$$E_{m(\text{compared building})} = 471380.50 \text{ MJ}$$

(2) Measurement of energy consumption in the architectural design and construction stage E_c

① Calculation of construction energy consumption

The construction equipment for the evaluated building and compared building mainly includes small cranes, cement mixers, small forklift trucks, and electric drills. The comparison of the construction energy consumption of the evaluated building and compared building is shown in Table 10.10.

Table 10.9 Energy consumption comparison between the evaluated building and compared building in the building material production stage

No	Building materials	Unit energy consumption	Evaluated building		Compared building	
			Usage amount	Energy consumption	Usage amount	Energy consumption
1	Steel	15.12 $MJ\cdot kg^{-1}$	3690 kg	55,792.80 MJ	3690 kg	55,792.80 MJ
2	Wood	48.345 $MJ\cdot m^{-3}$	277.6 m^3	13,420.60 MJ	276.69 m^3	13,376.58 MJ
3	Cement	3.28 $MJ\cdot kg^{-1}$	16,689.44 kg	54,741.36 MJ	16,689.44 kg	54,741.36 MJ
4	Flat glass	6.4 $MJ\cdot kg^{-1}$	138.6 kg	887.04 MJ	624.6 kg	3997.44 MJ
5	Hollow glass	2.52 $MJ\cdot m^{-2}$	38.88 m^2	97.98 MJ	0 m^2	0 MJ
6	Slag brick	1.4943 $MJ\cdot kg^{-1}$	157,188.93 kg	261,784.82 MJ	157,188.93 kg	261,784.82 MJ
7	Sand	13 MJ·m-3	31.83 m^3	413.79 MJ	31.83 m^3	413.79 MJ
8	Pottery	21.096 $MJ\cdot kg^{-1}$	3811.26 kg	80,402.34 MJ	3811.26 kg	80,402.34 MJ
9	Asphalt coil	6.446 $MJ\cdot m^{-2}$	135.18 m^2	871.37 MJ	135.18 m^2	871.37 MJ
10	Extruded polystyrene plate	115.65 $MJ\cdot kg^{-1}$	233.09 kg	26,956.86 MJ	0 kg	0 MJ

Table 10.10 Comparison of construction energy consumption between evaluated building and compared building

Construction machinery	Running time/h		Energy consumption/MJ	
	Evaluated building	Compared building	Evaluated building	Compared building
Small crane	14	14	176.4	176.4
Cement mixer	21	21	529.2	529.2
Small forklift truck	6	6	842.4	842.4
Electric drill	48	48	86.4	86.4

According to Formula (10.4), the construction energy consumption of the evaluated building and compared building are

$$E_{\text{pro(evaluated building)}} = 1614.6\ \text{MJ}$$
$$E_{\text{pro(compared building)}} = 1614.6\ \text{MJ}$$

② Calculation of energy consumption in building material transportation

In this case, the transportation means and distance of the evaluated building are the same as those of the compared building with the same building materials. The main transportation means of large building materials is a pickup truck with a load capacity of 2 t and a unit transportation energy consumption of 3.88 MJ/km.

Table 10.11 Comparison of building material transportation distance and energy consumption between evaluated building and compared building

Building materials	Transportation distance/km		Energy consumption/MJ	
	Evaluated building	Compared building	Evaluated building	Compared building
Slag brick reinforcement	50	50	15,714	15,714
Extruded polystyrene board, glass, exterior wall tiles	380	50	1474.4	194.0
Cement	70	70	2444.4	2444.4
Sand	5	5	329.8	329.8
Wood	5	5	1513.2	1513.2

The comparison of energy consumption in building material transportation between the evaluated building and compared building is shown in Table 10.11.

The energy consumption of the evaluated building and compared building in building material transportation are

$$E_{\text{trans(evaluated building)}} = 21475.8\ \text{MJ}$$
$$E_{\text{trans(compared building)}} = 20195.4\ \text{MJ}$$

According to Formula (10.3), the energy consumption of the evaluated building and compared building in the architectural design and construction stage are

$$E_{\text{c(evaluated building)}} = 23090.4\ \text{MJ}$$
$$E_{\text{c(compared building)}} = 21810.0\ \text{MJ}$$

The design energy consumption of farm buildings is usually negligible.

(3) Calculation of energy consumption in the building use and maintenance stage E_r

In this example, the evaluated building is a farmer's residence that is used as a farmhouse, with 7 air conditioners installed. According to the field investigation, the annual heating consumption of the evaluated building is 4200 kWh, the annual cooling consumption is 4725 kWh, and the annual total electricity consumption of the air conditioning is 8925 kWh. With the lighting power consumption, the annual energy consumption of the building is estimated to be 34,968.2 MJ.

To ensure reasonable comparison results, no energy-saving measures have been adopted in the compared building, and the thermal insulation is poor. Therefore, we calculated the total power consumption of air conditioning in the compared building at 4/3 times that of the evaluated building according to the survey, so the total power consumption of air conditioning per year is $8925 \times 4/3 = 11{,}900$ kWh.

The lighting electricity consumption of the compared building is the same as that of the evaluated building, as follows: 34,968.2–8925 × 3.6 = 2838.2 MJ.

The construction equipment energy consumption over the life cycle of the evaluated building and compared building is

$$E_{\text{use(evaluated building)}} = 1748412 \text{ MJ}$$
$$E_{\text{use(compared building)}} = 2283910 \text{ MJ}$$

It is assumed that both the evaluated building and the compared building carry out 1 major repair over the whole life cycle. According to the "engineering quantity analogy" method of building life-cycle maintenance energy consumption proposed by the author, the maintenance energy consumption of the building life cycle is calculated as 25% of the construction energy consumption when the building is newly built.

$$E_{\text{mat(evaluated building)}} = 129600.98 \text{ MJ}$$
$$E_{\text{mat(compared building)}} = 123506.77 \text{ MJ}$$

The energy consumption of the evaluated building and compared building in the building use and maintenance stage are

$$E_{\text{r(evaluated building)}} = E_{\text{use(evaluated building)}} + E_{\text{mat(evaluated building)}} = 1878026.84 \text{ MJ}$$
$$E_{\text{r(compared building)}} = E_{\text{use(compared building)}} + E_{\text{mat(compared building)}} = 2407479.23 \text{ MJ}$$

(4) Calculation of energy consumption in the building demolition stage E_{d}

In this example, both the evaluated building and compared building are existing residential buildings in use, so the energy consumption of the two buildings in the stage of demolition is estimated at 10% of their construction energy consumption.

The estimated energy consumption of the evaluated building and compared building in the demolition stage are

$$E_{\text{d(evaluated building)}} = E_{\text{c(evaluated building)}} \times 10\% = 2309.04\text{MJ}$$
$$E_{\text{d(compared building)}} = E_{\text{c(compared building)}} \times 10\% = 2181.00 \text{ MJ}$$

(5) Calculation of building total energy consumption over the life cycle E_{T}

It is assumed that the service time of both the evaluated building and compared building is calculated as 50 years. According to Formula (10.1), the total energy consumption of the evaluated building and compared building over their life cycles is

$$E_{\text{T(evaluated building)}} = 2398795.24 \text{ MJ}$$
$$E_{\text{T(compared building)}} = 2902850.73 \text{ MJ}$$

Table 10.12 Comparison of life-cycle energy consumption evaluation between evaluated building and compared building

Evaluation project	Evaluated building	Compared building	Energy-saving efficiency
Total life-cycle energy consumption/MJ	2,398,795.24	2,902,850.73	58.68%
Annual life-cycle energy consumption/MJ·a^{-1}	47,975.90	58,057.01	
Per unit area of annual life-cycle energy consumption/MJ·m^{-2}·a^{-1}	191.14	231.30	

(6) Comparison of energy-saving efficiency between evaluated building and compared building

The comparison of the life-cycle energy consumption evaluation between the evaluated building and compared building is shown in Table 10.12. The 3 sets of calculation results in the table show that the life-cycle energy use of the evaluated building is 58.68% that of the compared building, indicating the obvious energy-saving efficiency of the newly built energy-saving farmhouse.

2. Evaluation of the energy-saving efficiency of existing buildings

The evaluation of the energy-saving efficiency of existing energy-saving building in villages and towns should adopt the method of obtaining actual data through field testing. This evaluation takes the field test data of the energy-saving building of Yanhe Village Reception Centre over 6 years of use as an example.

1) Actual monitoring data

In 2009, the Yanhe Village Reception Centre was established and put into use. From June 2015 to June 2016, the author used an HC-T200 portable temperature data recorder to conduct annual continuous monitoring of indoor and outdoor temperature changes in the Reception Centre.

The annual monitoring results of the Reception Centre from June 2015 to June 2016 are shown in Fig. 10.7, the typical temperature analysis of low temperatures during the monitoring period is shown in Fig. 10.8, the typical temperature analysis of high temperatures during the monitoring period is shown in Fig. 10.9, and the typical temperature analysis of maximum temperature differences during the monitoring period is shown in Fig. 10.10.

2) Analysis of monitoring data

(1) Annual monitoring results of Reception Centre

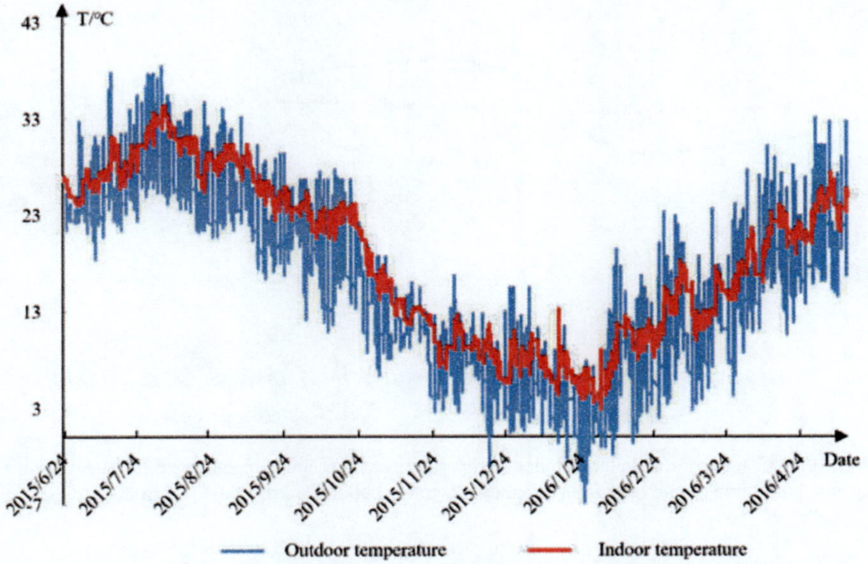

Fig. 10.7 Comparison of indoor-outdoor temperature changes and temperature differences in Yanhe Village Reception Centre from 2015 to 2016

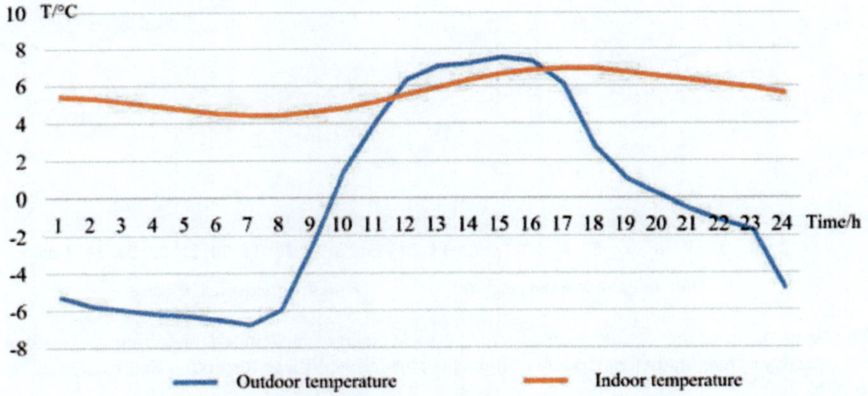

Fig. 10.8 Comparison of indoor-outdoor temperature changes and temperature differences in the Yanhe Village Reception Centre from 2015 to 2016

According to Fig. 10.7, after the energy-saving Reception Centre building was built and used for 6 years, the monitoring data analysis showed that the indoor temperature variation coefficient was 0.45 during the whole data collection period, which was less than the outdoor temperature variation coefficient of 0.56. That is, the indoor temperature variation range was far more stable than the outdoor range, indicating that the Reception Centre had good thermal insulation performance.

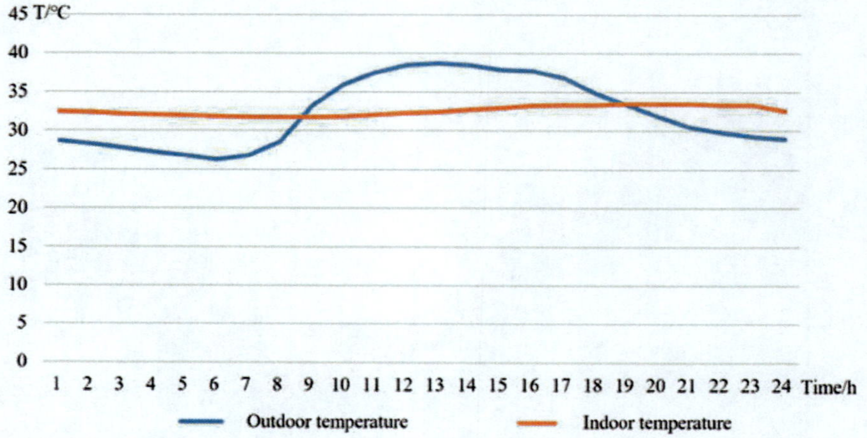

Fig. 10.9 Comparison of indoor-outdoor temperature changes and temperature differences during typical high-temperature days in the Yanhe Village Reception Centre from 2015 to 2016

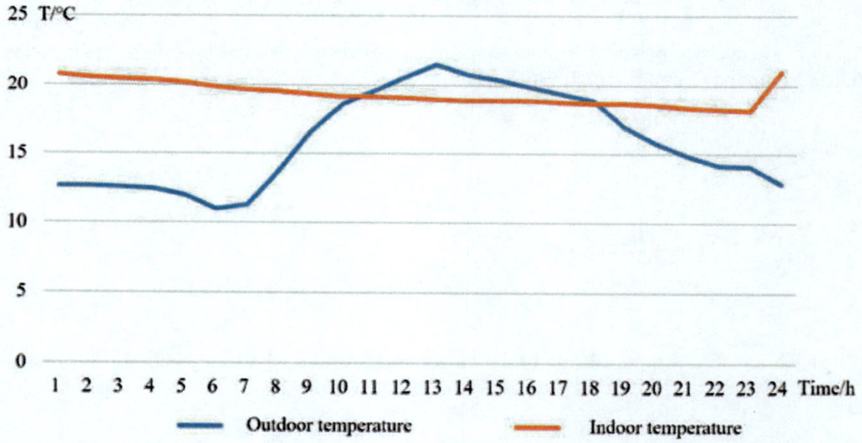

Fig. 10.10 Comparison of indoor-outdoor temperature changes and temperature differences during a typical day of maximum temperature difference in the Yanhe Village Reception Centre from 2015 to 2016

(2) Analysis of typical low-temperature data in monitoring time

Figure 10.8 shows that the lowest temperature was −6.8 °C on January 25, 2016, which was the lowest-temperature day in the data collection period. Therefore, in this evaluation, this day was selected as the typical low-temperature day to compare the indoor and outdoor temperatures of the Reception Centre. The 24-h monitoring data show that without any heating equipment, when the outdoor temperature was −6.8

°C, the indoor temperature reached 4.4, 11.2 °C higher than the outdoor temperature, and the 24-h indoor average temperature was 5.6 °C.

(3) Analysis of typical high-temperature data in monitoring time

Figure 10.9 shows that the highest temperature was 38.7 °C on August 3, 2015, which was the highest-temperature day in the data collection period. Therefore, in this evaluation, this day was selected as the typical high-temperature day to compare the indoor and outdoor temperatures of the Reception Centre. The 24-h monitoring data show that without any refrigeration equipment, when the outdoor temperature was 38.7 °C, the indoor temperature was 32.5, 5.2 °C lower than the outdoor temperature, and the 24-h indoor average temperature was 32.6 °C.

(4) Analysis of typical maximum difference temperature data in monitoring time

Figure 10.10 shows that the temperature difference between day and night was 10.5 °C on April 4, 2016, was the maximum temperature difference between day and night in the data collection period. Therefore, in this evaluation, this day was selected as the typical day of maximum temperature difference between day and night to compare the indoor and outdoor temperatures of the Reception Centre. The 24-h monitoring data show that the temperature difference between day and night outside was 22.53 °C, and without any refrigeration equipment, the maximum temperature difference inside the Reception Centre was only 3 °C.

(5) Conclusion of monitoring evaluation

The Yanhe Village Reception Centre has good energy-saving efficiency and thermal comfort after 6 years of use.

3. Comprehensive evaluation of building energy-saving efficiency

According to the evaluation method of village and town building energy-saving efficiency proposed in Sect. 10.3.1, farmhouse buildings in Yanhe Village were designed and constructed for energy savings. The efficiency of energy-saving buildings in this village has reached more than 50% of the energy-saving efficiency requirements stipulated by the local Hubei Provincial Government through the comprehensive evaluation. As a result, the evaluation of building energy-saving efficiency in Yanhe Village was awarded 1 point.

Figure 10.11 shows the actual energy-saving construction of the demonstration farmhouse in Yanhe Village.

Figure 10.12 shows the actual energy conservation construction of the public demonstration building, the Reception Centre of Yanhe Village.

Fig. 10.11 Exterior insulation and energy-saving construction of demonstration farmhouse in Yanhe Village

Fig. 10.12 Thermal insulation and energy-saving construction of demonstration public building, the Reception Centre

10.4 Evaluation of Village and Town Green Building Water-Saving Facility Efficiency

10.4.1 Evaluation Method of Village and Town Building Water-Saving Facility Efficiency

1. Evaluation content

1) Water quota of village and town buildings

At present, common rural buildings in China are mainly ordinary peasant houses and rural public buildings, such as village public offices, kindergartens, clinics, and farmer training centres.

According to *Standard for Water Saving Design of Civil Buildings* GB 50,555-2010 and the data of the on-site investigation, the recommended water quota of buildings is shown in Tables 10.13 and 10.14.

Table 10.13 Recommended quota for average domestic water usage and water savings of rural houses q_z

Category of rural houses		Standards for sanitary appliance settings	Water savings and water consumption quota q_z (L/people·d)		
			First area	Second area	Third area
Ordinary housing	I	Lavatory pan, washtub	80–100	50–70	45–65
	II	Lavatory pan washtub, wash bowl, washtub, washing machine, water heater and bath equipment	90–140	60–100	50–80

Notice 1. Water quotas of water savings in the chart are formulated according to GB 50,555-2010 (Medium-sized and small city quotas).
2. First area: Provinces of Hubei, Hunan, Jiangxi, Zhejiang, Fujian, Guangdong, Guangxi, Hainan, Shanghai, Jiangsu, Anhui, and Chongqing. Second area: Sichuan, Guizhou, Yunnan, Heilongjiang, Jilin, Liaoning, Beijing, Tianjin, Hebei, Shanxi, Henan, Shandong, Ningxia, Shanxi, Inner Mongolia east of Hetao and Gansu east of Yellow River. Third area: Xinjiang, Qinghai, Inner Mongolia west of Hetao and Gansu west of Yellow River.
3. Total water consumption is referred to in the chart, and dual water supply is not considered.Total water consumption is referred to in the chart, and dual water supply is not considered.

Table 10.14 Recommended quota for average domestic water usage and water savings of rural public buildings q_g

Types of buildings and standards for sanitary appliance settings	Water savings and water consumption quota q_g	Units
Office building	25–40	L/person·team
Nursery school	25–40	L/child·d
No accommodation	6–12	L/person·time
Medical staff	40–80	L/person·d

2) Water-use efficiency of domestic water appliances

At present, commonly used water appliances in rural buildings in China include water taps, showers, squatting pans and pedestal pans. The water quota and water-use efficiency grade of common water appliances are shown in Table 10.15. The "–" symbol in the table indicates that there is no need to fill it in.

3) Reasonable use of non-traditional water sources

Reasonable use of non-traditional water sources is an important step in saving water in villages and towns, especially in water-deficient areas.

The utilization ratio of non-traditional water sources in village and town buildings is calculated as follows:

Table 10.15 Water-use limit values and water-efficiency grades for common village and town water appliances

Water appliance	Evaluation content								Water-efficiency limit	Water savings
Water nozzle	Water-efficiency grade index			Level 1	Level 2	Level 3	–	–	Level 3	Level 2
	Average water consumption (L)			5.0	6.0	8.0	–	–		
Shower	Water-efficiency grade index			Level 1	Level 2	Level 3	–	–	Level 3	Level 2
	Flow (L/s)			0.08	0.12	0.15	–	–		
Squatting pan	Water-efficiency grade index			Level 1	Level 2	Level 3	–	–	Level 3	Level 2
	Water-efficiency grade index			Level 1	Level 2	Level 3	–	–		
Toilet	Water-efficiency rating index			Level 1	Level 2	Level 3	Level 4	Level 5	–	–
	Water consumption (L)	Single gear	Average value	4.0	5.0	6.5	7.5	9.0	Level 5	Level 2
		Double gear	Large gear	4.5	5.0	6.5	7.5	9.0	–	–
			Small gear	3.0	3.5	4.2	4.9	6.3	–	–
			Average value	3.5	4.0	5.0	5.8	7.2	–	–

$$R_u = \frac{W_u}{W_t} \times 100\% \tag{10.15}$$

where R_u is utilization ratio of non-traditional water sources, %;

W_u is actual consumption of non-traditional water sources, m^3/a; and

W_t is total water consumption, m^3/a.

4) Water metre allocation rate in built-up areas

Residences in built-up areas in towns should be equipped with water metres. According to *Guidelines for Water-saving Community Evaluation* GB/T 26,928-2011, the water metre allocation rate of communities should reach 100%.

2. Evaluation method

The evaluation of water-saving facilities in rural buildings adopts a scoring method from 0 to 1 point. If the evaluation indicator meets the current national/sector/industry standards or norms, 1 point is given; if not, 0 points are given. The selected index should be marked with "√" when filling in the form.

The evaluation data acquisition methods include on-site surveys and site detection, which should meet the current national standards. When site detection cannot be carried out (for example, in remote rural areas), the rank of rated water usage efficiency can be referred to.

The evaluation standards and methods of the water-saving facility efficiency of village and town buildings are shown in Table 10.16.

In the specific application, it is necessary to determine whether the evaluation object is a village or a town. In combination with the actual situation of the village or town, the corresponding evaluation index should be selected (and ticked √), and the indicators that do not apply to the evaluation are indicated with "–". Then, the "full score" items of the evaluation are counted, and the evaluation is carried out item by item. One point is allocated for standard items and 0 points for non-standard items. Because different villages have different types of buildings and different domestic water appliances, the full marks in Table 10.16 differ. If the score reaches 70% of the full score, then the facility efficiency is qualified.

10.4.2 Evaluation Case of Village and Town Building Water-Saving Facility Efficiency

For convenience of comparison, in this case, the same farmhouse considered in Sect. 9.4.1 was selected for building water-saving efficiency evaluation. It is a 2-story building with a building area of 251 m^2; it has 7 rooms and 2 halls and was completed in 2009. There are 3 faucets, 2 squatting pans, 2 solar shower systems and 1 washing machine. The permanent resident population is 3 people, and the annual

Table 10.16 Evaluation standards and methods of village and town building water-saving facility efficiency

Evaluation items	Evaluation indexes		Evaluation standard	Evaluation results			
				Qualified (1)	Sub-standard (0)	Score	Total score
Building water savings & usage quota (1–3 points)	(1) Normal housing	There are toilets and washbasins	District 1 < 100 (L/people·d) District 2 < 70 (L/people·d) District 3 < 65 (L/people·d)				
		There are toilets, washbasins, sinks and washing machines, water heaters and bathing equipment	District 1 < 140 (L/people·d) District 2 < 100 (L/people·d) District 3 < 80 (L/people·d)				
	(2) Public buildings	Office building	≤40 (L/people·d)				
		Kindergarten (no accommodation)	≤40 (L/child·d)				
		Medical staff (infirmary)	≤12 (L/people·time)				
		Training centre (set up public toilets and washrooms)	≤80 (L/people·d)				
Water efficiency of domestic water devices (6–8 points)	(3) Water nozzle	Average water consumption (L): Grade 1 (5.0), Grade 2 (6.0), Grade 3 (8.0)					
		Water-efficiency limit	Grade 3				
		Water-saving evaluation value	Grade 2				

(continued)

Table 10.16 (continued)

Evaluation items	Evaluation indexes		Evaluation standard	Evaluation results			
				Qualified (1)	Sub-standard (0)	Score	Total score
	(4) Shower	Water flow (L/s): Grade 1 (0.08), Grade 2 (0.12), Grade 3 (0.15)					
		Water-efficiency limit	Grade 3(<0.15 L/s)				
		Water-saving evaluation value	Grade 2				
	(5) Squatting pan	Average water consumption (L): Grade 1 (5.0), Grade 2 (6.0), Grade 3 (8.0)					
		Water-efficiency limit	Grade 3				
		Water-saving evaluation value	Grade 2				
	(6) Pedestal pan	Water consumption (L) Single level: Grade 1 (4.0), Grade 2 (5.0) Grade 3 (6.5), Grade 4 (7.5), Grade 5 (9.0) Double level: High-level grade 1 (4.5), grade 2 (5.0), grade 3 (6.5), grade 4 (7.5), grade 5 (9.0) Low-level grade 1 (3.0), grade 2 (3.5), grade 3 (4.2), grade 4 (4.9), grade 5 (6.3) Average: grade 1 (3.5), Grade 2 (4.0), Grade 3 (5.0), Grade 4 (5.8), Grade 5 (7.2)					
		Water-efficiency limit	Grade 5				
		Water-saving evaluation value	Grade 2				

(continued)

Table 10.16 (continued)

Evaluation items	Evaluation indexes	Evaluation standard	Evaluation results			
			Qualified (1)	Sub-standard (0)	Score	Total score
Rational use of non-traditional water resources (1 point)	(7) Utilization rate of non-traditional water source utilization	Used for indoor flushing, green irrigation, into the digester fertilizer, road flushing. The amount of water used from non-traditional sources should not be less than 50% of the total. See Formula (10.15) for the calculation				
Residential water metre allocation rate in built-up areas of the town (1 point)	(8) Water metre assembly rate in built-up areas of the town	100% (used only for township evaluation)				

Form-filling explanation:
(1) Located in ____________ (District 1/District 2/District 3), ____________ (Village/Town built-up area);
(2) Farmhouse type is ____________ (Residential building/Public building);
(3) Full score is ____________;
(4) Field test data from ____________ (Field test/Rated water-efficiency rating)

average number of tourists received is 1.6 people/day. Therefore, the evaluation is based on 5 permanent residents.

According to the above investigation and analysis, Table 10.16 is applied to evaluate the efficiency of the water-saving facilities of the evaluated building in Yanhe Village item by item, with 8 items 8 and a score of 7 points, accounting for 87.5% of the total score of 8 points. The score is more than 70% of the total; thus, the evaluation result is qualified. See Table 10.17 for the evaluation results of the building water-saving facility efficiency of Yanhe Village.

10.5 Evaluation of Village and Town Green Building Indoor Environmental Quality

10.5.1 Evaluation Method of Village and Town Building Indoor Environmental Quality

1. 1 Selection of evaluation methods

At present, there is no unified evaluation standard for indoor environmental quality in China. The commonly used evaluation methods can be divided into objective evaluation, subjective evaluation and standard evaluation.

(1) Objective evaluation. The representative pollutants are selected as the evaluation index, real pollutant data are obtained through on-site testing or laboratory testing after sampling, and quantitative analysis is carried out with the current national standards to evaluate the indoor environmental quality.
(2) Subjective evaluation. The environment is described and evaluated by using the sense organs of the human body through on-site investigation and questioning of indoor personnel.
(3) Standard evaluation. An objective evaluation method that adopts single-factor evaluation is used. The standard-reaching evaluation method requires that all tested items meet *Indoor Air Quality Standard* GB/T 18,883-2002.

Combined with the characteristics of rural China, this evaluation uses a standard-reaching evaluation method for building rural indoor environmental quality evaluations.

2. Content and evaluation indexes

This evaluation mainly considers two indexes of green village and town building indoor environmental quality: indoor air quality and acoustic environmental quality.

Table 10.17 Evaluation results of the water-saving facility efficiency of a building in Yanhe Village

Evaluation items	Evaluation indexes		Evaluation standard	Evaluation result			
				Qualified (1)	Sub-standard (0)	Score	Total score
Building water savings & usage quota (1–3 points)	(1) Normal housing	There are toilets and washbasins	District 1 < 100 (L/people·d) District 2 < 70 (L/people·d) District 3 < 65 (L/people·d)	–		1	7
		There are toilets, washbasins, sinks and washing machines, water heaters and bathing equipment	District 1 < 140 (L/people·d)√ District 2 < 100 (L/people·d) District 3 < 80 (L/people·d)	1			
	(2) Public buildings	Office building	≤40 (L/people·d)	–			
		Kindergarten (no accommodation)	≤40 (L/child·d)	–			
		Medical staff (infirmary)	≤12 (L/people·time)	–			
		Training centre (set up public toilets and washrooms)	≤80 (L/people·d)	–			
Water efficiency of domestic water devices (6–8 points)	(3) Water nozzle	Average water consumption (L): Grade 1 (5.0), Grade 2 (6.0), Grade 3 (8.0)√				6	
		Water-efficiency limit	Grade 3	1			

(continued)

Table 10.17 (continued)

Evaluation items	Evaluation indexes		Evaluation standard	Evaluation result			
				Qualified (1)	Sub-standard (0)	Score	Total score
		Water-saving evaluation value	Grade 2	1			
	(4) Shower	Water flow (L/s): Grade 1 (0.08), Grade 2 (0.12), Grade 3 (0.15)√					
		Water-efficiency limit	Grade 3(<0.15 L/s)	1			
		Water-saving evaluation value	Grade 2	1			
	(5) Squatting pan	Average water consumption (L): Grade 1 (5.0), Grade 2 (6.0), Grade 3 (8.0) √					
		Water-efficiency limit	Grade 3	1			
		Water-saving evaluation value	Grade 2	1			
	(6) Pedestal pan	Water consumption (L) Single level: Grade 1 (4.0), Grade 2 (5.0) Grade 3 (6.5), Grade 4 (7.5), Grade 5 (9.0) Double level: High-level grade 1 (4.5), grade 2 (5.0), grade 3 (6.5), grade 4 (7.5), grade 5 (9.0) Low-level grade 1 (3.0), grade 2 (3.5), grade 3 (4.2), grade 4 (4.9), grade 5 (6.3) Average: grade 1 (3.5), Grade 2 (4.0), Grade 3 (5.0), Grade 4 (5.8), Grade 5 (7.2)					

(continued)

Table 10.17 (continued)

Evaluation items	Evaluation indexes		Evaluation standard	Evaluation result			
				Qualified (1)	Sub-standard (0)	Score	Total score
		Water-efficiency limit	Grade 5	–			
		Water-saving evaluation value	Grade 2	–			
Rational use of non-traditional water resources (1 point)	(7) Utilization rate of non-traditional water source utilization		Used for indoor flushing, green irrigation, into the digester fertilizer, road flushing. The amount of water used from non-traditional sources should not be less than 50% of the total. See Formula (10.15) for the calculation.√		0	0	
Residential water metre allocation rate in built-up areas of the town (1 point)	(8) Water metre assembly rate in built-up areas of the town		100% (used only for township evaluation)	–			

(continued)

Table 10.17 (continued)

Evaluation items	Evaluation indexes	Evaluation standard	Evaluation result			
			Qualified (1)	Sub-standard (0)	Score	Total score
Form-filling explanation: (1) Located in District 1 (District 1/District 2/District 3), Village (Village/Town built-up area); (2) Farmhouse type is Residential building (Residential building/Public building); (3) Full score is 8; (4) Field test data from Field test (Field test/Rated water-efficiency rating)						

1) Description of evaluation index

(1) If it is necessary to test and evaluate the total number of indoor bacteria, this evaluation can be performed after obtaining the relevant data by on-site sampling laboratory tests.
(2) Only "indoor air environmental quality" can be evaluated for the indoor environmental quality of village and town buildings.

2) Evaluation contents

(1) Indoor air quality

The indoor air environmental quality evaluation of rural houses contains 9 items: formalin, benzene, ammonia gas, toluene, xylene, TVOC, PM_{10}, temperature and relative humidity.

The evaluation index of village and town building indoor air quality is shown in Table 10.18.

Table 10.18 Evaluation index of village and town building indoor air quality

Evaluation items	Evaluation indexes	Evaluation Contents		Evaluation Standard
Indoor air quality	Indoor temperature	22–28 °C	Summer air conditioning	Indoor air-quality standard
		16–24 °C	Winter heating	
	Relative humidity	40–80%	Summer air conditioning	
		30–60%	Winter heating	
	Formaldehyde	Concentration $\leq$ 0.10 mg/m^3	Average value in one hour	
	Benzene	Concentration $\leq$ 0.11 mg/m^3	Average value in one hour	
	Ammonia gas	Concentration $\leq$ 0.20 mg/m^3	Average value in one hour	
	Toluene	Concentration $\leq$ 0.20 mg/m^3	Average value in one hour	
	Xylene	Concentration $\leq$ 0.20 mg/m^3	Average value in one hour	
	TVOC	Concentration $\leq$ 0.60 mg/m^3	Average value in eight hours	
	PM_{10}	Concentration $\leq$ 150 μg/m^3	Average value in 24 h	
	$PM_{2.5}$	Concentration $\leq$ 75 μg/m^3	Average value in 24 h	Refer to ambient air-quality standards

(2) Indoor acoustic environmental quality

The acoustic environmental quality concerns the noise level (A level).

The evaluation index of village and town building indoor acoustic environmental quality is shown in Table 10.19.

3. Evaluation technical route

The village and town building indoor environmental quality evaluation technical route is shown in Fig. 10.13.

4. Evaluation sample collection and monitoring

1) Distribution of sample

The method of point distribution can be diagonal or quincunx so that all points are distributed in the room. The sampling point should avoid opening air ducts and vents, and the distance from the wall and large furniture should be more than 1 m.

2) Height of sample

The sampling height is 0.7–1.5 m from the ground.

3) Number of samples

The number of sampling points should be determined according to the indoor area and site conditions. Generally, 1–3 sampling points should be set for rooms below 50 m^2, 3–5 sampling points for rooms between 50 and 100 m^2, and at least 5 sampling points for rooms above 100 m^2.

Sampling in the living room should be carried out under conditions of people's normal activities. The room should be monitored for at least one day and sampled twice a day without opening doors and windows.

Table 10.19 Evaluation index of village and town building indoor acoustic environmental quality

Evaluation items	Evaluation indexes	Evaluation Contents	Evaluation Standard
Acoustic environmental quality	Noise level (A Level)	Bedroom allows noise level $\leq$ 45 dB (daytime)/ $\leq$ 37 dB (night)	Code for design of sound insulation for civil buildings
		Noise level in living room and office space $\leq$ 45 dB	

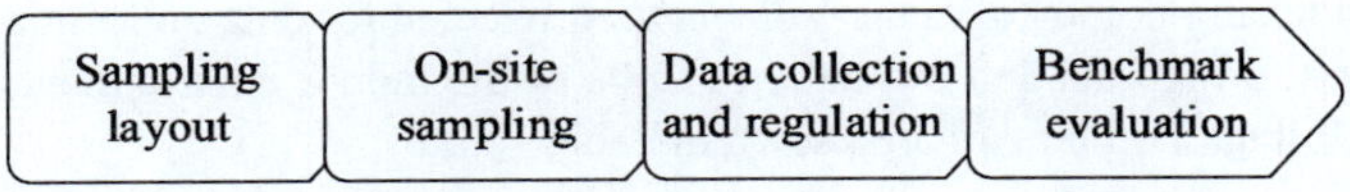

Fig. 10.13 Village and town building indoor environmental quality evaluation technical route

4) Monitoring methods

The monitoring methods for village and town building indoor environmental quality should be in accordance with the current national standards.

5. Evaluation method of village and town building indoor environmental quality

This evaluation method uses the standard evaluation method to evaluate village and town building indoor environmental quality. If the evaluation index reaches the relevant national evaluation standard, 1 point is given; if it fails to meet the standard, 0 points are given. There are 11 evaluation indexes in the village and town building indoor environmental quality evaluation system: 10 indoor air quality evaluation indexes and 1 acoustic environmental quality evaluation index. The full score of the village and town building indoor environmental quality evaluation is 11 points.

If only the indoor air environmental quality of the building is evaluated, the full score is 10 points.

10.5.2 Evaluation Case of Village and Town Building Indoor Environmental Quality

1. Selection of evaluation monitoring points

Three monitoring points were selected from representative farmhouses in Yanhe Village: Dushanju farmhouse (bedroom) S_{11}, Ginkgo Mountain Villa (hotel room) S_{12}, and Reception Centre (office room) S_{13}.

S_{11} was built and put into use in 2008. Every bedroom area is 18 m^2, and there is a monitoring point.

S_{12} was built in the 1970s as a warehouse building. In 2005, it was transformed into rural tourist accommodations with a room area of 20 m^2, and there is a monitoring point.

S_{13} was completed and put into use in 2009. The conference room area is 204 m^2, and there are 5 monitoring points.

The distribution of indoor environmental quality monitoring points of farmhouses in Yanhe Village is shown in Fig. 10.14.

The actual situation of indoor environmental quality testing of farmhouses in Yanhe Village is shown in Fig. 10.15.

2. Monitoring data and evaluation

Three monitoring points in Yanhe Village were selected for data collection.

The monitoring data and evaluation results of the indoor environmental quality of Yanhe Village farmhouses are shown in Table 10.20.

3. Analysis of evaluation results

1) The indoor environmental quality evaluation of the buildings in Yanhe Village was generally excellent, with an average score of 9.7 points.

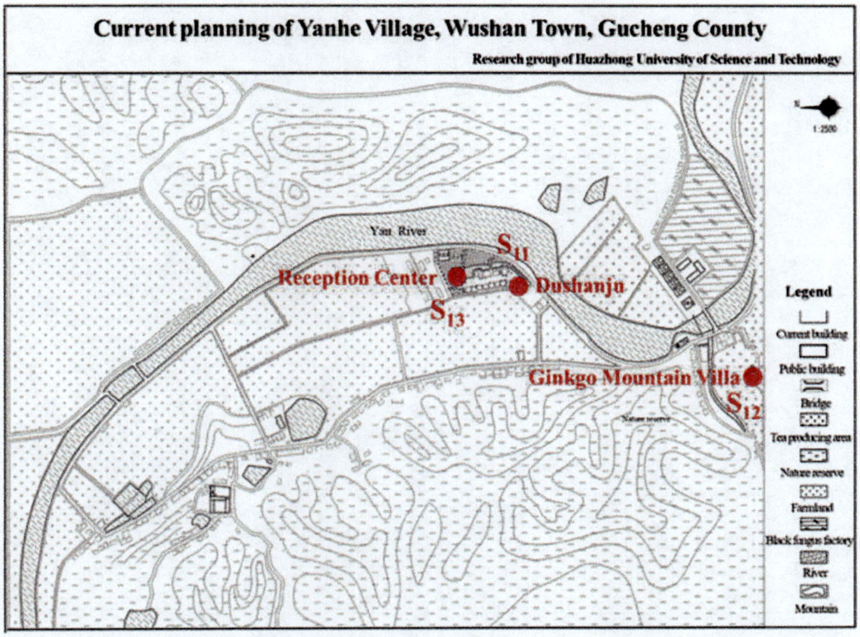

Fig. 10.14 Distribution of indoor environmental quality monitoring points of farmhouses in Yanhe Village

2) The indoor sound level at the 3 sample points slightly exceeded the standards.
3) The formaldehyde content at sample point S_{13} slightly exceeded the standard.
4) Problem Analysis

(1) Indoor noise during the day did not meet the standards. The reason is that the leading industry of Yanhe Village is rural tourism, and there are many tourists and vehicles during the day, which leads to a slight excess of indoor noise.
(2) The indoor formaldehyde level of sample S_{13} exceeded the standard. The reason is that the conference room is always closed except during meetings, and the indoor ventilation is insufficient; thus, the formaldehyde concentration in the furniture exceeds the standards.

10.6 Evaluation of Low-Carbon Operational Management of Village and Town Green Buildings

10.6.1 Evaluation Method of Village and Town Building Low-Carbon Operational Management

1. Basic concept

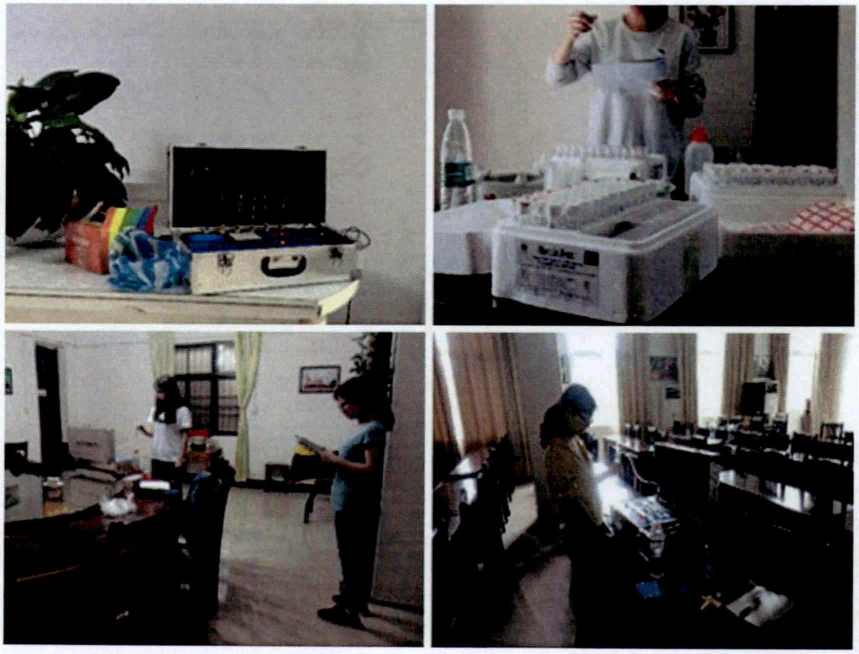

Fig. 10.15 Actual situation of indoor environmental quality testing of farmhouses in Yanhe Village

(1) Carbon source refers to the process, activity or mechanism of releasing carbon dioxide into the atmosphere.
(2) Carbon sink refers to the process and mechanism by which plants absorb carbon dioxide from the atmosphere and water in the soil through photosynthesis to fix carbon dioxide in plants and turn it into carbon [3].
(3) Carbon sink of buildings refers to the amount of carbon dioxide absorbed and stored by greenery and vegetation from the air within the scope of the designated building.
(4) Low-carbon community refers to controlling and reducing greenhouse gas emissions throughout the planning, construction and use of the community through the comprehensive plan of energy, resources, transportation, land use, construction, management and resident participation to meet or exceed the standards for energy conservation and emissions reductions set by national or local governments.

2. Evaluation contents of village and town building low-carbon operational management of

The evaluation target is the low-carbon operational status of village and town buildings and communities.

Table 10.20 Monitoring data and evaluation results of the indoor environmental quality of Yanhe Village farmhouses

Evaluation items	Evaluation indexes	S_{11}				S_{12}				S_{13}			
		Measured value	Evaluation results			Measured value	Evaluation results			Measured value	Evaluation results		
			Qualified (1)	Sub-standard (0)	Total score		Qualified (1)	Sub-standard (0)	Total score		Qualified (1)	Sub-standard (0)	Total score
Indoor air quality	Room temperature	25	1		10	25	1		10	26	1		9
	Relative humidity	73	1			68	1			72	1		
	Formaldehyde	0.1	1			0.05	1			0.15		0	
	Benzene	Not detected	1			Not detected	1			Not detected	1		
	Ammonia	Not detected	1			Not detected	1			Not detected	1		
	Toluene	Not detected	1			Not detected	1			Not detected	1		
	Xylene	Not detected	1			Not detected	1			Not detected	1		
	TVOC	Not detected	1			Not detected	1			Not detected	1		
	PM_{10}	18	1			6	1			9	1		
	$PM_{2.5}$	9	1			3	1			5	1		
Noise environmental quality	Noise level (A sound level)	46.1		0		46.2		0		45.1		0	

Note The score standard for this table is shown in Tables 10.18 and 10.19

1) Evaluation of low-carbon operation in village and town buildings

Building carbon emissions are the main carbon emission source in villages and towns. Energy conservation and emissions reduction of rural houses are key measures to effectively reduce energy consumption in villages and towns and control carbon emissions.

(1) Improving the low-carbon environment of village and town buildings includes making full use of external natural conditions to ensure good sunshine, daylight, and natural ventilation in rural areas; making full use of surrounding road networks and public service facilities; and adhering to a compact layout and land resource conservation.
(2) Adopting low-carbon energy equipment in villages and towns includes making full use of solar energy, wind energy, geothermal energy, biogas and other clean and renewable energy; prohibiting the use of low-quality coal, straw burning and other high-polluting and high-emissions household energy; encouraging the use of energy-saving air conditioning and heating facilities; and strictly controlling indoor building temperatures. The heating temperature in winter should be kept at 18–22 °C, and the cooling temperature in summer should be kept at 22–27 °C.

2) Evaluation of low-carbon operation in community environmental facilities

The low-carbon operation of village community environmental facilities should make full use of and continuously improve the natural ecological environment in the community and effectively control and reduce carbon emissions from the operation of environmental facilities.

(1) Community water resource low-carbon utilization facilities include installing water-saving facilities in all village and town buildings; promoting water-saving technologies such as landscape and car-wash water; using non-traditional water sources such as rainwater and river water for non-drinking water usage; and using permeable bricks for paving materials.
(2) Community garbage low-carbon collection and treatment facilities include classified collection of household waste; sealed collection and transportation of domestic garbage; 100% harmless treatment of domestic garbage; and assigning full-time staff to be responsible for community garbage collection and transfer management.
(3) Community greening and landscape facility maintenance includes a green area rate of the community in the township of not less than 30%, and a greening rate of the green space itself of more than 70%; reasonable use of roof greening and vertical landscaping of building facades to increase green coverage; protection of trees and green plants according to local conditions; and manual maintenance and pruning to manage green plants and reduce carbon emissions from the use of fuel machinery.

3) Evaluation of low-carbon operations in traffic and water supply and drainage facilities

Village and town transportation and water supply and drainage facilities should use low-carbon technologies such as low-impact development, reclaimed water utilization, and low-carbon energy utilization to save resources, reduce energy consumption, reduce pollution, and improve the community environment.

(1) Low-carbon transportation facilities and low-carbon travel include prioritizing walking, cycling, and public transportation; prioritizing permeable surfaces for non-motorized roads, ground parking and other hard paving; and using clean-energy vehicles for public transportation to encourage low-carbon travel.
(2) Low-carbon water supply and drainage facility operation It includes establishing classified water supply systems and diverting rainwater and wastewater and using small dispersion and low-cost maintenance techniques such as detention ponds, rainwater wetlands, rainwater infiltration and filtration collection facilities, and rainwater plantation tank trenches.

4) Evaluation of village and town building low-carbon operational management

Establishing a low-carbon operational plan, implementing a low-carbon demonstration project and establishing a social participation mechanism of low-carbon operational management are necessary to achieve the goal of low-carbon building operation.

(1) A low-carbon building operational plan and demonstration project includes developing a low-carbon operational management plan for rural houses or communities; demonstration projects for the low-carbon operation of rural houses that have been built or are under construction (refer to 1)-4) above for evaluation content); and a relatively complete monitoring system for the energy consumption of rural houses and communities.
(2) Social participation in low-carbon operational management of rural houses includes encouraging community residents and experts to actively participate in the whole process of preparation and implementation of low-carbon operational management plans; establishing low-carbon development and management bulletin boards, display boards and brochures; and conducting regular low-carbon education and training.

3. Evaluation method of village and town low-carbon operational building management

The evaluation of village and town building low-carbon operational management adopts a scoring method from 0 to 1 point. If the evaluation indicator meets the standard, 1 point is given; if not, 0 points are given.

The full score is 9 points. If the score exceeds 7 points, it is qualified as meeting the technical requirements for the low-carbon operation of green houses.

In the actual evaluation, if there are multiple requirements in a certain scoring standard, 1 point can be given as long as one of them is satisfied. For the four types of evaluation content, one of the scoring standards is selected only once in principle to avoid repeated evaluation.

The village and town building low-carbon operational management evaluation standards and methods are shown in Table 10.21.

During the actual evaluation operation, the actual scores of each sub item of the evaluated village or town should be filled into the corresponding "evaluation score" column in the table to obtain the evaluation results directly.

10.6.2 Evaluation Case of Village and Town Building Low-Carbon Operation and Management

According to the above investigation and analysis, Table 10.21 is applied to evaluate the low-carbon operation and management of Yanhe Village item by item. The result is a score of 8 points, accounting for 88.9% of the total score of 9 points. The score is more than 70% of the total, so the evaluation result is qualified. The evaluation results of the building low-carbon operational management in Yanhe Village are shown in Table 10.22.

Table 10.21 Evaluation standards and methods of village and town building low-carbon operational management

Evaluation items	Evaluation indexes	Evaluation standard	Evaluation result			
			Qualified (1)	Sub-standard (0)	Score	Total score
Low-carbon operation of village and town buildings (2 points)	(1) Improve the low-carbon use environment of buildings	Ensure that farmhouses have good conditions of sunshine, lighting and natural ventilation; make full use of the surrounding road network and public service facilities; compact layout and save land resources				
	(2) Adopt low-carbon energy equipment in village and town buildings	Use solar energy, wind energy, geothermal energy, biogas and other clean and renewable energy; prohibit the use of low-quality coal, straw burning and other high-polluting and high-emissions household energy; use energy-saving air conditioning and heating facilities; and strictly control the indoor temperature of buildings				
Low-carbon operation of community environmental facilities (3 points)	(3) Community water resources low-carbon utilization facility	Use water-saving facilities in all buildings; promote water-saving technologies such as landscape and car-wash water; use non-traditional water sources such as rainwater and river water for non-drinking water usage; and use permeable bricks for paving materials				

(continued)

Table 10.21 (continued)

Evaluation items	Evaluation indexes	Evaluation standard	Evaluation result			
			Qualified (1)	Sub-standard (0)	Score	Total score
	(4) Community garbage low-carbon collection and treatment facility	Classified collection of household waste; sealed collection and transportation of domestic garbage; 100% harmless treatment of domestic garbage; assigning full-time staff to be responsible for community garbage collection and transfer management				
	(5) Community greening and landscape facility maintenance	The green area rate of the community in the township should be not less than 30%, and the greening rate of the green space itself should be greater than 70%; use of roof greening and vertical landscaping of building facades choose local greening plants; assign special personnel to maintain greening and landscape facilities				
Traffic and water supply and drainage facilities for low-carbon operation (2 points)	(6) Low-carbon transportation facilities and low-carbon travel	Prioritize the use of public transport, walking and bicycling; guide residents to low-carbon travel				

(continued)

Table 10.21 (continued)

Evaluation items	Evaluation indexes	Evaluation standard	Evaluation result			
			Qualified (1)	Sub-standard (0)	Score	Total score
	(7) Low-carbon water supply and drainage facility operation	Establish classified water supply systems and divert rainwater and wastewater; use small dispersion and low-cost maintenance techniques such as detention ponds, rainwater wetlands, rainwater infiltration and filtration collection facilities, and rainwater plantation tank trenches				
Low-carbon operational capacity of building (2 points)	(8) Low-carbon operational plan and demonstration building project	Have low-carbon operational plan, implementing or implying low-carbon demonstration projects; establish building and community energy monitoring and assessment systems				
	(9) Social participation in low-carbon operational management of rural houses	Encourage community residents and experts to actively participate in the whole process of preparation and implementation of low-carbon operational management plans; set up low-carbon development and management bulletin boards, display boards and brochures; and conduct regular low-carbon education and training				

Table 10.22 Evaluation results of building low-carbon operational management in Yanhe Village

Evaluation items	Evaluation indexes	Evaluation results			
		Qualified (1)	Sub-standard (0)	Score	Total score
Low-carbon operation of village and town buildings (2 points)	(1) Improve the low-carbon use environment of buildings	1		2	8
	(2) Adopt low-carbon energy equipment in village and town buildings	1			
Low-carbon operation of community environmental facilities (3 points)	(3) Community water resources low-carbon utilization facility	1		3	
	(4) Community garbage low-carbon collection and treatment facility	1			
	(5) Community greening and landscape facility maintenance	1			
Traffic and water supply and drainage facilities for low-carbon operation (2 points)	(6) Low-carbon transportation facilities and low-carbon travel		0	1	
	(7) Low-carbon water supply and drainage facility operation	1			
Low-carbon operational capacity of building (2 points)	(8) Low-carbon operational plan and demonstration building project	1		2	

(continued)

Table 10.22 (continued)

Evaluation items	Evaluation indexes	Evaluation results			
		Qualified (1)	Sub-standard (0)	Score	Total score
	(9) Social participation in low-carbon operational management of rural houses	1			

References

1. Zhong P (2005) Study on energy consumption and environmental impact of building life cycle. Dissertation, Sichuan University
2. Hua H, Li W, Wang XM, et al (2017) Comparison of energy consumption measurement and energy-saving efficiency of adobe building. J Civ Eng Manag 34(6):17–22+28
3. Dong HY, Yun JF, Wang GZ (2012) Carbon sink profile. Science Press, Beijing

Chapter 11
Comprehensive Evaluation of Green Village and Town Planning and Construction

Abstract The comprehensive evaluation of green village and town planning and construction refers to the overall inspection and evaluation of the implementation and completion status of green village and town planning and construction, which plays an important role in ensuring and optimizing the realization of the planning and construction goals. This chapter puts forward the comprehensive evaluation content, establishes a comprehensive evaluation method, conducts a comprehensive case evaluation, and builds a comprehensive evaluation platform of green village and town planning and construction.

Based on the implementation evaluation methods for the plan and the environmental infrastructure construction and green village and town green building construction proposed in Chaps. 8–10, the sub-item comprehensive evaluation and overall comprehensive evaluation methods for green village and town planning and construction are established in this chapter.

11.1 Comprehensive Evaluation Content of Green Village and Town Planning and Construction

1. Comprehensive evaluation index

According to the theoretical system and evaluation method of green village and town construction in this book, the comprehensive evaluation index system of green village and town planning and construction is composed of three types of secondary evaluation indexes: plan evaluation, infrastructure construction evaluation, and green building construction evaluation. This evaluation index system has 18 three-level indexes and 390 evaluation items. Each index is based on the current relevant national laws, regulations and standards.

The comprehensive evaluation index of green village and town planning and construction are shown in Table 11.1.

X. Wang and H. Hua, *Green Village and Town Construction in China*,
https://doi.org/10.1007/978-981-16-2098-0_11

Table 11.1 Comprehensive evaluation index of green village and town planning and construction

Number	Primary indexes	Number	Secondary indexes	Number	Tertiary indexes
A	Comprehensive evaluation of planning and construction	A_1	Evaluation of plan	A_{11}	Evaluation of site selection and functional division optimization
				A_{12}	Evaluation of plan niche suitability
				A_{13}	Evaluation of water resource utilization and sewage disposal plan
				A_{14}	Evaluation of waste disposal plan
				A_{15}	Evaluation of clean energy utilization plan
				A_{16}	Evaluation of plan implementation and eco-environmental restoration
		A_2	Evaluation of infrastructure construction	A_{21}	Evaluation of drinking water source protection and water-saving facility construction
				A_{22}	Evaluation of domestic sewage treatment facility construction
				A_{23}	Evaluation of waste classified collection and harmless treatment facility construction
				A_{24}	Evaluation of clean energy facility construction
				A_{25}	Evaluation of greening and landscape facility construction

(continued)

Table 11.1 (continued)

Number	Primary indexes	Number	Secondary indexes	Number	Tertiary indexes
				A_{26}	Evaluation of disaster prevention, mitigation and early-warning facility construction
		A_3	Evaluation of green building construction	A_{31}	Evaluation of green building construction selection
				A_{32}	Evaluation of green building construction
				A_{33}	Evaluation of building energy-saving efficiency
				A_{34}	Evaluation of building water-saving facility efficiency
				A_{35}	Evaluation of building indoor environmental quality
				A_{36}	Evaluation of building low-carbon operational management

2. Evaluation content and evaluation method

The content and method of comprehensive evaluation for green village and town planning and construction are shown in Fig. 11.1.

3. Evaluation process

The working process of comprehensive evaluation of green village and town planning and construction is divided into three stages: investigation and work plan formulation, demonstration and evaluation, and report compilation.

The working process of the comprehensive evaluation of green village and town planning and construction is shown in Fig. 11.2.

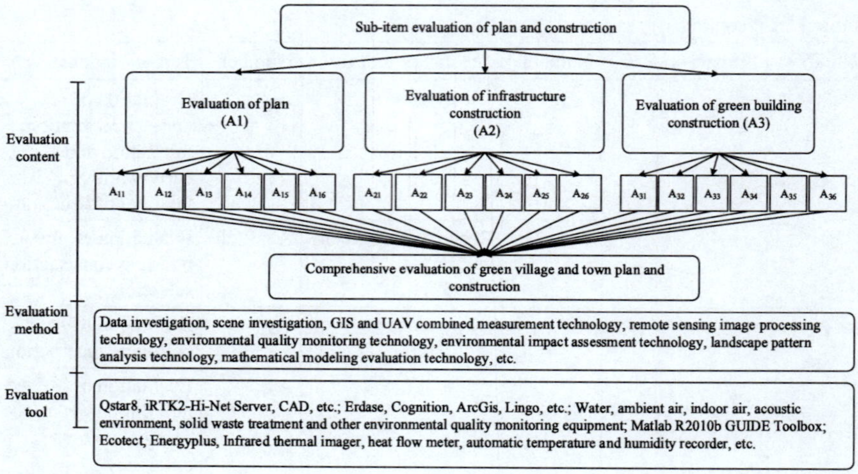

Fig. 11.1 Content and method of comprehensive evaluation of green village and town planning and construction

11.2 Comprehensive Evaluation Method of Green Village and Town Planning and Construction

1. Evaluation procedure

The comprehensive evaluation of green village and town planning and construction has two parts: sub-item evaluation and comprehensive evaluation.

1) Sub-item evaluation

According to the comprehensive evaluation index system of green village and town planning and construction established in Sect. 11.1, 18 sub-evaluation items, such as site selection and functional optimization evaluation of land, proposed in Table 11.1 are used to analyse and evaluate various evaluation indexes of the evaluated village and town, and the corresponding evaluation results are obtained. See Chaps. 8–10 for each sub-item evaluation method and evaluation case.

When conducting sub-item evaluation, we should combine the key points and characteristics of green village and town construction and development in different regions and select some of the above items for key evaluation rather than forcing all sub-items to be evaluated individually.

2) Comprehensive evaluation

(1) The results of the sub-item evaluation are processed by data normalization.
(2) According to the normalized data, combined with the weight of the evaluation indexes at all levels, the comprehensive evaluation of green village and town planning and construction is carried out.

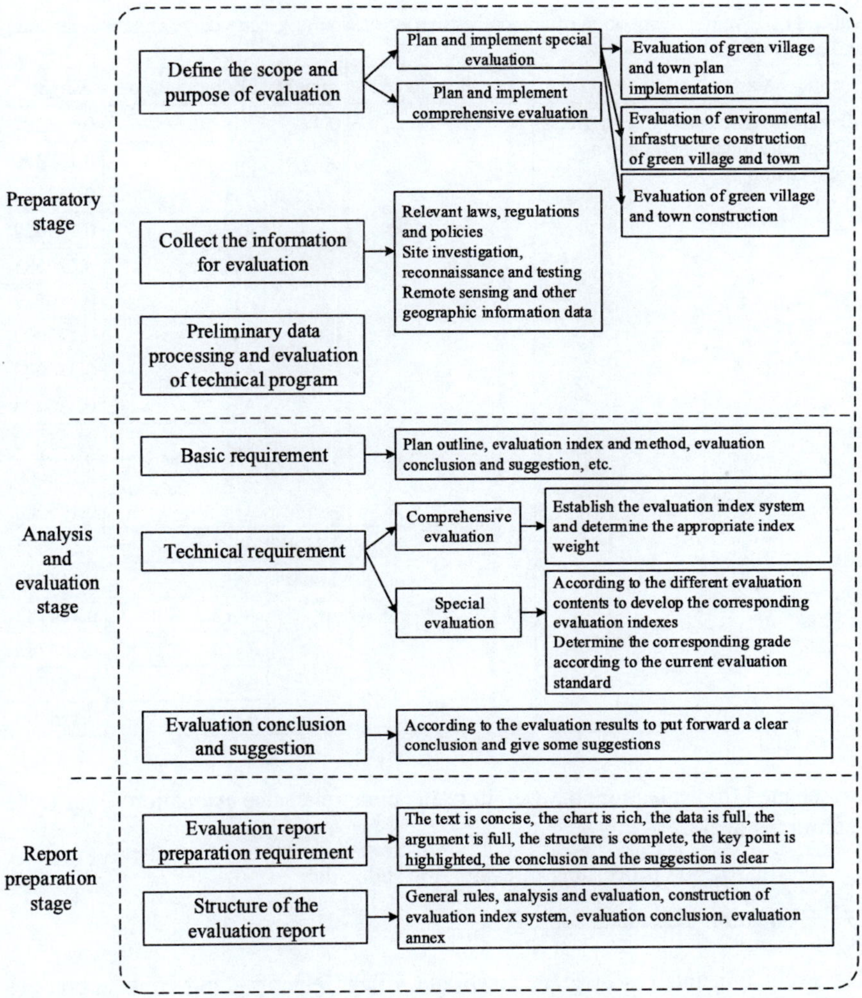

Fig. 11.2 Comprehensive evaluation process of green village and town planning and construction

2. Evaluation weight

There are many methods of determining the weight of a comprehensive evaluation. To adapt to the characteristics of simple operation required by the evaluation of villages and towns, this evaluation adopts a method combining the expert evaluation method and analytic hierarchy process to determine the weight of the comprehensive evaluation.

The weight distribution of the comprehensive evaluation of green village and town planning and construction is shown in Table 11.2. When only some of the indicators

Table 11.2 Weight distribution of comprehensive evaluation of green village and town planning and construction

Primary indexes	Secondary indexes	Weight	Tertiary indexes	Weight
Comprehensive evaluation of green village and town planning and construction A	A_1	0.4	A_{11}	0.156382
			A_{12}	0.139320
			A_{13}	0.221157
			A_{14}	0.197029
			A_{15}	0.175533
			A_{16}	0.110579
	A_2	0.4	A_{21}	0.125000
			A_{22}	0.250000
			A_{23}	0.250000
			A_{24}	0.125000
			A_{25}	0.125000
			A_{26}	0.125000
	A_3	0.2	A_{31}	0.164623
			A_{32}	0.164623
			A_{33}	0.184783
			A_{34}	0.184783
			A_{35}	0.184783
			A_{36}	0.116406

are selected for evaluation, the weight of the comprehensive evaluation needs to be determined again.

3. Normalization processing of evaluation index data

1) Evaluation index data source

The evaluation indexes of green town and village planning and construction are obtained by using the qualitative and quantitative analysis methods of each special evaluation technology. See Chaps. 8–10 for details.

2) Normalization method of evaluation index data

The comprehensive evaluation indexes of green village and town planning and construction have the characteristics of extensive professional evaluation and strong professional technology. The final evaluation results and expression methods of each index are different. Therefore, before the comprehensive evaluation of green village and town planning and construction, the evaluation results of each sub-index should be expressed in the same way, that is, the normalized treatment of the evaluation index data.

Normalization is a way to simplify the calculation, and the results do not change the meaning and conclusions drawn from the evaluation results.

Normalization processing of the plan data and construction evaluation focuses on the three-level index. To simplify the calculation, two normalization processing methods are mainly adopted in this book for the final evaluation results of the three-level indicators: one is the data mapping method, and the other is the evaluation level assignment method.

(1) Data mapping method

The simple data mapping method normalizes the evaluation results data between 0 and 1. In this evaluation, the 0–1 scoring method was used to score multiple evaluation indexes. The evaluation results were normalized by the data mapping method. The calculation method is

$$M = \frac{n_i}{\sum n_i}$$

where M is the normalized processing result;
n is individual evaluation results; and
i is the ith evaluation indicator.

(2) Evaluation level assignment method

The evaluation results are classified into certain levels, such as excellent, good, average, generally poor and very poor, or A, B, C, D, E. This approach adopts the evaluation level assignment method to normalize the evaluation results.

The evaluation results assignment method is shown in Table 11.3.

4. Comprehensive evaluation results and grading

1) Calculation procedure of comprehensive evaluation

The calculation steps of the comprehensive evaluation of green village and town planning and construction are shown in Fig. 11.3.

2) Classification of comprehensive evaluation

Table 11.3 Assignment method of evaluation results

Evaluation level	Excellent/A	Good/B	Average/C	Generally poor/D	Very poor/E
Value	5	4	3	2	1

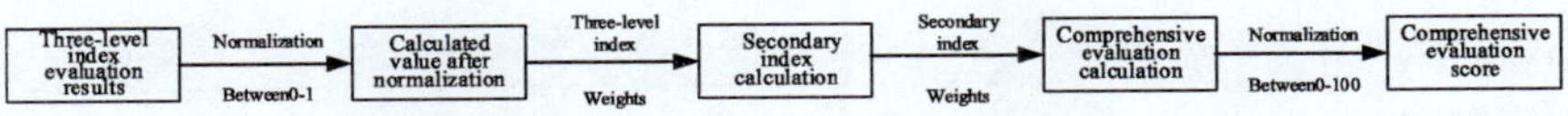

Fig. 11.3 Calculation steps of comprehensive evaluation of green village and town planning and construction

Table 11.4 Grading of comprehensive evaluation of green village and town planning and construction

Comprehensive evaluation score	100–90	89–80	79–50	49–20	19 or less
Evaluation level	I	II	III	IV	V
Evaluation results	Excellent	Good	Average	Generally poor	Very poor

The 18 evaluation indexes for the comprehensive evaluation of planning and construction are divided into five levels. The comprehensive evaluation grade standards are set according to the actual green village and town construction. Among them, the excellent rate and good rate are both 10%, the average rate is 30%, and the generally poor rate and very poor rate are 30% and 20%, respectively.

The classification of the comprehensive evaluation of green village and town planning and construction is shown in Table 11.4. For convenience of calculation, 5 points is recommended. For example, at the B level, 80% indicates that the B level has been achieved, and 85% indicates that the preferred item or bonus item has been achieved.

3) Final results of comprehensive evaluation

(1) Evaluation level

The final result of the comprehensive evaluation of the green village and town planning and construction is expressed as the evaluation level. According to the level from high to low, level I is the highest evaluation level, and level V is the lowest evaluation level.

(2) The warning level

The evaluation levels of qualified villages and towns in the comprehensive evaluation of planning and construction are marked with green marks. When the comprehensive score is below 50, the evaluated village or town will receive a non-qualification warning, which can be issued according to the three levels of yellow, orange and red.

The early-warning levels of the comprehensive evaluation of green village and town planning and construction are shown in Fig. 11.4.

5. Content of comprehensive evaluation report

The compilation of the evaluation report on the green village and town planning and construction should include the following contents:

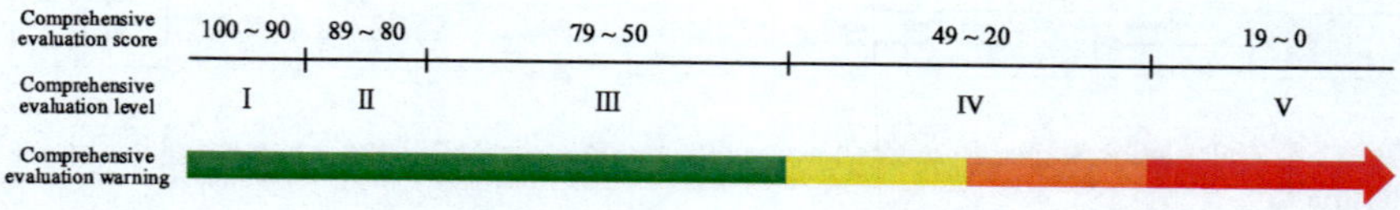

Fig. 11.4 Early-warning level division of comprehensive evaluation for green village and town planning and construction

(1) Evaluation overview. Summarize the origin of the task, clarify the evaluation basis, the purpose and principle of the evaluation, and the focus of the evaluation.
(2) Project overview. Analyse the design scheme or project status of each planning and construction project.
(3) Data analysis. Identify the impact factors related to the green ecological construction of the village or town, and make a preliminary analysis of the site survey and test data.
(4) Comprehensive evaluation. According to the evaluation impact factors related to the construction project, according to the relevant construction and evaluation standards, construct the evaluation index system and use appropriate evaluation methods for the construction project evaluation. In the comprehensive evaluation of green village and town construction, carry out weight analysis and calculation for each evaluation index, and then carry out comprehensive evaluation.
(5) Evaluation conclusion. Summarize the work results of the planning and construction evaluation, and put forward clear evaluation conclusions and improvement suggestions.
(6) Evaluation attachment. Attach the necessary evaluation-related charts and tables, such as current status survey scope, monitoring point distribution charts and tables, and professional testing reports.

11.3 Comprehensive Evaluation Case of Green Village and Town Planning and Construction

1. Field investigation and survey of village and town planning and construction

From 2014 to 2017, the authors conducted a total of eight field surveys and on-site investigations in Yanhe Village, Hubei Province, with three major categories and 18 indicators, and carried out all of the separate sub-evaluations.

The specific content and data required for comprehensive evaluation are shown in the relevant cases from Chaps. 5–10 of this book.

2. Comprehensive evaluation of village and town planning and construction

1) Summarizing the sub-item evaluation results of green village and town planning and construction

According to the comprehensive evaluation index of green village and town planning and construction established in Table 11.1, according to the evaluation content and evaluation method proposed in Fig. 11.1, 18 evaluation items of the three categories of green village and town planning and construction in Yanhe Village can be evaluated. For evaluation cases, please refer to Chaps. 5–10 of this book. See Table 11.5 for a summary of the relevant evaluation results.

Table 11.5 Method of sub-item evaluation results summary and data normalization of Yanhe Village

Number	Sub-item evaluation results	Normalization method
A_{11}	The comprehensive value of the spatial fit of the site selection plan is 0.815 The comprehensive value of the spatial coincidence of the functional zoning plan is 0.914	–
A_{12}	The niche suitability is 0.69, which is suitable	–
A_{13}	Water resource utilization plan: full score 13 points, score 10 points, and score rate 76.9% Sewage treatment plan: full score 4 points, score 4 points, and score rate 100%	Data mapping method
A_{14}	Full score 9 points, score 9 points, and score rate 100%	Data mapping method
A_{15}	The evaluation result is grade B, and the score rate is 80%	Grade assignment method—Data mapping method
A_{16}	The ecological environmental protection effect is good, and the score rate is 80%	Grade assignment method—Data mapping method
A_{21}	Full score 11 + 1 points, score 10 + 1 points, and score rate 91.7%	Data mapping method
A_{22}	Full score 1 point, score 1 point, and score rate 100%	–
A_{23}	Full score 7 points, score 7 points, and score rate 100%	Data mapping method
A_{24}	Grade A, score 1 points, and score rate 100%	Grade assignment method—Data mapping method
A_{25}	Full score 16 points, score 16 points, and score rate 100%	Data mapping method
A_{26}	Full score 14 points, score 12 points, and score rate 85.7%	Data mapping method
A_{31}	Full score 21 points, score 21 points, and score rate 100%	Data mapping method
A_{32}	Full score 19 points, score 16 points, and score rate 84.2%	Data mapping method
A_{33}	Full score 1 points, score 1 points, and score rate 100%	–
A_{34}	Full score 8 points, score 7 points, and score rate 87.5%	Data mapping method
A_{35}	Full score 11 points, score 9.7 points, and score rate 88.2%	Data mapping method
A_{36}	Full score 9 points, score 8 points, and score rate 88.9%	Data mapping method

2) Normalization of the data of each sub-item evaluation result

The data of each sub-item of the Yanhe Village planning and construction evaluation were normalized by using the data mapping method and evaluation level assignment method.

The methods of sub-item evaluation results summary and data normalization of Yanhe Village are shown in Table 11.5.

3) Comprehensive evaluation

According to the comprehensive evaluation method of green village and town planning and construction established in Sect. 11.2. Firstly, the evaluation data of 18 sub-items in three categories should be implemented normalized data processing. After normalization, the calculated value is multiplied by the three-level index weight. After addition and averaging, three second-level indexes are calculated. Then, the score of the first-level index is obtained through comprehensive evaluation.

The comprehensive evaluation results of the green village planning and construction in Yanhe Village are shown in Table 11.6.

Table 11.6 Comprehensive evaluation results of the green village planning and construction in Yanhe Village

Primary indexes	Point	Secondary indexes	Weight	Point	Tertiary indexes	Weight	Calculated value after normalization
A	91.3	A_1	0.4	85.3	A_{11}	0.156382	86.5
					A_{12}	0.139320	69.0
					A_{13}	0.221157	88.5
					A_{14}	0.197029	100
					A_{15}	0.175533	80.0
					A_{16}	0.110579	80.0
		A_2	0.4	97.2	A_{21}	0.125000	91.7
					A_{22}	0.250000	100
					A_{23}	0.250000	100
					A_{24}	0.125000	100
					A_{25}	0.125000	100
					A_{26}	0.125000	85.7
		A_3	0.2	91.6	A_{31}	0.164623	100
					A_{32}	0.164623	84.2
					A_{33}	0.184783	100
					A_{34}	0.184783	87.5
					A_{35}	0.184783	88.2
					A_{36}	0.116406	88.9

4) Analysis of evaluation results

The comprehensive evaluation conclusions of the planning and construction of Yanhe Village are as follows:

(1) Sub-item evaluation results are excellent

Table 11.6 shows that the evaluation score of plan implementation in Yanhe Village is 85.3 points, and the evaluation grade is Class II. The evaluation score of environmental infrastructure construction is 97.2 points, the evaluation score of green building construction is 91.6 points, and the evaluation grades all reached Class I.

All 3 sub-categories of the planning and construction of Yanhe Village met the national requirements, and the evaluation results reached excellent standards.

(2) Comprehensive evaluation results are excellent

The comprehensive evaluation score of the green village planning and construction of Yanhe Village was 91.3 points, and the comprehensive evaluation grade was Class I. The overall evaluation results were excellent.

11.4 Comprehensive Evaluation Platform of Green Village and Town Planning and Construction

To make the evaluation more convenient and more efficient, the authors' research team established a comprehensive evaluation platform for the green village and town planning and construction. The platform consists of four parts: the database of evaluation and monitoring, the retrieval method of evaluation and monitoring, the evaluation platform of planning and construction, and the application of the comprehensive evaluation platform of green village and town planning and construction.

1. Database of evaluation and monitoring of green villages and towns

1) Contents of evaluation and monitoring database of green villages and towns

The database consists of two parts: the evaluation technology database and the monitoring technology database.

(1) The evaluation technology database consists of three parts: an evaluation technology database of green village and town plans, an evaluation technology database of green village and town infrastructure construction, and an evaluation technology database of green village and town farmhouse construction.
(2) The monitoring technology database consists of two parts: a monitoring technology database of green village and town plan and implementation and a monitoring technology database of green village and town farmhouse construction.

2) Database development technology

The database uses Microsoft Access to realize integrated storage and management of plan, construction, evaluation and monitoring technical attribute data. The WinFrom application program was developed by using the C# programming language and Visual Studio tool, which integrates the Microsoft Access database.

3) Database management

(1) System login

Users must open the monitoring system, enter a username and password (*******), and click the login button to enter the main interface of the system.

The main evaluation platform system interface of green village and town planning and construction is shown in Fig. 11.5.

(2) Database management and usage

Let us take the monitoring database of green village and town planning and construction as an example to illustrate the management and usage method.

Users need to find "monitoring and evaluation technology" in the menu bar of the main interface and click on the sub-menu "monitoring technology" or "evaluation technology".

Fig. 11.5 Evaluation platform system interface of green village and town planning and construction

For example, to use the monitoring technology database, users need to click "monitoring technology". The interface includes mainly the directory area, function area and display area.

Directory area: In the system design, a tree structure is used to display the two categories—the monitoring technology of the plan and implementation, and the monitoring technology of farmhouse construction—and five sub-monitoring technologies—the monitoring technology of land use and function division, the monitoring technology of environmental quality influencing factors of ecological carrying capacity, the monitoring technology of energy-saving efficiency of farm buildings, the indoor environmental quality testing technology of farmhouses, and the safety testing technology of existing farm buildings. A double click of the tree structure node will display the corresponding monitoring technology in the display area.

Display area: A double click of "monitor technology" in the table will display the specific introduction and operational method.

Function area: Through the link with the database, "search by name" and "search by keyword" can be applied to all monitoring technologies. The search results can be displayed simultaneously in the display area.

2. Retrieval method of evaluation and monitoring of green villages and towns

Through the link with the database, "search by name" and "search by keyword" can be applied to all evaluation monitoring technologies, and the search results can be displayed simultaneously in the display area.

Let us take an inquiry about NO_X detection technology in air quality as an example.

Users need to enter "Air" In The "monitoring Technology" interface search box and select "search by keyword"; then, all air monitoring-related technologies will be displayed in the display area. Then, users need to select "NO_X detection" and double click to check detailed information such as basic principles, experimental instruments, and operating procedures.

The interface "NO_X detection" of the indoor environmental quality search display is shown in Fig. 11.6.

3. Evaluation platform of green village and town planning and construction

The comprehensive evaluation platform for green village and town planning and construction consists of three parts: indicator area, function area and conclusion area. The instructions for users are as follows:

(1) Main interface: users need to find "plan evaluation" in the menu bar of the main interface and click to jump to the monitoring and evaluation platform interface, which includes the indicator area, function area and conclusion area.
(2) Indicator area: According to the system setting, 3 categories and 18 indicators are set. Users need to input the indicator value according to the actual situation. It should be noted that the index value should not be empty and should not exceed the limit (0–100); otherwise, the system will respond with an error message.

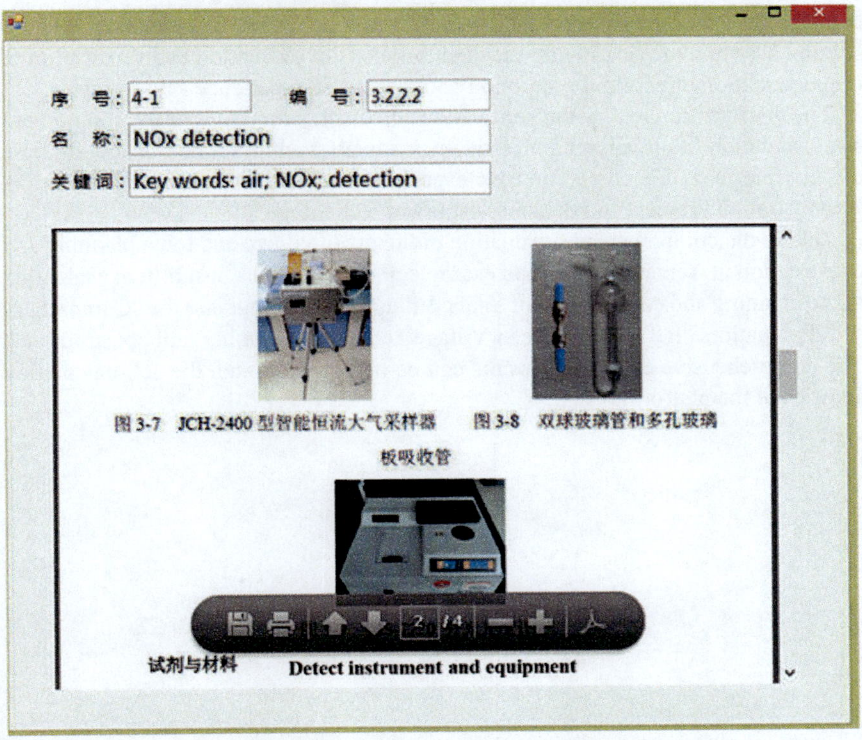

Fig. 11.6 "NO_x detection" interface of indoor environmental quality search display

(3) Function area: After all the indicators are input, users can click the "submit" button, and the system will then calculate the evaluation results automatically and display the score. Based on the analysis of the score and the specific indicator values, the system can automatically generate reference conclusions for decision makers to refer to. In the process, if users make a mistake, they can click "clear" to re-enter information.

(4) Conclusion area: After users click the "submit" button, the detailed evaluation report will be displayed synchronously in the conclusion area. The evaluation report can be printed so that the data can be stored and analysed for improvement.

4. Application of comprehensive evaluation platform of green village and town planning and construction

The green village and town planning and construction is a long-term and dynamic development process, and it is necessary to establish a demonstration base that has the dual functions of evaluating and verifying the implementation effect and promoting and demonstrating experience achievements. It is necessary to establish a demonstration base and provide it with the dual functions of evaluation and verification of implementation effects and promotion and display of experience achievements.

The platform can realize the verification and display functions of the comprehensive evaluation intuitively and effectively. It mainly includes a demonstration base display interface, demonstration base overview, demonstration project distribution, demonstration projects and detailed display of technologies.

Taking the comprehensive evaluation of the green village and town planning and construction of Yanhe Village as an example, the results of each sub-item evaluation of the planning and construction of Yanhe Village will be input into the "Comprehensive Evaluation Platform for Green Village and Town Planning and construction". The comprehensive evaluation results can be obtained through the self-calculation abilities of the platform.

Correction to: Evaluation of Green Village and Town Environmental Infrastructure Construction

Correction to:
Chapter 9 in: X. Wang and H. Hua, *Green Village and Town Construction in China*,
https://doi.org/10.1007/978-981-16-2098-0_9

In the original version of the book, author-provided belated corrections have been incorporated in Table 9.1 of Chapter 9.

The updated version of this chapter can be found at
https://doi.org/10.1007/978-981-16-2098-0_9

X. Wang and H. Hua, *Green Village and Town Construction in China*,
https://doi.org/10.1007/978-981-16-2098-0_12